Benchmark Papers in Acoustics

Series Editor: R. Bruce Lindsay
Brown University

Volume

1 UNDERWATER SOUND / *Vernon M. Albers*
2 ACOUSTICS: Historical and Philosophical Development / *R. Bruce Lindsay*
3 SPEECH SYNTHESIS / *James L. Flanagan and Lawrence R. Rabiner*
4 PHYSICAL ACOUSTICS / *R. Bruce Lindsay*
5 MUSICAL ACOUSTICS, PART I: Violin Family Components / *Carleen M. Hutchins*
6 MUSICAL ACOUSTICS, PART II: Violin Family Functions / *Carleen M. Hutchins*
7 ULTRASONIC BIOPHYSICS / *Floyd Dunn and William D. O'Brien, Jr.*
8 VIBRATION: Beams, Plates, and Shells / *Arturs Kalnins and Clive L. Dym*
9 MUSICAL ACOUSTICS: Piano and Wind Instruments / *Earle L. Kent*
10 ARCHITECTURAL ACOUSTICS / *Thomas D. Northwood*
11 SPEECH INTELLIGIBILITY AND SPEAKER RECOGNITION / *Mones E. Hawley*
12 DISC RECORDING AND REPRODUCTION / *H. E. Roys*
13 PSYCHOLOGICAL ACOUSTICS / *Earl D. Schubert*
14 ACOUSTIC TRANSDUCERS / *Ivor D. Groves, Jr.*
15 PHYSIOLOGICAL ACOUSTICS / *Juergen Tonndorf*
16 ACOUSTICAL MEASUREMENTS: Methods and Instrumentation / *Harry B. Miller*

A Benchmark® Books Series

ACOUSTICAL MEASUREMENTS Methods and Instrumentation

Edited by

HARRY B. MILLER

Naval Underwater Systems Center

Hutchinson Ross Publishing Company

Stroudsburg, Pennsylvania

Benchmark Papers in Acoustics, Volume 16
Library of Congress Catalog Card Number: 81–7175
ISBN: 0–97933–415–0

84 83 82 1 2 3 4 5
Manufactured in the United States of America.

LIBRARY OF CONGRESS CATALOGING IN PUBLICATION DATA
Main entry under title:
Acoustical measurements.
(Benchmark papers in acoustics; 16)
Includes indexes.
1. Sound—Measurement—Addresses, essays, lectures.
I. Miller, Harry B. (Harry Bernard), 1916– . II. Series.
QC243.A24 534'.42 81-7175
ISBN 0-87933-415-0 AACR2

Distributed world wide by Academic Press,
a subsidiary of Harcourt Brace Jovanovich,
Publishers.

CONTENTS

PART II: ELECTROACOUSTIC MICROPHONES AND CLOSED-CAVITY CALIBRATIONS

PART III: DIRECT VISUAL OBSERVATION OF SOUND WAVES

PART V: SOUND ABSORPTION COEFFICIENT MEASUREMENTS AND ACOUSTIC IMPEDANCE MEASUREMENTS

PART VI: PHASE DISTORTION AND TRANSIENT DISTORTION IN ELECTROACOUSTIC SYSTEMS

SERIES EDITOR'S FOREWORD

The "Benchmark Papers in Acoustics" constitute a series of volumes that make available to the reader in carefully organized form important papers in all branches of acoustics. The literature of acoustics is vast in extent and much of it, particularly the earlier part, is inaccessible to the average acoustical scientist and engineer. These volumes aim to provide a practical introduction to this literature since each volume offers an expert's selection of the seminal papers in a given branch of the subject—that is, those papers that have significantly influenced the development of that branch in a certain direction and introduced concepts and methods that possess basic utility in modern acoustics as a whole. Each volume provides a convenient and economical summary of results as well as a foundation for further study for both the person familiar with the field and the person who wishes to become acquainted with it.

Each volume has been organized and edited by an authority in the area to which it pertains. Each volume provides an editorial introduction summarizing the technical significance of the field being covered. Each article is accompanied by editorial commentary, with necessary explanatory notes, and an adequate index is provided for ready reference. Articles in languages other than English are either translated or abstracted in English. It is the hope of the publisher and editor that these volumes will constitute a working library of the most important technical literature in acoustics of value to students and research workers.

The present volume, *Acoustical Measurements: Methods and Instrumentation*, has been edited by Harry B. Miller, consulting electroacoustics engineer at the Naval Underwater Systems Center in New London, Connecticut. Mr. Miller is an authority in the field of acoustical measurements and instrumentation, having served previously as head of the Acoustic Engineering Department of the Brush Development Company and as head of the Advanced Development Electroacoustics Laboratory of the Electronics Division of General Dynamics. In its 45 carefully selected seminal articles, this volume covers thoroughly the development of acoustical instrumentation and measurement methods during the last 125 years, with particular emphasis on the work of the past 75 years. Each of the six groups of articles into which the volume is divided is provided not only with an introduction but also with commentaries on the significance of the papers in the group and their relation to other papers in the volume. There is ample emphasis on the historical background but without neglect of the very important more recent contributions. Both in its coverage and its method of presentation this volume should command the enthusiastic attention of all workers in acoustics.

R. BRUCE LINDSAY

PREFACE

Anyone who has had occasion to repeatedly consult the standard English and American treatises on sound has probably been struck by the repeated appearance of the same references showing up again and again in all the treatises. These referenced writings, ground-breaking papers usually written in English, German or French, were never on hand at the moment I wanted them, often not readily obtainable at all and certainly not available in an English translation. Here, then, is a collection of some of the famous referenced writings on acoustical measurements. They are all presented in English, some translated especially for this volume.

This collection of source papers has turned out to be the equivalent of a textbook on applied acoustics by way of the case-history method. Most of the writers are good teachers as well as gifted experimentalists; however, many of the papers, especially the pre-1930 papers, often call out for clarification. Therefore, as a special feature, this volume includes the editor's notes (not listed in the table of contents) which provide commentaries, often extensive, on nearly every paper or group of papers in the volume.

One aim of the commentaries is to transform "a quaint paper, maybe worth skimming through" into the powerful paper it was perceived to be by its author's contemporaries. Two obstacles that often stand in the way of the reader are that the paper is out of his or her area of expertise, and that the time gap is too great. The commentaries try to remove these two obstacles, often by translating the author's approach into an alternative way of looking at the problem. The alternative approach is usually one that would appeal to an electrical engineer. Thus problems in heat transfer, mechanical amplification, acoustic impedance, and so on, are discussed via parallel examples from electrical engineering, such as skin effect phenomena, coupled-circuit theory, electrical transmission line behavior, and so forth. In addition, following each set of commentaries are detailed references for further reading, each giving the specific pages that pertain to the narrow problem under discussion. Before closing, let me urge the reader to have within arm's reach as many as possible of the nine referenced works listed at the close of the Introduction.

I would like to thank R. Bruce Lindsay for his helpful suggestions and for his translations from the German of Papers 16, 17, and 19.

Finally, I would like to dedicate this book to Vivian Miller, my wife, without whose unflagging support the book would not have become a reality.

HARRY B. MILLER

CONTENTS BY AUTHOR

ACOUSTICAL MEASUREMENTS

INTRODUCTION

This volume is concerned with applied acoustics, specifically with acoustical measurement instruments and methods. Well over one hundred selections were studied and considered for inclusion in this volume, but more than half had to be rejected, reluctantly, for lack of space. Accordingly we present the survivors, papers by approximately forty gifted experimentalists. The papers deal with measurement problems in various branches of applied acoustics. They are mostly journal articles, but some are extracts from books.

At the end of this introduction a list of essential references is given. These references should preferably be within arm's reach of the reader when he or she uses this volume. Of these references, one is considered indispensable: the *Journal of the Acoustical Society of America*, or *JASA*. This is the foremost archival journal in the field of acoustics, covering the entire field. Since this journal is widely accessible, important papers that appeared in *JASA* are sometimes only briefly sampled in this volume. By this device it has been possible to reprint a number of non-*JASA* papers that would otherwise have remained unretrieved in the *Physical Review, American Institute of Electrical Engineers Proceedings*, and other journals.

The volume is divided into six parts. In Part I, "Nonelectroacoustic Oscilloscopes, Wave Analyzers, and Microphones," we can trace a continuous path from Lissajous to Helmholtz (observing vibrations of mechanical bodies), then to Webster, to D. C. Miller, and to E. T. Paris (observing, indirectly, vibrations of a gas). This era could be called the "pre-electronic era."

Then suddenly E. C. Wente and the electronic era appeared, as presented in Part II, "Electroacoustic Microphones and Closed Cavity Calibrations." The era of Webster, Miller, and Paris gradually disappeared. One is tempted to refer to the pre-Wente era, the early-Wente era, and the post-Wente era. Part II is concerned with the early-Wente era, which coexisted with the Webster-Miller-Paris era.

Almost simultaneous with Lissajous was A. Toepler, as shown in Part III, "Direct Visual Observation of Sound Waves." Toepler's passion was to try to see, by direct observation, a sound wave travel through the air; he did it. Toepler called his method the "schlieren method." This was a third path that coexisted with the paths of Part I and Part II. This path led from Toepler to Dvorak and others, and then, after World War II, to F. Zernike and others of the early Fourier-optics era.

Part IV is entitled "Free-Field Calibration Methods and the Development of Reciprocity Concepts." We have chosen to give special emphasis in this volume to the technology that evolved from the Wente era (electroacoustic transducers). Hence, Part IV is almost wholly concerned with the calibration problems that pertain to electroacoustic transducers.

Part V is called "Sound-Absorption Coefficient Measurements and Acoustic-Impedance Measurements." Simultaneously with the activities of Webster and Miller and Wente and Fletcher, the architectural acousticians were fighting their own measurement problems. Wallace Sabine had made good use of the schlieren method (actually, Dvorak's shadow method) in his ray-tracing studies of reverberation, using room models. However, many more problems needed solutions. The most fruitful answers have come from the study of sound-absorption coefficient measurements, and acoustic-impedance measurements, which is treated in Part V. The reader searching for unsolved problems will find a gold mine in this field. A good method for measuring the sound-absorption coefficient of material insonified by an obliquely incident wave still needs to be invented—that is, impedance tube devices still measure only normal-incidence sound, and the motional-impedance method for measuring acoustic impedance is still waiting for a transducer whose mechanical resonance can be quickly and easily retuned over a broad bandwidth.

Part VI is entitled "Phase Distortion and Transient Distortion in Electroacoustic Systems." The steady improvement in the design of loudspeakers, microphones, and electroacoustic systems in general, which began in the 1930s, took a sudden leap forward in the 1960s as a result of the availability of large computers and of the Fast Fourier Transform (FFT). The subsequent explosion in the technology of computer-aided signal processing, in conjunction with moderately priced small computers, revolutionized the whole acoustic-measurement field. Part VI begins with some background papers that preceded this revolution and then presents some papers that depend upon the FFT.

With improved methods for measuring transient distortion, to cite a specific example, great improvements began to be seen (or heard) in the design of acoustic components. Loudspeakers and amplifiers are two prominent examples. Design upgrading, therefore, is also the concern of Part VI, which clearly shows how improvements in measurement methods and instrumentation lead directly to improvements in the design and production of acoustic components and systems.

Following each paper or group of papers is a section of Editor's Notes. The comments address themselves to the question: "What is going on, where is the author leading us?" The editor's approach is sometimes to present a simplified version of the author's problem, or perhaps an analogous problem from another field, or sometimes to focus attention onto the one component in an equation that is really the controlling factor.

On two occasions the editor felt constrained to intrude more deeply. Each intrusion takes the form of an original paper. Paper 11 grew out of A. G. Webster's unfulfilled promise to explain his double resonator in a future paper. Paper 33 grew out of a desire to show how W. Schottky's reciprocity method fits into its proper position between the Helmholtz-Rayleigh method and the MacLean-Cook method.

The annotated bibliographies allow the interested reader to pursue a subject more deeply but very narrowly, if he so desires, via specific page numbers in the referenced material.

BIBLIOGRAPHY

Beranek, L. L., 1949, *Acoustic Measurements*, Wiley, New York.
Bobber, R. J., 1970, *Underwater Electroacoustic Measurements*, Naval Research Laboratory, Washington, D. C.
Crandall, I. B., 1926, *Theory of Vibrating Systems and Sound*, Van Nostrand, New York.
Lindsay, R. B., ed., 1972, *Acoustics: Historical and Philosophical Development*, Dowden, Hutchinson & Ross, Stroudsburg, Pa., 465 p.
Morse, P. M., 1948, *Vibration and Sound*, 2nd ed., Van Nostrand, New York.
Olson, H. F., 1947, *Elements of Acoustical Engineering*, 2nd ed., Van Nostrand, New York.
Stewart, G. W., and R. B. Lindsay, 1930, *Acoustics*, Van Nostrand, New York.
Strutt, J. W. (Lord Rayleigh), 1945, *Theory of Sound*, Dover Publications, New York.
Wood, A. B., 1941, *A Textbook of Sound*, 2nd ed., Bell, London.

Part I

NONELECTROACOUSTIC OSCILLOSCOPES, WAVE ANALYZERS, AND MICROPHONES

Editor's Comments on Paper 1

1 **LISSAJOUS**
Excerpt from *Memoir on the Optical Study of Vibratory Motion*

In 1857, Jules Lissajous published his famous paper on a means for making visible the vibrations of one or more vibrating bodies and a means for the optical compounding of these vibrations. Blackburn's pendulum and Wheatstone's kaleidophone had previously traced out ellipses and other figures qualitatively. However, Lissajous was interested in a quantitative instrument, and he achieved his major goal: to tune an unknown vibrator approximately to unison with a standard tuning fork by the use of the eye alone, without the help of the ear, and to record the sharpness or flatness of the unknown's pitch quantitatively, in cycles per second (or hertz in the present notation).

Lissajous at first compounded two vibrations moving in the same direction. He was thereby able to count the beats between two slightly mistuned tuning forks (see Fig. 5). He then tried to spread out this one-dimensional pattern into a two-dimensional pattern that would make use of a time axis. Even in his work with a single vibrator, however, he could not get a linear sawtooth sweep for his time axis. For example, his rotating mirror, projecting light onto a flat screen, gave a distortion reminiscent of a Mercator-projection map of Greenland (see Fig. 3). Lissajous's solution to the sweep problem is shown in Figure 10. Instead of combining two vibrations along the same path and displaying the resultant amplitude versus time where time is an explicit variable, he combined the two vibrations along two perpendicular paths, plotting amplitude A_1 against amplitude A_2 and using time as an implicit variable. This gave him a nearly stationary pattern, still distorted but having uniform brightness, suitable either for direct viewing by the microscope or for projection onto a screen. He was thereby able to measure frequency relationships and phase differences with great accuracy. In addition, Lissajous supplied a simple rule for decoding even complicated patterns based on counting the vertical cusps and the horizontal cusps.

The Lissajous vibration patterns, shown in Plate II, are still widely enough renowned to appear in the larger Webster dictionaries, but his less publicized work, important to Helmholtz and others and seen in part in Plate I, is usually overlooked nowadays. He came very close to inventing the Helmholtz vibration microscope, falling short because electro-magneto-mechanical drive systems were not yet perfected (Helmholtz generously gave him full credit for the invention).

The original French paper is eighty-five pages long. Our English presentation has been shortened considerably, while yet trying to retain the flavor of Lissajous's style. The result, Paper 1, is a combination of strict translation and paraphrase. (All material in brackets has been interpolated by the volume editor.) Thus when Lissajous's own words sound repetitious or even naive, remember that an enthusiastic pioneer is describing how he views things.

Editor's notes and an annotated bibliography are given at the end of the translated paper in order to clarify some points that a modern reader might find unclear.

Jules A. Lissajous gave many public lecture-demonstrations of his method, projecting vibration patterns onto a screen (as did Dayton C. Miller sixty years later) in order to display certain attributes of vibrating rods. His work was funded, modestly, by the French Academy of Sciences. (One of his reviewers was Babinet.)

1

MEMOIR ON THE OPTICAL STUDY OF VIBRATORY MOTION

J. Lissajous

This excerpt was translated expressly for this Benchmark volume by Harry B. Miller, Naval Underwater Systems Center, from "Mémoire sur l'étude optique des mouvements vibratoires," in Ann. Chim. Phys. *ser. 3,* ***51****:147–231 (1857)*

[*Editor's Note:* The superscripts in the text refer to the Editor's Notes at the end of the translation.]

The memoir that I have the honor of submitting to the Academy has for its object the development of an optical method suitable for the study of vibratory motion. This method, based on the persistence of visual sensations, and on the compounding of two or more simultaneous vibrations, allows—without the help of the ear—the study of every kind of vibratory motion, and consequently of every kind of sound.

CHAPTER ONE: OUTLINE OF THE METHOD

I. Means for Making Visible the Vibratory Motion of Bodies

I have been led to conceive this method by the desire to make vibratory phenomena visible to a large audience. Struck by the advantage offered, for the teaching of acoustics, by the use of projection methods, which Paul Desains has made so much use of in his course of physics at the Faculty of Sciences of Paris, I have wanted to try to make visible the oscillatory motion of vibrating bodies, without having recourse to the mechanical tracing of the vibration, as did Young, Savart, Duhamel, and Wertheim—all very successfully.

I first worked with the tuning fork, that little instrument being of all vibrating bodies the one that is most convenient to handle. This work was merely the particular application of a principle whose generality will stand out clearly from what will follow.

To make visible, either directly or by projection, the vibration of a tuning fork, I fasten to the end of one of the branches on the convex face (Fig. 1, Plate I) a small metallic plane mirror M. The other branch carries a counterweight M', in order that the added mass shall be the same on the two branches, which is necessary in order for the tuning fork to vibrate easily and for a long time. I observe in this mirror the reflected image of a candle placed a few meters away; then I make the tuning fork vibrate. Immediately I see the image become larger in the direction of the length of the branches. If I then make the tuning fork turn about its axis, the appearance changes, and I perceive in the mirror a bright sinusoidal line, whose undulations indicate by their shape the amplitude, greater or lesser, of the vibration.

If we wish to operate via projection in a dark chamber, we let a beam of sunlight fall on the mirror. Then the reflected ray throws onto the wall or onto a screen a tracing that enlarges in the direction of the vibration as soon as we excite the tuning fork, and this tracing is transformed into a sinusoidal line as soon as we make the tuning fork turn about its axis.

The same procedure holds for [large] vibrating bodies that by virtue of their weight and their shape, do not lend themselves easily to rapid turning. It suffices indeed, instead of making the body turn, to receive the ray reflected by the first mirror on a second mirror that at the same time is made to turn more or less quickly around an axis perpendicular to the mean direction of the reflected ray and located in the same plane where this ray executes its vibrations. We thus see, either directly in the movable mirror or by projection onto a screen, the sinusoidal line that demonstrates the existence of the vibratory motion.

If one wishes to make the phenomenon sharper and brighter, the following arrangements can be used:

1. *For direct observation:* You take as a source of light a lamp surrounded by an opaque chimney in which a small hole has been pierced, and you observe the phenomenon with the aid of a telescope, previously focused, so as to clearly see the reflected image before any motion has been given to the apparatus.
2. *For projection:* You take for a source of light the sun or an electric light, whose rays are allowed to pass through an opening of circular form O in a thin diaphragm, Figure 2. The beam that has come through this opening is caused to reflect onto mirrors M and m. It then passes through a lens L (achromatic if possible), which is placed in such a way as to form on the screen EF as sharp as possible an image I of the circular opening O.

The explanation of these various phenomena is very simple. The pencil of light reflected by the mirror M must oscillate as soon as the mirror itself oscillates. The tip of this pencil must describe, in the eye or on the screen, a straight line G-H, Figure 3, the various points of which remain illuminated for a certain time because of the persistence of the sensation of vision in the eye. If the duration of the sensation is much greater than the duration of a complete oscillation, the line appears illuminated throughout its entire extent, in a permanent fashion. It is necessarily more luminous at the extremities H and G, Figure 4, where the movement of the image is slower, because there the sum of the stimuli received by the eye in a given time is greater. If one adds to the oscillatory motion of the mirror a rotational motion around the mirror axis, then the tip of the reflected pencil traces a sinusoidal line S, Figure 3, either in the eye or upon the screen.[1]

The different points of this line become visible successively, but as the sensation produced in the eye does not cease immediately, this line appears to be illuminated over an extent that depends on the duration, great or small, of the sensation.

This principle can furthermore be put into play in many different ways, which reduce to the following elements: giving a sharp blow, by any means at all, to one of the points of the vibrating body so that the motion of this point is clearly distinguished from the motion of the others, then combining the oscillatory motion of the point with a continuous displacement directed perpendicularly to the direction of the vibration.[2] It is thus, for example, that one can make visible the sinusoidal line that represents oscillatory motion—by displacing the eye rapidly in a direction perpendicular to the vibration.

These first experiments are nothing but the preliminaries of the method. They are the development and the application, in another form, of ideas already put into effect in other apparatus, notably in the ingenious kaleidophone of Mr. Wheatstone. I must acknowledge the share of inspiration that I owed at the time of my first researches to the remarkable work with rapidly rotating mirrors done by Wheatstone, Arago, and Foucault.

What follows seems to me to result completely from the former proceedings. It comprises, to say it properly, an optical method whose general application seems to me to be realizable today, as one will be able to judge by the details given in the course of this memoir.

The problem whose solution I have undertaken is the following: to compare, without the help of the ear, the vibrations of two bodies so as to know the exact relation of the number of vibrations that they execute in the same time, as well as all the circumstances that, for the duration of the phenomenon, characterize their relative motion.

This problem can be solved by obtaining, by means of a special trick, the "optical compounding"—if I may express myself thus—of two vibratory motions following either the same direction or perpendicular directions.

II. Optical Compounding of Two Vibrations Moving in the Same Direction

The compounding of vibrations along the same direction has a useful application only in the case where the two bodies are in unison or near unison.

To obtain this compounding we place, facing one another, two tuning forks equipped with mirrors. The light leaving point O, Figure 5, falls on the first mirror along a direction nearly normal to the surface M; from there onto the second mirror M'; and from there to the eye or upon the screen E (in this last case the beam must pass through a lens L in order to produce a clear image of the point O).[3] The axes of the tuning forks are vertical and the mirrors practically parallel.

Supposing, first, the tuning forks to be in close agreement, let us excite only the first one; the image elongates as we have seen before. If we now excite the two tuning forks at the same time, the elongation of the image becomes larger or smaller according to whether there is or is not agreement between the simultaneous motions communicated [via superposition] to the image by the vibration of the two tuning forks.

If the two tuning forks pass at the same time and in the same direction through their point of equilibrium, the elongation of the image reaches its maximum. If on the contrary they pass through it at the same time but in opposite directions, the elongation reaches its minimum and can even equal zero if the amplitudes of vibration communicated to the light beam reflected by the two mirrors are exactly equal. Finally, the elongation of the image takes on values more or less large, according to whether the elapsed time is more or less long between the precise moments when the two tuning forks pass through their points of equilibrium.

It is the relation between this time [the beat period] and the duration of one [tuning fork] period that constitutes what we shall call "phase difference."

In the case where the tuning forks are in exact agreement, the luminous curve experiences merely a progressive decrease in height due to the extinguishing of the vibrations. But in the case where the agreement is slightly off, the luminous curve experiences periodic variations in its elongation;[4] and in the same time that the ear hears the beating that indicates the flaw in agreement, the eye perceives in the clearest fashion the accompanying pulsations of the image.

[*Editor's Note:* The author then discusses an experimental arrangement, shown in Figure 7, to demonstrate linear superposition of two vibrations.[5] He refers to the "law of infinitely small vibrations of a pendulum" as a prelude to showing that the resulting simple sine waves can be a basis for synthesizing a complex wave. He likes to emphasize that these laws apply to a material particle as well as to a spot of light. He then shows how two rotating vectors can combine to trace out two superposed traveling waves (Fig. 9). In the first case the two vectors rotate at the same angular velocity. In the second case they have two slightly different angular velocities. This leads to a discussion of the visual observation of beats. He then attacks the same problem analytically and concludes:]

The identity of the equations we have set up [for a mathematical point, as in Fig. 9], with those that express the motion of a particle of air subjected to the simultaneous action of two undulations that differ in intensity and in period, indicates well enough the intimate relation existing between this kind of "optical beating," and acoustic beating. [*Editor's Note:* The author is here jumping from the vibrations of two tuning forks in a vacuum, observed optically, to the vibration of a particle of air acted on by these two tuning forks, and observed acoustically. Thus he listens to the beating of the sound waves in air and simultaneously observes, optically, the "beating" of the two sine waves of light.

The author then enters into a raging dispute of the period: will two vibrating bodies, situated near each other, always interact and hence always produce beats? Savart apparently was claiming that this mutual reaction, or coupling, was always present. Lissajous says that his experiments show otherwise. To say it in other terms, if two identical vibrators of resonance frequency f_1 are tightly coupled by some means, they will produce no f_1, but rather $f_1 + \Delta f$ and $f_1 - \Delta f$; and these two vibrations will radiate two sound waves that will beat (Savart's thesis). But if the coupling coefficient is reduced to zero, Δf becomes zero and we get f_1 along with another f_1 and the two vibrations will radiate a single-frequency sinusoidal wave. There will be no beats (Lissajous's thesis). Both results, of course, can be demonstrated experimentally. Lissajous now stops discussing this whole line of investigation for awhile, and turns to his most famous contribution:]

III. Optical Compounding of Two Vibrations Moving in Perpendicular Directions

[*Editor's Note:* The author refers to Figure 10 and explains the path of the doubly reflected pencil of light that emerges from point O and is observed by the telescope L, and he continues:] If we make the tuning fork A vibrate, the image oscillates rapidly in the horizontal direction and appears as a luminous line directed horizontally. If we make only the vertical tuning fork B vibrate, the image elongates in the vertical direction. But if we make the two tuning forks vibrate at the same time, the image takes on a complex motion due to the combination of the two movements. It traces out in space a more or less complicated curve, whose form depends on the relation between the number of vibrations executed in the same time by the two tuning forks. [*Editor's Note:* The author then starts with: (1) *Tuning forks in unison. Relation of numbers of vibration: one to one.* He refers to Plate II, Series 1:1, points out the straight line and the ellipses, and discusses them briefly. He reminds us that a given shape will be preserved throughout its transient duration only if the amplitudes of the two component vibrations decrease by the same percentage. But if the damping is not the same for the two vibrators,] then the axis of the ellipse approaches the direction of the tuning fork most strongly excited, and at the same time the ellipse tends toward a straight line.

If the tuning forks are not exactly in tune, the initial phase difference is not maintained, and the curve passes through all its variations. It seems to be animated by a sort of swinging, all the more rapid the more the tuning forks are out of tune. And when the curve has passed through all its distinct forms to return to the initial form, we can be sure that one of the tuning forks has executed one full period more than the other.

[Then the author proceeds to the Series 1:2.] The figure perceived at the first instant preserves the same form, if the agreement is perfect. If there is a small difference in tuning, the figure passes through all the indicated transformations; and it has passed through all the possible forms when the "sharper" tuning fork has accomplished one period more than the number it would execute if the agreement were perfect. [*Editor's Note:* Observe that it is possible to achieve a stationary pattern for frequencies 1 vs. 2 or 1 vs. 3 or 3 vs. 4, and so on; and to "freeze" any phase differences you wish. The numerator "one" serves merely as the time axis or distance axis, which is here folded back onto itself; and the denominator "three" is a traveling wave traveling along this time axis or distance axis.[6] So we see that the term *phase difference* does have meaning here, even though the two frequencies are different. Note also that this is not the paradox of trying to produce a stationary pattern from two standing waves on a string, the waves having, for example, frequencies 1 and 3. Lissajous is dealing with two traveling waves. The reader might wish to imagine a bent coat hanger slowly rotating around a horizontal axis.

The author then refers to: 1:3, 2:3, and 3:4, and he adds a valuable observation.] The complication of the figures increases as the two terms [numerator and denominator] . . . each take on a higher value. It would then be impossible to recognize the relation we were dealing with, if the curve did not carry within itself the precise information that is needed. . . . All around the curve there are a number of peaks situated on the four sides of the rectangle. The rectangle's dimensions are precisely equal to the amplitudes of the horizontal and vertical vibrations. Count the number of peaks located on the upper part, then the number of peaks located on one of the vertical sides of the rectangle, e.g., the right side: these two numbers are the numerator and the denominator of the fraction that expresses the relation of the two numbers of vibrations executed by these two bodies. The numerator corresponds to the vertical tuning fork [refer to the Series 1:3 for a simple example], the denominator to the horizontal. [For] example, Row 5, Series 3:4, first picture: we have three peaks on the top side, four on the right side of the curve, the relation of the number of vibrations is 3:4. The vertical tuning fork is a [musical] fourth down from the horizontal tuning fork.

CHAPTER TWO: GEOMETRIC THEORY OF THE COMPOUNDING OF TWO PERPENDICULAR VIBRATIONS

[*Editor's Note:* The author then launches into a geometrical discussion of the addition of vectors, referring to Figures 12, 14, 15, and 16. He tells how he wanted to build a good tracing machine, probably a harmonograph, but] I have had to stop in the face of the prospect of expenses greater than the resources at my disposal. [Then a footnote:] I hope soon to fill this gap with the help of funds that the Academy of Sciences has been willing to place at my disposal. [*Editor's Note:* The author then discusses an interim device he has built to shorten the manual labor of constructing a curve.]

On a level brass plaque I traced two circles AA' and BB', Figure 13, whose diameters were in the ratio 2 to 3 (I could have used any other ratio); I drew through the common center O of the two circles, two rectangular lines; I then divided each quadrant of each side into 8 equal parts (although I could have chosen a different number). [Observe the faint marks on the circumferences.] I then joined, two by two, the points of division of the circle AA' by lines parallel to BB' and then drew the tangents at A and A'. I then operated in the same way on the circle BB' by means of connecting lines parallel to AA'. The totality of these lines, prolonged up to their mutual meeting, formed a cross-ruled rectangle presenting a number of points of intersection equal to 17 x 17 or 289.

I had this whole operation accomplished by a professional engraver. We then drew on paper a number of trials of this cross-ruled rectangle. Each of these trial papers allowed us, as will be seen, to trace rapidly and accurately the various types of our figures.

Suppose that it is a matter of tracing the figure that corresponds to the relation two to three: two for the vertical tuning fork, three for the horizontal tuning fork. We know in advance the phase difference, call it 5/32 [of the period of the numerator; refer to the 3/16 picture in Row 4 of Plate II for a closely related example], and we know that the horizontal tuning fork leads the vertical.

We count, beginning from the origin O, 5 divisions on the radius OA [Fig. 13]. We label the point thus obtained [number zero]. This point belongs to the desired curve, for it corresponds to a vertical displacement of zero, and to a horizontal displacement . . . corresponding to 5/32 of a period [where 8/32 corresponds to a 90° lead angle and 5/32 is close to 60°].

Once this point is indicated, we obtain the next one by advancing 3 divisions horizontally and 2 vertically, and we arrive thus at the second point on the curve. . . . Continuing in the same manner, we will obtain new points on the curve. We will take care, every time we reach the boundaries (horizontal or vertical), to reverse our direction so as to complete the number of divisions we are required to traverse [for this increment, say 3].

We will continue thus until we have returned to the starting point. We will then join up all the various points by a continuous tracing that will be more or less exact according to whether the points are less spaced or more spaced apart. We will additionally increase the precision of the drawing by noting that the curve must be tangent to the 4 sides of the rectangle. [*Editor's Note:* The author notes that all the curves in Plate II were traced out in this manner. Then he says:] Thus the various figures corresponding to a given relation in numbers of vibrations are nothing other than the different appearances that would be presented by the same curve traced, not on a plane surface but in space on a [transparent] cylinder, if one made the cylinder turn around on its axis.[7] In this case the eye of the observer would be placed on a perpendicular to the axis situated in the plane of the mean circumference [located halfway up the cylinder].

Each figure, we have just proved, can be considered as the projection of a sinusoidal line that has been traced onto a [transparent] cylinder. . . .

The three-dimensional appearance of these curves can be presented to the eyes in a very simple manner with the aid of a stereoscopic viewer, as we have mentioned in [another paper]. . . . [*Editor's Note:* After a little more discussion of the behavior of two tuning forks, the author proceeds to a discussion of one tuning fork plus a more general vibrator.]

CHAPTER THREE

I. First Section: Generalization of the Method

The optical method of which we have availed ourselves first to study the motion of tuning forks can be generalized and applied to the study of every kind of vibratory motion. It is merely necessary to compound at right angles the motion of the body one wishes to study with that of another body, which serves as a standard.

The first condition to fulfill is not to alter the motion of the body one is studying when adding a mirror or any other optical apparatus. We have succeeded in this by the use of an apparatus to which we have given the name "vibration comparator."[8]

This apparatus, whose disposition can vary according to the applications one wishes to make, is composed essentially of a microscope whose ocular and objective are independently mounted [see Fig. 20] yet located with a spacing suitable for producing their usual effect. The ocular is mounted on a fixed support; the objective is attached to one of the branches of a tuning fork whose other branch carries a counterweight. The plane of the two branches of the tuning fork is perpendicular to the axis of the microscope. It results from this that every point located at a distance from the objective suitable for being seen clearly in the microscope furnishes at the focus of the ocular an image that oscillates rapidly as soon as the tuning fork is caused to vibrate; and so the spot is transformed into a straight line perpendicular to the branches of the tuning fork.

Let us now take the vibrating body we wish to study, and let us make one of its points much brighter than the others by some simple means, which can easily be imagined in each particular case and whose usage produces no perceptible change in the body. Let us place this body facing the microscope so that the point in question is clearly visible, and let us orient the two pieces of apparatus so that the oscillation of the point takes place in a plane parallel to the objective and perpendicular to the line along which the objective vibrates. Then the image of the point [see Fig. 25] seen through the microscope will be animated simultaneously by two oscillatory movements perpendicular to each other. The compounding of these two movements will furnish one of the luminous curves studied above and will allow us to understand precisely the motion by which the body is animated.

Since the tuning fork of the comparator can be tuned to any tone whatsoever, we can, for example, tune it by eye against a given standard [for example, D in Fig. 20]. It will suffice for this, to mark a dot on the end of one of the branches, using a sharp point. This dot, with the help of suitable illumination, will appear in the field of the microscope as a bright spot. We will then be able to substitute for the standard tuning fork the body we wish to study and see if it produces exactly the same number of vibrations. It is in this manner that we will succeed, without the help of the ear, in adjusting as many tuning forks as we like against a given standard, or equally, in verifying the differences that can exist if the tuning forks have been standardized by another method. We can also, by the same means, tune tuning forks to the third, the fifth, the octave, etc. of a given standard.

Let us note in passing that we can achieve great precision in establishing between the standard and the comparator a tiny difference, which can be estimated by the swinging of the curve.[9] It will be necessary, in order that the other tuning forks conform to the primary standard, that they produce in a given time exactly the same number of swingings of the curve.

Figure 20 shows the apparatus: O, ocular; a, objective; M, tuning fork of the comparator; D, tuning fork under study.

Our method allows equally well the study of the vibration of gases. It suffices for this effect to glue onto a membrane a little mirror of very light weight. As soon as the membrane vibrates, the mirror oscillates and communicates its motion to a luminous ray that is made to reflect from its surface. We have thus been able, in the amphitheater of the Sorbonne, to make visible by projection the oscillation of a membrane excited by a trumpet situated several meters away. [*Editor's Note:* The presentation would be similar to Figure 4 rather than Figure 3.]

II. Second Section: Diverse Applications

[*Editor's Note:* The author discusses diverse applications, accomplished or projected. Under the heading *Optical Study of Beats,* he repeats his belief that Savart has assigned more insistence on the mutual reaction of two bodies situated near each other than experiment bears out. This is a coupled-circuits problem, as commented on previously. Under the heading *Phase Stability of Bowed Tuning Forks,* the author anticipates Helmholtz in showing that two carefully bowed tuning forks preserve their freely oscillating phase relation. He summarizes: "The bow, suitably used, sustains the vibration without modifying its phase." The author then tells us, under the heading *And Other Things, Including Determination of Absolute Frequencies,* what a source of difficulty it is that the freely oscillating tuning forks "extinguish" so quickly and sometimes extinguish (or decay) at different rates, thus, for example, causing the rotation of an ellipse. He tells us of the blessing Foucault has brought with his electromagnetic excitation of tuning forks. He shows his own implementation (Fig. 24). He discusses excitation at a subharmonic, which would be similar to bowing a string with a a staccato excitation—that is, a periodic pulse excitation of a resonant system equivalent to a Class-C oscillator. The author states:]

It is especially for the case of two tuning forks intended for determining beats, that the prolongation of the vibrations is useful. . . . I have succeeded in prolonging indefinitely these vibrations by means of the ingenious mercury interrupter invented recently by Léon Foucault. . . . If this last condition is fulfilled [tuning the buzzer to a frequency two octaves below the tuning fork], the electromagnets of the tuning fork impress upon the two branches periodic kicks whose rhythm agrees with that of the tuning fork. At every four vibrations of the tuning fork, its motion is aided by the kick which the electromagnet gives it, and it begins to vibrate as if under the influence of a kind of electric bow whose action restores to it, periodically, the kinetic energy lost through communication of its vibratory motion to the surroundings. . . .[10]

[*Editor's Note:* The author then goes on to describe some beautiful experiments he did with a monochord, using the vibration microscope shown in Figure 25. Sometimes the harmonics are not in proper relation to the fundamental.] This defect in tuning between harmonic and fundamental furnishes a very sensitive means for discovering defects in the homogeneity of a string. . . . [*Editor's Note:* The author predicts that his method will be used in physiological optics as well as in electricity (to measure the duration of an electric spark). He continues:] But its greatest advantage, in my eyes, is to furnish to acousticians a means for checking, every time they think it necessary, the results obtained by the ear. . . . Thus this is perhaps the surest means of terminating experimentally the interminable argument between physicists and musicians on the subject of sharps and flats and intervals of the scale. [*Editor's Note:* This may refer to the nonlinear relation between pitch (subjective) and frequency (objective), which Lissajous was not aware of, as well as the problem of tuning a piano to the well-tempered scale, which calls for intervals equal to the twelfth root of 2.] Using apparatus whose vibratory motion would be measured optically, one could produce sounds whose estimation would be entrusted to musical ears; and then when the musicians would have chosen for the notes of the scale those sounds that seemed to them most fitting, the optical study of these sounds would establish in an incontestable way their numerical relations. One would thus give to each organ its natural share: to the ear, the care of estimating the correctness of sounds; to the eye, the care of counting exactly the number of their vibrations. [*Editor's Note:* The author then proceeds to the final chapter, which is really an appendix.]

CHAPTER FOUR

Calculation of the General Equation of the Curves Obtained in the Compounding of Perpendicular Vibratory Motions

[*Editor's Note:* The author derives equations for some of the curves shown in Plate II, including the ellipse in Row 1 and the parabola in Row 2. He concludes by saying that he had to work out at least a few such cases in order to add analytical verification to the results he had obtained first by purely geometric considerations, "although in fact analysis can add only numerical data to the general results already found by geometry alone."]

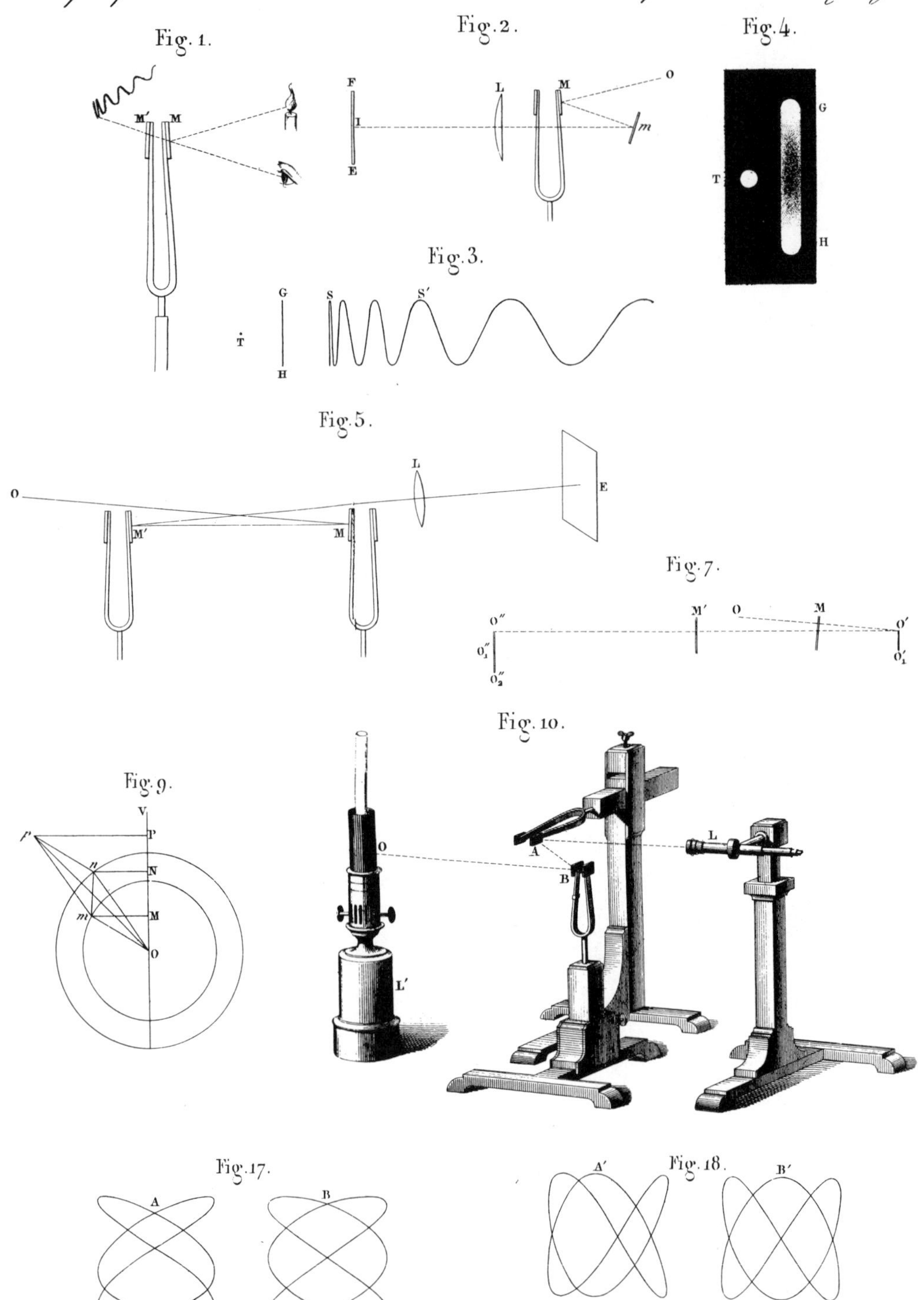
Etude optique des Mouvements vibratoires, par M. Lissajou
Fig. 1.
M′
M
Fig. 2.
F
I
E
L
M
O
m
Fig. 4.
G
T
H
Fig. 3.
G
T
H
S
S′
Fig. 5.
O
L
E
M′
M
Fig. 7.
O″
M′
O
M
O′
Fig. 10.
Fig. 9.
V
P
N
M
O
p
n
m
O
A
B
L
L′
Fig. 17.
A
B
Fig. 18.
A′
B′

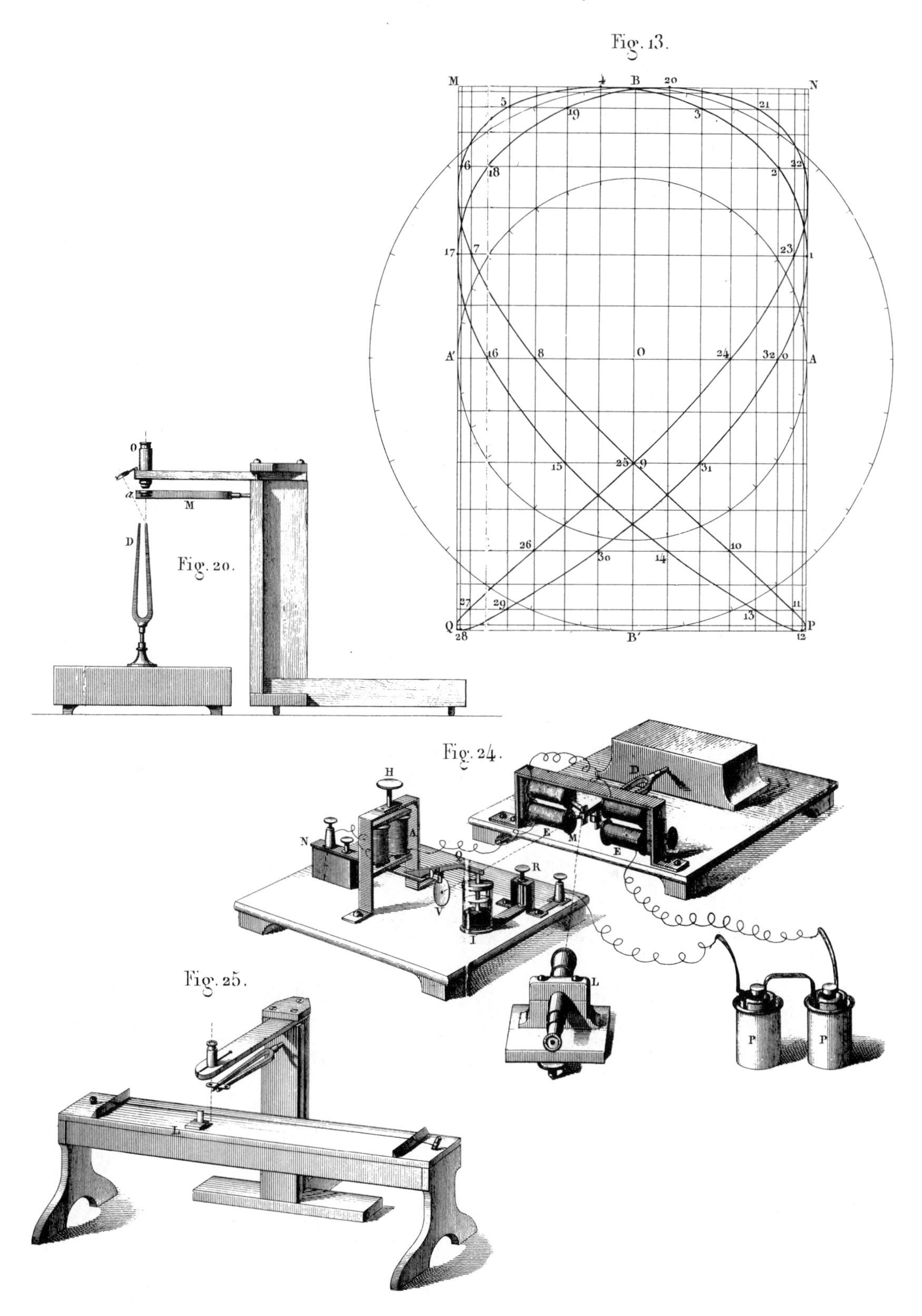

Fig. 13.

Fig. 20.

Fig. 24.

Fig. 25.

Etude optique des mouvements vibratoires; par M. Lissajous.

Courbes obtenues par la composition optique de deux mouvements vibratoires de directions rectangulaires. (*Fig.* 11.)

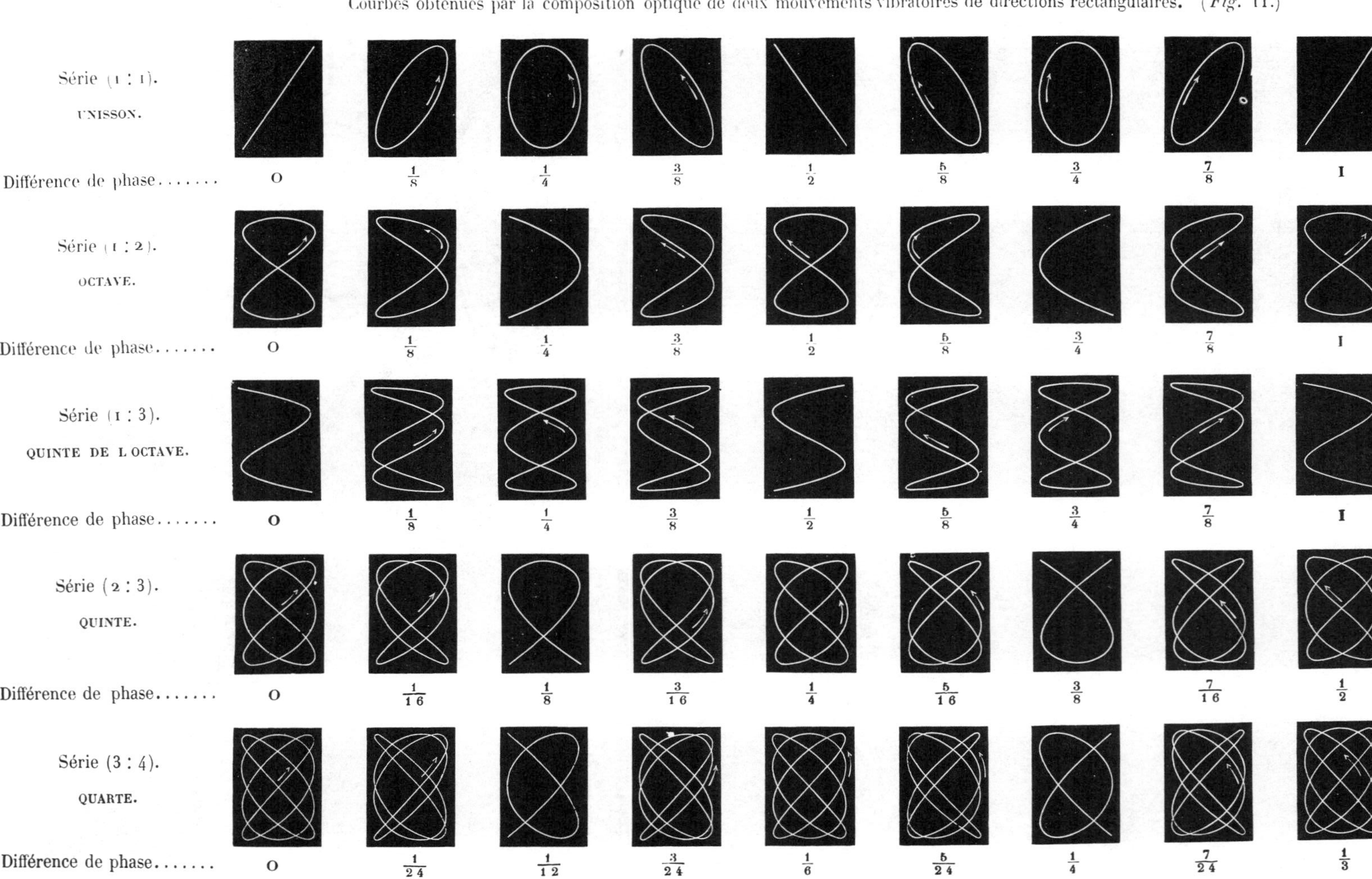

EDITOR'S NOTES

1. The sinusoidal line S–S′ of Figure 3 is distorted because it is associated with a tuning fork or mirror undergoing a constant rotational velocity instead of a constant translational velocity. In the projection case, the most compressed portion (near S) occurs when the light strikes the mirror head-on; and this expands (near S′) as the tuning fork rotates by, say, 45° because the projection is onto a plane. In direct viewing by the eye (Fig. 1), this distortion is the other way around: the most expanded portion occurs at the head-on position, whereas the most compressed portion occurs when the mirror rotates by, say, 45°, for the viewer imagines he is looking at a sine wave painted onto the inside of a cylinder. Hence, the sketch in Figure 1 is misleading.
2. This implies a sawtooth type of sweep (translation instead of rotation). This would indeed give an undistorted sine wave. Unfortunately, people in Lissajous's time were unable to generate a high-velocity, large-displacement mechanical sawtooth wave. We could do it today with a large-excursion loudspeaker.
3. Note that the author is still able to do his demonstrations before a large audience, but he has given up rotational velocity for this demonstration. Hence he can display only "a breathing line," as in Figure 4.
4. The audience must here imagine the decaying "beats curve" solely from the changing vertical line. [Excellent oscillograms of such a curve can be found in Morecroft, *Principles of Radio Communication*, 2d ed., 1927, Wiley, New York, pp. 279–281; and Chaffee, Amplitude Relations in Coupled Circuits, *Proc. I.R.E.* **4**:283 (1916).]
5. In Figure 5 the author has been assuming linear superposition of the two light-beam traces reflected from the two vibrating tuning forks. He justifies this linear behavior in Figure 7, where he says that for infinitesimal displacements, the off-center mirrors of Figure 5 are equivalent to the on-center mirrors of Figure 7. Note that a tiny vertical rotation of the first mirror M imparts an almost pure translation to the beam of light that hits it. Then the reflected beam hits the second mirror M′ which imparts a second pure translation to the beam. This behavior, which we summarize as $R \sin \Delta\theta \simeq R\Delta\theta$, is referred to by him as "the law of infinitely small vibrations of a pendulum."
6. What Lissajous has accomplished is the following. Since he could not generate the desired sawtooth displacement for his

time base, he gave up the explicit use of the time domain and instead made two synchronized tuning forks work together in a different way (see, for example, the frequency relation 3:4). This gave him an accurate "wave analyzer" or frequency analyzer. However, the seeds of an accurate oscilloscope are still present when he uses the relation 1:3 or 1:4, and so forth. Thus, if the reader will take one cycle of a sine wave tracing, mark it halfway along its time axis, and fold it back on itself at the halfway point, he will see that it forms a crude ellipse. This folding would occur in practice with a (synchronous) time-base generator that generates an isosceles triangle. If now we replace the isosceles triangle with its approximation, a sine wave (−90° to +90° to −90°), as the synchronous time-base generator, the two sharp cusps where the sine wave tracing has reversed, are smoothed out. Moreover, the synchronous time-base generator is already at hand: a second tuning fork. The smoothed-out figure is a true ellipse, as seen in Row 1 of Plate II. Then in Row 3 (rotate the sheet), second and third pictures, we clearly see a sine wave tracing, three cycles long, folded back on itself and compressed at the left and right edges. Helmholtz picked this up, as we shall see in the Helmholtz extract, and extrapolated it to an accurate oscilloscope.

7. This is discussed in more detail by Helmholtz, as we shall see.
8. This is a brilliant modification of the technique used in Figure 10. There, each fresh unknown tuning fork B would have to be temporarily equipped with mirrors, and among other things, this would destroy any chance for the mass-production adjusting of tuning forks. Instead, however, Lissajous now merely marks a bright dot on the top of one prong (of D, Fig. 20). Moreover, a moving lens is as good as a moving mirror to establish one axis (say, the abscissa). Helmholtz took up Lissajous's vibration comparator (giving full credit to Lissajous), improved it, and called it a "vibration microscope."
9. The sense of these swingings (or the apparent rotation of the figure), either clockwise or counterclockwise, can tell you if your unknown tuning fork is too sharp or too flat.
10. If Lissajous had been able to combine this electric-bow arrangement with his vibration comparator of Figure 10 or Figure 25, he would have had Helmholtz's vibration microscope in every detail. Probably he was unable to obtain small yet powerful electromagnets and so was forced to work with the complex shown in Figure 24; and this was too awkward to be combined with a microscope.

ANNOTATED BIBLIOGRAPHY

Cohen, I. B., 1946, Lissajous Figures, *Acoust. Soc. Am. J.* **17**:228–230.
Discusses W. C. Sabine's interest in the subject.

Hales, W. B., 1946, Recording Lissajous Figures, *Acoust. Soc. Am. J.* **16**: 137–146.
Good curves, good history.

Rapport sur un Mémoire de M. Lissajous, 1857, *Compt. Rend.* **45**:48–52.
This is an excellent summary of Lissajous's memoir, prepared for the French Academy of Sciences by a three-man commission that included Babinet. The report recommends that Lissajous continue to be funded, although his method "strikes one more by its oddness than by its importance."

Trendelenburg, F., 1950, *Einführung in die Akustik*, 2nd ed., Springer Verlag, Berlin, pp. 14–19.
Good discussion of Lissajous's work.

Editor's Comments on Paper 2

2 HELMHOLTZ
Excerpts from *On the Sensations of Tone*

Within three or four years after J. Lissajous published his memoir (Lissajous, 1857), Hermann L. F. Helmholtz picked up Lissajous's vibration comparator and developed it into his own vibration microscope. One of Helmholtz's major aims was to work out the details of how a bowed violin string appeared in both a time presentation and a space presentation. We quote from Rayleigh's *Theory of Sound* (Strutt, 1894, vol. 1, p. 208):

> 138. There is one problem relating to the vibrations of strings which we have not yet considered, but which is of some practical interest, namely, the character of the motion of a violin (or cello) string under the action of the bow. In this problem the *modus operandi* of the bow is not sufficiently understood to allow us to follow exclusively the *a priori* method: the indications of theory must be supplemented by special observation. By a dexterous combination of evidence drawn from both sources, Helmholtz has succeeded in determining the principal features of the case, but some of the details are still obscure.

Rayleigh later says: "The remainder of the evidence on which Helmholtz's theory rests was derived from direct observation with the vibration microscope."

As Paper 2, we present first Helmholtz's discussion of the bowed violin string, including his Appendix VI that concludes by showing what the space pattern (shape) of the bowed string must be (Fig. 63). His treatment of the plucked string (given more fully in Lindsay, 1973) follows because the plucked string sheds some needed light on the motion of the bowed string.

REFERENCES

Lindsay, R. B., 1973, *Acoustics: Historical and Philosophical Developments*, Dowden, Hutchinson & Ross, Stoudsburg, Pa., pp. 324–350.
Lissajous, J., 1857, Mémoire sur l'étude optique des mouvements vibratoires, *Ann. Chim. Phys.* ser. 3, **51**:147–242.
Strutt, J. W., (Lord Rayleigh). 1894, *Theory of Sound*, Macmillan, London, p. 208.

ANNOTATED BIBLIOGRAPHY

Benade, A. H., 1976, The Oscillations of a Bowed String, *Fundamentals of Musical Acoustics*, Oxford University Press, New York, pp. 505–526.
Excellent discussion and extension of Helmholtz's work.
Morse, P. M., 1940, Plucked Strings, *Vibration and Sound*, 2nd ed., McGraw-Hill, New York, pp. 78–81.
Good analysis of the time patterns as well as the space patterns.
Richardson, E. G., 1940, Plucked and Bowed Strings, *Sound*, 3rd ed., Arnold, London, pp. 85–98.
See Wood, below.
Strutt, J. W. (Lord Rayleigh), 1945, Transverse Vibrations of Strings *Theory of Sound*, vol. 1, Dover Publications, New York, pp. 184–191, 208–213, 230–231.
The standard English-language reference. Similar to Helmholtz's Appendix VI but more difficult.
Wood, A., 1940, Plucked and Bowed Strings, *Acoustics*, Blackie & Son, London, pp. 377–384.
Wood and Richardson together make Helmholtz's exposition even clearer.

2

Reprinted from pages 80–83, 384–387 and 53–54 of *On the Sensations of Tone,* 2nd ed., A. J. Ellis, trans., Longmans, Green, London, 1930, 575 p.

ON THE SENSATIONS OF TONE

H. L. F. Helmholtz

[*Editor's Note:* In the original, material precedes this excerpt.]

4. *Musical Tones of Bowed Instruments.*

No complete mechanical theory can yet be given for the motion of strings excited by the violin-bow, because the mode in which the bow affects the motion of the string is unknown. But by applying a peculiar method of observation, proposed in its essential features by the French physicist *Lissajous*, I have found it possible to observe the vibrational form of individual points in a violin string, and from this observed form, which is comparatively very simple, to calculate the ¶ whole motion of the string and the intensity of the upper partial tones.

Look through a hand magnifying glass consisting of a strong convex lens, at any small bright object, as a grain of starch reflecting a flame, and appearing as a fine point of light. Move the lens about while the point of light remains at rest, and the point itself will appear to move. In the apparatus I have employed, which is shown in fig. 22 opposite, this lens is fastened to the end of one prong of the tuning-fork G, and marked L. It is in fact a combination of two achromatic lenses, like those used for the object-glasses of microscopes. These two lenses may be used alone as a doublet, or be combined with others. When more magnifying power is required, we can introduce behind the metal plate A A, which carries the fork, the tube and eye-piece of a microscope, of which the doublet then forms the object-glass. This instrument may be called a *vibration microscope.* When it is so arranged that a fixed luminous point may be clearly seen through it, and the fork is set in vibration, the doublet L moves periodically up and down in pendular vibration. The observer, however, appears to see the luminous point itself vibrate, and, since the separate vibrations succeed each other so rapidly that the impression on the eye cannot die away during the time of a whole vibration, the path of the luminous point appears as a fixed straight line, increasing in length with the excursions of the fork.*

The grain of starch which reflects the light to be seen, is then fastened to the resonant body whose vibrations we intend to observe, in such a way that the grain moves backwards and forwards horizontally, while the doublet moves up and down vertically. When both motions take place at once, the observer sees the real horizontal motion of the luminous point combined with its apparent vertical motion, and the combination results in an apparent curvilinear motion. The field of vision in the microscope then shows an apparently steady and unchangeable bright

Fig. 22.

curve, when either the periodic times of the vibrations of the grain of starch and of the tuning-fork are exactly equal, or one is exactly two or three or four times as great as the other, because in this case the luminous point passes over exactly the same path every one or every two, three, or four vibrations. If these ratios of the vibrational numbers are not exactly perfect, the curves alter slowly, and the effect to the eye is as if they were drawn on the surface of a transparent cylinder which slowly revolved on its axis. This slow displacement of the apparent curves is not disadvantageous, as it allows the observer to see them in different positions. But if the ratio of the pitch numbers of the observed body and of the fork differs too

* The end of the other prong of the fork is thickened to counterbalance the weight of the doublet. The iron loop B which is clamped on to one prong serves to alter the pitch of the fork slightly; we flatten the pitch by moving the loop towards the end of the prong. E is an electro-magnet by which the fork is kept in constant uniform vibration on passing intermittent electrical currents through its wire coils, as will be described more in detail in Chapter VI.

much from one expressible by small whole numbers, the motion of the curve is too rapid for the eye to follow it, and all becomes confusion.

If the vibration microscope has to be used for observing the motion of a violin string, the luminous point must be attached to that string. This is done by first blackening the required spot on the string with ink, and when it is dry, rubbing it over with wax, and powdering this with starch so that a few grains remain sticking. The violin is then fixed with its strings in a vertical direction opposite the microscope, so that the luminous reflection from one of the grains of starch can be clearly seen. The bow is drawn across the strings in a direction parallel to the prongs of the fork. Every point in the string then moves horizontally, and on setting the fork in motion at the same time, the observer sees the peculiar vibrational curves already mentioned. For the purposes of observation I used the *a′* string, which I tuned a little higher, as *b′*♭, so that it was exactly two Octaves ¶ higher than the tuning fork of the microscope, which sounded *B*♭.

In fig. 23 are shown the resulting vibrational curves as seen in the vibration microscope. The straight horizontal lines in the figures, a to a, b to b, c to c

Fig. 23.

¶

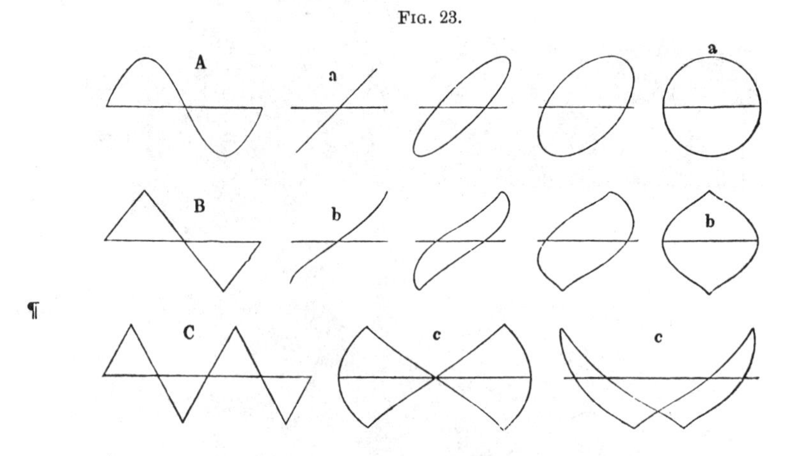

show the apparent path of the observed luminous point, before it had itself been set in vibration; the curves and zigzags in the same figures, show the apparent path of the luminous point when it also was made to move. By their side, in A, B, C, the same vibrational forms are exhibited according to the methods used in Chapters I. and II., the lengths of the horizontal line being directly proportional to the corresponding lengths of *time*, whereas in figures a to a, b to b, c to c, the horizontal lengths are proportional to the *excursions* of the vibrating microscope. ¶ A, and a to a, show the vibrational curves for a tuning-fork, that is for a simple pendular vibration; B and b to b those of the middle of a violin string in unison with the fork of the vibration microscope; C and c, c, those for a string which was tuned an Octave higher. We may imagine the figures a to a, b to b, and c to c, to be formed from the figures A, B, C, by supposing the surface on which these are drawn to be wrapped round a transparent cylinder whose circumference is of the same length as the horizontal line. The curve drawn upon the surface of the cylinder must then be observed from such a point, that the horizontal line which when wrapped round the cylinder forms a circle, appears perspectively as a single straight line. The vibrational curve A will then appear in the forms a to a, B in the forms b to b, C in the forms c to c. When the pitch of the two vibrating bodies is not in an exact harmonic ratio, this imaginary cylinder on which the vibrational curves are drawn, appears to revolve so that the forms a to a, &c., are assumed in succession.

It is now easy to rediscover the forms A, B, C, from the forms a to a, b to b, and c to c, and as the former give a more intelligible image of the motion of the string than the latter, the curves, which are seen as if they were traced on the surface of a cylinder, will be drawn as if their trace had been unrolled from the cylinder into a plane figure like A, B, C. The meaning of our vibrational curves will then precisely correspond to the similar curves in preceding chapters. When four vibrations of the violin string correspond to one vibration of the fork (as in our experiments, where the fork gave *B*♭ and the string *b*′♭, p. 82*a*), so that four waves seem to be traced on the surface of the imaginary cylinder, and when moreover they are made to rotate slowly and are thus viewed in different positions, it is not at all difficult to draw them from immediate inspection as if they had been rolled off on to a plane, for the middle jags have then nearly the same appearance on the cylinder as if they were traced on a plane.

The figures 23 B and 23 C (p. 82*b*), immediately give the vibrational forms for the middle of a violin string, when the bow bites well, and the prime tone of the ¶ string is fully and powerfully produced. It is easily seen that these vibrational forms are essentially different from that of a simple vibration (fig. 23, A). When the point is taken nearer the ends of the string the vibrational figure is shown in fig. 24, A, and the two sections $\alpha\beta$, $\beta\gamma$, of any wave, are to one another as the two sections of the string which lie on either side of the observed point. In the figure

Fig. 24.

A

α β γ

B

this ratio is 3 : 1, the point being at ¼ the length of the string from its extremity. Close to the end of the string the form is as in fig. 24, B. The short lengths of line in the figure have been made faint because the corresponding motion of the ¶ luminous point is so rapid that they often become invisible, and the thicker lengths are alone seen.*

These figures show that every point of the string between its two extremities vibrates with a constant velocity. For the middle point, the velocity of ascent is equal to that of descent. If the violin bow is used near the right end of the string descending, the velocity of descent on the right half of the string is less than that of ascent, and the more so the nearer to the end. On the left half of the string the converse takes place. At the place of bowing the velocity of descent appears to be equal to that of the violin bow. During the greater part of each vibration the string here clings to the bow, and is carried on by it; then it suddenly detaches itself and rebounds, whereupon it is seized by other points in the bow and again carried forward.†

* [Dr. Huggins, F.R.S., on experimenting, finds it probable that under the bow, the relative velocity of descent to that of the rebound of the string, or ascent, is influenced by the tension of the hairs of the bow.—*Translator.*]

† These facts suffice to determine the complete motion of bowed strings. See Appendix VI. A much simpler method of observing the vibrational form of a violin string has been given by Herr Clem. Neumann in the *Proceedings* (*Sitzungsberichte*) *of the I. and R. Academy* at Vienna, mathematical and physical class, vol. lxi. p. 89. He fastened bits of wire in the form of a comb to the bow itself. On looking through this grating at the string the observer sees a system of rectilinear zigzag lines. The conclusions as to the mode of motion of the string agree with those given above.

[*Editor's Note:* Material has been omitted at this point.]

APPENDIX VI.

ANALYSIS OF THE MOTION OF VIOLIN STRINGS.

(See p. 83*a*.)

ASSUME the lens of the vibration microscope to make horizontal vibrations, then vibrational curves will be observed like those represented in fig. 23, p. 82*b*, *c*. Call the vertical ordinate y and the horizontal x; then y is directly proportional to the displacement of the vibrating point, and x to that of the vibrating lens. The latter performs a simple pendular vibration. If the number of its vibrations be n ¶ and the time t, we have generally

$$x = A \,.\, \sin\,(2\pi nt + c)$$

where A and c are constant.

Now if y also makes n vibrations, x and y are both periodic and have the same periodic time. Hence, at the end of each period, x and y have the same values as before, and the observed point is at exactly the same place as at the beginning of the period. This holds for every point in the curve and for every fresh repetition of the vibratory motion, so that the curve appears stationary.

Suppose a vibrational form of the kind depicted in figs. 5, 6, 7, 8, 9, pp. 20 and 21, in which the horizontal abscissa is directly proportional to the time, to be wrapped round a cylinder, of which the circumference is equal to a single period of those curves, so that the time t is now to be measured along the circumference of the cylinder. Call x the distance of a point from a plane drawn through the axis of the cylinder. Then in this case also

¶

$$x = A \,.\, \sin\,(2\pi nt + c),$$

where $A \,.\, \sin c$ is the value of x for $t = 0$, and A is the radius of the cylinder. Hence, if the curve drawn upon the cylinder be viewed by an eye at an infinite distance in the line $x = 0$, $y = 0$, the curve has exactly the same appearance as in the vibration microscope.

If x and y have not exactly the same period; if, for example, y makes n vibrations while x makes $n + \Delta\, n$, where $\Delta\, n$ is a very small number, the expression for x may be written

$$x = A \,.\, \sin\,[2\pi nt + (c + 2\pi t\,\Delta\, n)].$$

In this case, then, c which was formerly constant, increases slowly. But c represents the angle between the plane $x = 0$ and the point in the drawing for which $t = 0$. In this case, then, the imaginary cylinder round which the drawing is supposed to be wrapped, revolves slowly.

Since a magnitude which is periodic after the period π, may be also considered as periodic after the periods 2_π, or 3_π, or ν_π, where ν is any whole number, these remarks apply also for the case when the period of y is an aliquot part of the period of x, or conversely, or both are aliquot parts of the same third period, that is, for the case when the tones of the tuning-fork and of the observed body stand in any consonant ratio; the only limitation is that the common period of vibration must not exceed the time required for a luminous impression to become extinct in the eye. [See Donkin's *Acoustics*, pp. 36-44.]

From the observed curves, fig. 23 B, C, p. 82*b*, and fig. 24 A, B, p. 83*b*, it follows that all points of the string ascend and descend alternately, that the ascent is made with a constant velocity, and also the descent with a constant velocity, which is however different from the velocity of ascent. When the bow is drawn across a node of one of the upper partials of the string, the motion takes place in all nodes of the same tone precisely in the manner described. For other points of the string, little crumples are perceptible in the vibrational figure, but they do not prevent us from clearly recognising the principal motion. ¶

FIG. 62.

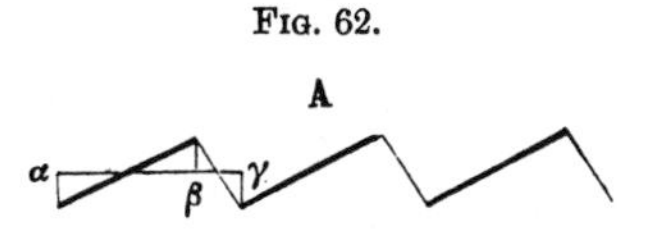

If in fig. 62 we reckon the time from the abscissa of the point α, so that for α, $t = 0$, and further for the point β put $t = \tau$, and for the point γ put $t = T$, so that the last represents the length of a whole period; then

$$\left.\begin{array}{l} y = ft + h, \text{ from } t = 0 \text{ to } t = \tau; \\ y = g\,(T - t) + h, \text{ from } t = \tau \text{ to } t = T, \end{array}\right\} \qquad (1)$$

whence for $t = \tau$, it results that

$$f\tau = g\,(T - \tau). \qquad ¶$$

Now suppose y to be developed in one of Fourier's series:

$$y = A_1 \,.\, \sin\frac{2\pi t}{T} + A_2 \,.\, \sin\frac{4\pi t}{T} + A_3 \,.\, \sin\frac{6\pi t}{T} \;\&c.$$

$$+ B_1 \,.\, \cos\frac{2\pi t}{T} + B_2 \,.\, \cos\frac{4\pi t}{T} + B_3 \,.\, \cos\frac{6\pi t}{T} \;\&c.$$

then it results from integration that

$$A_n \,.\, \int_0^T \sin^2\frac{2n\pi t}{T} \,.\, dt = \int_0^T y \,.\, \sin\frac{2n\pi t}{T} \,.\, dt$$

$$B_n \,.\, \int_0^T \cos^2\frac{2n\pi t}{T} \,.\, dt = \int_0^T y \,.\, \cos\frac{2n\pi t}{T} \,.\, dt, \qquad ¶$$

and this gives the following values for A_n and B_n:

$$A_n = \frac{(g + f) \,.\, T}{2n^2\pi^2} \,.\, \sin\frac{2n\pi t}{T}$$

$$B_n = -\frac{(g + f) \,.\, T}{2n^2\pi^2} \,.\, \left\{1 - \cos\frac{2n\pi t}{T}\right\}$$

and y may then be written in the form

$$y = \frac{(g + f) \,.\, T}{\pi^2} \,.\, \Sigma_{n=1}^{n=\infty} \left\{\frac{1}{n^2} \,.\, \sin\frac{\pi n\tau}{T} \,.\, \sin\frac{2\pi n}{T}\left(t - \frac{\tau}{2}\right)\right\} \qquad (2)$$

In equation (2), y denotes merely the distance of any determinate point of the string from its position of rest. If x denotes the distance of this point from the beginning of the string, and L the length of the string, then the general form of y, as in equation (1b) of App. III., p. 575b, is

$$y = \Sigma_{n=1}^{n=\infty} \left\{ C_n \cdot \sin \frac{n\pi x}{L} \cdot \sin \frac{2\pi n}{T} \left(t - \frac{\tau}{2} \right) \right\}$$
$$+ \Sigma_{n=1}^{n=\infty} \left\{ D_n \cdot \sin \frac{2\pi x}{L} \cdot \cos \frac{2\pi n}{T} \left(t - \frac{\tau}{2} \right) \right\} \ldots\ldots\ldots\ldots \quad (3)$$

By comparing equations (2) and (3) we find immediately that all D's vanish, or

$$D_n = 0, \qquad \text{and}$$
$$C_n \cdot \sin \frac{n\pi x}{L} = \frac{g + f}{\pi^2} \cdot \frac{T}{n^2} \cdot \sin \frac{n\pi\tau}{T} \ldots\ldots\ldots\ldots \quad (3a)$$

¶ Here $g + f$ and τ are independent of x, but not of n. On taking the equations for $n = 1$ and $n = 2$, and then dividing one by the other, there results

$$\frac{C_2}{C_1} \cdot \cos \frac{\pi x}{L} = \frac{1}{4} \cdot \cos \frac{\pi\tau}{T}.$$

From which it follows that for $x = \frac{L}{2}$, τ is also $= \frac{T}{2}$, as observation shews. But if $x = 0$, then according to observations τ is also $= 0$. Hence

$$C_2 = \frac{1}{4} \cdot C_1, \text{ and } \frac{x}{L} = \frac{\tau}{T} \ldots\ldots\ldots\ldots \quad (3b)$$

so that $g + f$ is independent of x. Let v be the amplitude of the vibration of the point x in the string, then

¶
$$f\tau = g(T - \tau) = 2v,$$
$$g + f = \frac{2v}{\tau} + \frac{2v}{T - \tau} = \frac{2vT}{\tau(T - \tau)} = \frac{2vL^2}{Tx(L - x)}.$$

And since $g + f$ is independent of x, we must have

$$v = 4V \cdot \frac{x(L - x)}{L^2}$$

where V is the amplitude in the middle of the string. From equation (3b) it follows that the sections $\alpha\beta$ and $\beta\gamma$ of the vibrational figure, fig. 62, p. 385b, must be proportional to the corresponding parts of the string on both sides of the observed point. Hence we have finally

¶
$$y = \frac{8V}{\pi^2} \cdot \Sigma_{n=1}^{n=\infty} \left\{ \frac{1}{n^2} \cdot \sin \frac{n\pi x}{L} \cdot \sin \frac{2\pi n}{T} \left(t - \frac{\tau}{2} \right) \right\} \ldots\ldots \quad (3c)$$

for the complete expression of the motion of the string.

If we put $t - \frac{\tau}{2} = 0$, y will $= 0$ for all values of x, and hence all parts of the string pass through their position of rest simultaneously. From that time the velocity f of the point x is

$$f = \frac{2v}{\tau} = \frac{8V(L - x)}{LT}.$$

But this velocity lasts only during the time τ, where $\tau = \frac{Tx}{L}$. Hence after the time t, and

$$\text{as long as } t < \frac{Tx}{L}, \text{ we have } y = ft = \frac{8V}{LT} \cdot (L - x)t \ldots\ldots\ldots \quad (4)$$
$$\text{and hence } y < \frac{8V}{L^2} \cdot x(L - x).$$

From that point y returns with the velocity

$$g = \frac{2v}{T - \tau} = \frac{8Vx}{LT}.$$

And hence after the time $t = \tau + t_1$,

$$y = \frac{8V}{L^2} \cdot x(L - x) - \frac{8Vx}{Lt} \cdot t_1.$$

And since $$L - x = \frac{T - \tau}{T} \cdot L$$

we find $$y = \frac{8Vx}{LT} \cdot \left\{ T - (\tau + t_1) \right\}$$

$$= \frac{8Vx}{LT} \cdot (T - t) \quad \text{.................................} \quad (4a)$$ ¶

The deflection on one part of the string is therefore given by the equation (4), and on the other part by the equation (4a). Both equations shew that the form of the string is a straight line, which in (4) passes through $x = L$, and in (4a) through $x = 0$. These are the two extremities of the string. The point where these straight lines intersect is given by the condition

$$y = \frac{8V}{LT} \cdot (L - x)\,t = \frac{8V}{LT} \cdot x\,(T - t).$$

Whence $$(L - x)\,t = x\,(T - t)$$
and $$Lt = xT.$$

Hence the abscissa x of the point of intersection increases in proportion to the time. This point of intersection, which is at the same time the point of the string most remote from the position of rest, passes, therefore, with a constant velocity from one end of the string to the other, and during this passage describes a ¶ parabolic arc, for which

$$y = v = \frac{8V}{L^2} \cdot x\,(L - x).$$

Hence the motion of the string may be briefly thus described. In fig. 63 the foot d of the ordinate of its highest point moves backwards and forwards with a constant velocity on the horizontal line ab,

Fig. 63.

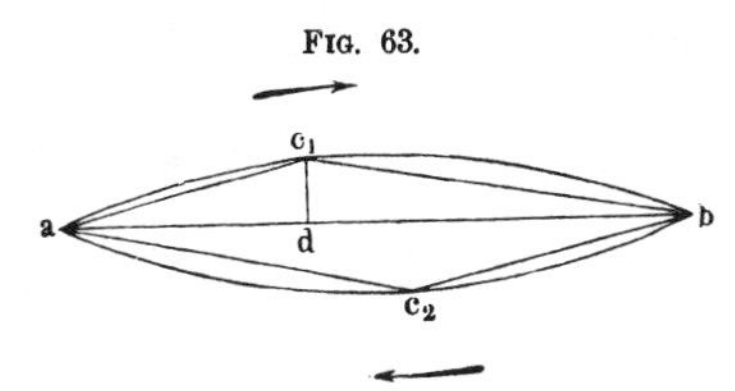

¶

while the highest point of the string describes in succession the two parabolic arcs ac_1b and bc_2a, and the string itself is always stretched in the two lines ac_1 and bc_1 or ac_2 and bc_2. [See Donkin's *Acoustics*, pp. 131-138.]

The small crumples of the vibrational figures which are so frequently observed, fig. 25, p. 84*b*, probably arise from the damping and disappearance of those tones which have nodes at the point bowed or in its immediate neighbourhood, and are consequently either unexcited or but slightly excited by the bow. When the bow is drawn across the string in a node of the *m*th partial tone situate near to the bridge, the vibrations of this *m*th, and further of the 2*m*th, 3*m*th, &c., tone have no influence on the motion of the point in the string touched by the bow, and they may consequently disappear, without changing the effect of the bow upon the string, and this really explains the crumples observed in the vibrational figure. [See also App. XX. sect. N. No. 5, on Prof. Mayer's Harmonic Curves.] I have not been able to determine by observation what happens when the bow is drawn across the string between two nodal points.

[*Editor's Note:* Material has been omitted at this point.]

The compound tone of a plucked string is also a remarkably striking example ¶ of the power of the ear to analyse into a long series of partial tones, a motion which the eye and the imagination are able to conceive in a much simpler manner. A string, which is pulled aside by a sharp point, or the finger nail, assumes the form, fig. 18, A (p. 54 *a*), before it is released. It then passes through the series of forms, fig. 18, B, C, D, E, F, till it reaches G, which is the inversion of A, and then returns, through the same, to A again. Hence it alternates between the forms A and G. All these forms, as is clear, are composed of three straight lines, and on expressing the velocity of the individual points of the strings by vibrational curves, these would have the same form. Now the string scarcely imparts any perceptible portion of its own motion directly to the air. Scarcely any audible tone results when both ends of a string are fastened to immovable supports, as metal bridges, which are again fastened to the walls of a room. The sound of the string reaches the air through that one of its extremities which rests upon a bridge standing on an elastic sounding board. Hence the sound of the string essentially depends on the motion of this extremity, through the pressure which it exerts on the sounding board. The magnitude of this pressure, as it alters periodically with the time, is shown in fig. 19, where the height of the line h h corresponds to the amount of pressure exerted on the bridge by that extremity of the string when the string is at rest. Along h h suppose lengths to be set off corresponding to consecutive intervals of time, the vertical ¶ heights of the broken line above or below h h represent the corresponding augmentations or diminutions of pressure at those times. The pressure of the string on the sounding board consequently alternates, as the figure shows, between a higher and a lower value. For some time the greater pressure remains unaltered ; then the lower suddenly ensues, and likewise remains for a time unaltered. The letters a to g in fig. 19 correspond to the times at which the string assumes the forms A to G in fig. 18. It is this alteration between a greater and a smaller pressure which produces the sound in the air. We cannot but feel astonished that a motion produced by means so simple and so easy to comprehend, ¶ should be analysed by the ear into such a complicated sum of simple tones. For the eye and the understanding the action of the string on the sounding board can be figured with extreme simplicity. What has the simple broken line of fig. 19 to do with wave-curves, which, in the course of one of their periods, show

FIG. 18.

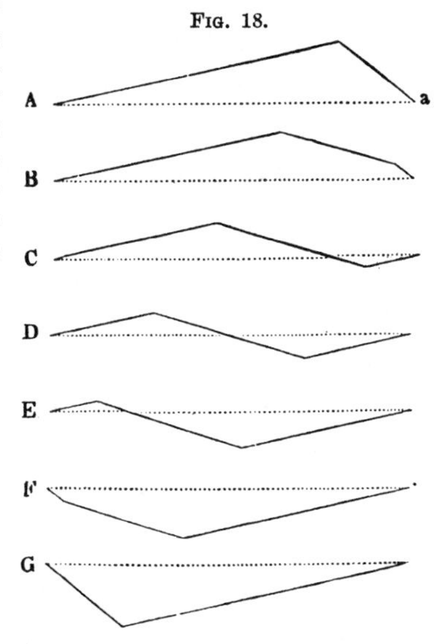

FIG. 19.

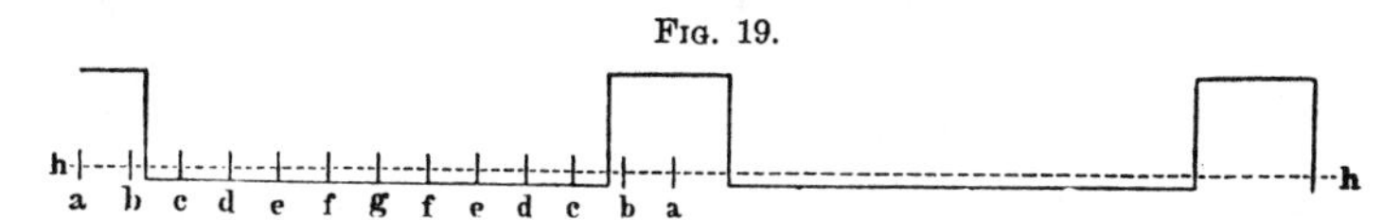

3, 4, 5, up to 16, and more, crests and troughs ? This is one of the most striking examples of the different ways in which eye and ear comprehend a periodic motion.

[*Editor's Note:* In the original, material follows this excerpt.]

EDITOR'S NOTES

Page 82: In Figure 23, the curves A, B, and C are calculated oscilloscope presentations using a linear time axis, whereas curves a, b, and c are experimental Lissajous-type presentations of the same motion. If we fold the two-cycle wave, C, back on itself at midpoint and then compress it at the left and right edges, we get the middle picture c. (See also Editor's Note 6 to Paper 1.) This is clearly similar to the first picture in Row 2 of Lissajous's Plate II (Paper 1).

Page 83, line 10: Since the tuning-fork time-base generator mainly compresses only the left and right edges, the more cycles we display, the better—that is, the center portion is almost undistorted. Thus Figure 23C (the middle picture) is almost sawtooth oscilloscopic and would be more so with four cycles displayed. Hence Helmholtz had no need to calculate the curves "rolled off onto a plane"; he could draw them directly.

Line 13: We will always interpret the phrase *vibrational form* as *oscilloscopic presentation* (to be distinguished later from space pattern or configuration). Note that the bowing point is always near the bridge, far from the middle of the string, but that the observation point can be at any point on the string, including the middle.

Line 17: *The point* means the observation point. Figure 63 shows that the shape or space pattern of the bowed string always traces the same figure as time progresses, no matter where the bow is applied. (This is quite otherwise for the plucked string.) Hence the ratios discussed with reference to Figure 24 are concerned entirely with the *observation point* in its time pattern presentation. Thus if we choose to make our observations at the middle of the string, the oscilloscopic form (time pattern) will be an isosceles triangle (as in Figure 23C) and by definition the middle marks off two equal lengths L/2. Likewise, in Figure 24B, if we choose our observation point close to the bridge, the time pattern will be a sawtooth triangle of time lengths 0.9τ and 0.1τ when the string lengths are 0.9L and 0.1L. However, the bowed *shape* itself will not have deviated from following its two parabolic arcs (Fig. 63)—that is, the string always forms a triangle. The apex of the triangle shoots back and forth like the shuttle of a loom; only instead of following, as a trajectory, a straight line, it follows two parabolic arcs. Note that if the bowing point is close to b, the highest displacement value (at the center) is greater than the displacement at the bowing point. This is quite otherwise for the plucked string.

Page 54: For the plucked string, the highest displacement value occurs at the plucking point (for example, just to the left of a in Fig. 18), and the trajectory of the apex is a parallelogram formed by using a common abscissa for Figures 18A and 18G. Only now we have *two* shuttles, shooting in opposite directions away from the apex and along the parallelogram. If we observe one point very near a, we will see a displacement time pattern the same as the force response shown in Figure 19, a periodic rectangle rich in even and odd harmonics. If we observe a point at the center, however, we will see a pattern containing only odd harmonics (for example, a periodic trapezoid approximating a square wave).

Last sentence: We might say that the eye perceives in the time domain [thus $f(t)$], whereas the ear perceives in the frequency domain [thus $F(\omega)$]. But either domain is merely the Fourier transform of the other [$f''(t) \leftrightarrow (j\omega)^2 F(\omega)$].

By means of the vibration microscope, Helmholtz could directly observe the time pattern of displacement at different observation points. Some of these time patterns are shown in Figure 24, and a force response versus time, proportional to a displacement response, is shown in Figure 19. Helmholtz was then able to work backward and confirm the space patterns he had calculated and guessed at, as shown in Figure 18 and Figure 63.

If we pluck near a (Fig. 18) but observe the displacement at the middle of the string, all the even harmonics will have apparently dropped out, and we will see as our time pattern an approximation to a periodic square wave. However, of course we still *hear* the even as well as odd harmonics because we hear what is present at the bridge, which is at point a. What we hear is a function of the force response shown in Figure 19.

If now we change our plucking point—for example, plucking the middle of the string—our envelope parallelogram (formed by using a common abscissa for Figs. 18A and G) becomes a symmetrical rhombus. All the even harmonics are now *everywhere* forbidden. Thus, even if our observation point is away from the middle of the string, we see a time pattern having only the odd harmonics 1, 3, 5, 7, and so on.

If now we were to depict seven snapshots of the bowed string (as we have depicted seven snapshots of the plucked string) by expanding on Figure 63, and take our bowing point about ¼″ to the left of b, the first snapshot would show a triangle with apex close to a. The second snapshot would show the apex moved to c_1. The displacement at the bowing point near b meanwhile is growing larger. The apex moves along the envelope parabola coming closer to the bowing point, whose displacement increases until it meets the apex right on the parabola. The apex then suddenly swings around b, dragging the string below the *x*-axis to a negative value (following the parabolic arcs to get there).

All this time the bow has been steadily sawing away, at a point ¼″ away from b (referring to Fig. 63). If we choose as an observation point the middle of the string and observe the time pattern, we will see a simple isosceles triangle, with no even harmonics, as shown in Figures 23B and C. This apparent suppression of the even harmonics also occurred in plucking a string off-center and observing at the midpoint.

Rayleigh and Helmholtz both derived identical equations for the motion of the bowed string and again for the plucked string. Helmholtz's presentations, however, are in this case more helpful to the reader than Rayleigh's.

Editor's Comments on Papers 3 Through 11

From about 1890 to about 1922, Arthur G. Webster of Clark University was continually working to improve the absolute measurement of sound by means of his nonelectroacoustic wave analyzers. Concurrently, Dayton C. Miller of the Case School of Applied Science (now Case Western Reserve University), from about 1909 to about 1916, was working to improve his broadband oscilloscope, another nonelectroacoustic instrument (H. B. Miller, 1977). Somewhat in parallel with these two, in England at the Signals Experimental Establishment (Woolwich, England), E. T. Paris also was working on his own nonelectroacoustic wave analyzers.

We present D. C. Miller's work first. Paper 3 consists of extracts from his 1916 book, *The Science of Musical Sounds,* which was adapted from a series of public lectures, and Paper 4 is excerpted from his 1937 book, *Sound Waves: Their Shape and Speed.* Miller's photographs of the *time-domain* response of musical instruments and of vowel sounds are famous and were an important factor in introducing many young scientists into the serious study of acoustics. We also present a one-page report, Paper 5, that Miller did jointly with Webster for the National Research Council.

We follow Miller with his friendly competitor, A. G. Webster, (Papers 6-9). Webster was primarily interested in the *frequency domain*. He was not only a first-class (and famous) mathematical physicist but a first-rate instrument designer as well. He was also something of a pixie, as can be seen from his informal remarks during meetings of learned societies. For example, he kept promising to write up this oral paper or that oral paper, but he rarely did so.

Webster's contribution to this volume comprises a group of papers: two complete papers and three extracts. We present first Paper 6, the 1922 *Nature* paper, then Papers 7A and 7B, the 1919 AIEE paper with a bit of the discussion period. These three papers together more than cover the material give in the much referred to paper (Webster, 1919) in the *Proceedings of the National Academy of Sciences* and do it better.

We follow with Paper 8, an extract from a 1920 National Academy of Sciences paper, wherein Webster showed some experimental results obtained with his phonometer. We complete our Webster sampling with Paper 9, written with P. Sabine, an extract from the famous November 1922 *Bulletin of the National Research Council.*

We proceed with Paper 10, one of E. T. Paris's various papers on the theory of the double resonator. Paris routinely used such a resonator in conjunction with the hot-wire microphone of W. S. Tucker in order to achieve the detection and measurement of very faint sounds. Paris's paper is of special interest in this book because of the clear light he sheds (indirectly) on Webster's wave analyzer, helping us to see the design criteria probably used by Webster in designing his phonometer. After a study of Paper 10, the progenitor of this paper (Paris, 1922) can be read with more ease. See especially the last three pages. Further light is shed by Paris's (1926) paper on the theory of the Boys double resonator.

We conclude with a brief paper by the volume editor (Paper 11). Webster's work is so ingenious, and yet so scantily explained by Webster, that the editor felt it desirable to present a brief analysis in the format of a paper, using electrical-equivalent circuits and steady-state excitation.

REFERENCES

Miller, H. B., 1977, Acoustical Measurements and Instrumentation, *Acoust. Soc. Am. J.* **61**:274–282.

Paris, E. T., 1922, On Doubly-Resonated Hot-Wire Microphones, *R. Soc. London Proc.* A**101**:391–410.

Paris, E. T., 1926, The Theory of the Boys Type of Double Resonator, *Philos. Mag.* **7**:751–769.

Webster, A. G., 1919, A Complete Apparatus for Absolute Acoustical Measurements, *Nat. Acad. Sci. Proc.* **5**:173–179.

ANNOTATED BIBLIOGRAPHY

Benade, A. H., 1976, The Brass Wind Instruments, *Fundamentals of Musical Acoustics*, Oxford University Press, New York, pp. 391–429.
A fine extension of Helmholtz's writings on the subject.

Boys, C. V., 1890,The Double Resonator, *Nature* **42**:604.
A little too brief explanation of his discovery.

Crandall, I. B., 1926, Resistance Coefficients for Conduits, *Theory of Vibrating Systems and Sound*, Van Nostrand, New York, pp. 229–236.
Good derivation of the resistance of an orifice.

Helmholtz, H. L. F., 1945, Theory of Pipes (Appendix VII), *Sensations of Tone*, 2nd ed., Dover Publications, New York, pp. 388–397.
Good discussion of conical versus cylindrical wind instruments.

Jeans, J., 1937, Orchestral Wind Instruments, *Science and Music*, Macmillan, New York, pp. 147–151.
Good summary of some of D. C. Miller's work.

Morse, P. M., 1948, Propagation of Sound in Horns, *Vibration and Sound*, 2nd ed., McGraw-Hill, New York, pp. 283–287.
Good discussion of resonant horns and pipes.

Olson, H. F., 1947, Acoustic Resistance, *Elements of Acoustical Engineering*, 2nd ed., Van Nostrand, New York, pp. 86–89.
Good collection of formulas.
Stewart, G. W., 1920, *Phys. Rev.* **16**:313.
Discussion of experiments on conical horns.
Stewart, G. W., and R. B. Lindsay, 1930, Distributed Acoustic Impedance, *Acoustics*, Van Nostrand, New York, pp. 132–149.
Good discussion of resonant horns.

3

Reprinted from pages 78, 79, 81, 85, 87–88, 142–143, 147–149, and 156–164 of *The Science of Musical Sounds,* Macmillan, New York, 1916, 286 p.

THE SCIENCE OF MUSICAL SOUNDS

D. C. Miller

[*Editor's Note:* In the original, material precedes these excerpts. We begin in the middle of Lecture III, from which Figures 67 and 68 have been omitted. They show photographs of a phonodeik used for photographing oscillograms. The details involved in this photography have also been omitted.]

THE PHONODEIK

For the investigation of certain tone qualities referred to in Lecture VI, the author required records of sound waves showing greater detail than had heretofore been obtained. The result of many experiments was the development of an instrument which photographically records sound waves, and which in a modified form may be used to project such waves on a screen for public demonstration; this instrument has been named the "Phonodeik," meaning to show or exhibit sound.[42]

The sensitive receiver of the phonodeik is a diaphragm, *d*, Fig. 66, of thin glass placed at the end of a resonator horn *h*; behind the diaphragm is a minute steel spindle mounted in jeweled bearings, to which is attached a tiny mirror *m*; one part of the spindle is fashioned into a small

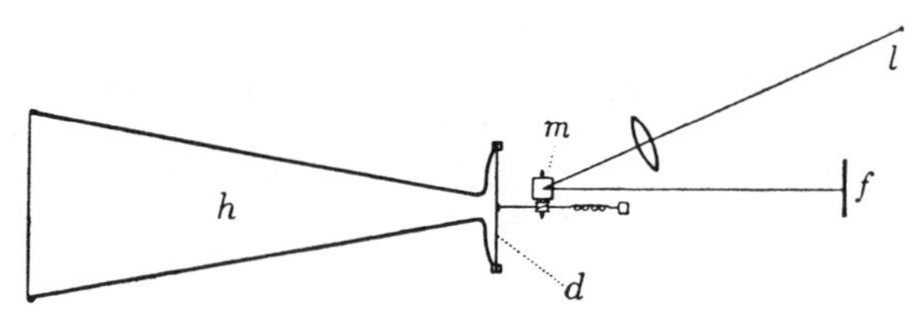

FIG. 66. Principle of the phonodeik.

pulley; a few silk fibers, or a platinum wire 0.0005 inch in diameter, is attached to the center of the diaphragm and being wrapped once around the pulley is fastened to a spring tension piece; light from a pinhole *l* is focused by a lens and reflected by the mirror to a moving film *f* in a special camera. If the diaphragm moves under the action of a sound wave, the mirror is rotated by an amount proportional to the motion, and the spot of light traces the record of the sound wave on the film, in the manner of the pendulum shown in Fig. 11, page 12.

In the instrument made for photography, Fig. 67, the usual displacement of the diaphragm for sounds of ordinary loudness is about half a thousandth of an inch, resulting in an extreme motion of one thousandth of an inch, which is magnified 2500 times on the photograph by the mirror and light ray, giving a record $2\frac{1}{2}$ inches wide; the film commonly employed is 5 inches wide, and the record is sometimes wider than this. The extreme movement of the diaphragm of a thousandth of an inch must include all the small variations of motion corresponding to the fine details of wave form which represent musical quality. Many of the smaller kinks shown in the photographs, such as Figs. 110 and 169, are produced by component motions of the diaphragm of less than one hundred-thousandth of an inch; the phonodeik must faithfully reproduce not only the larger and slower components, but also these minute vibrations which have a frequency of perhaps several thousand per second.

The fulfillment of these requirements necessitates unusual mechanical delicacy; the glass diaphragm is 0.003 inch thick, and is held lightly between soft rubber rings, which must make an air-tight joint with the sound box; the steel staff is designed to have a minimum of inertia, its mass is less than 0.002 gram (less than $\frac{1}{32}$ grain); the small mirror, about 1 millimeter (0.04 inch) square, is held in the axis of rotation; the pivots must fit the jeweled bearings more accurately than those of a watch; there must be no lost motion, as this would produce kinks in the wave, which when magnified would be perceptible in the photograph; there must be no friction in the bearings.

The phonodeik responds to 10,000 complete vibrations (20,000 movements) per second, though in the analytical work so far undertaken it has not been found necessary to investigate frequencies above 5000.

[*Editor's Note:* Material, including Figure 69 showing a photograph of a "demonstration phonodeik," has been omitted at this point.]

THE DEMONSTRATION PHONODEIK

The vibrator of the phonodeik employed in research is very minute and delicate, and its small mirror reflects too little light to make the waves visible to a large audience. For purposes of demonstration, a phonodeik has been especially constructed, Fig. 69, which will clearly exhibit the principal features of "living" sound waves.[43] The sound from a voice or an instrument is produced in front of the horn; the movements of the diaphragm with its vibrating mirror cause a vertical line of light which, falling upon a motor-driven revolving mirror, is thrown to the screen in the form of a long wave; the movements of the diaphragm are magnified 40,000 times or more, producing a wave which may be 10 feet wide and 40 feet long.

With this phonodeik a number of experiments may be made in further explanation of the principles of simple harmonic motion and wave forms. When the revolving mirror is kept stationary, the spot of light on the screen moves in a vertical line as the diaphragm vibrates; though these movements are superposed, their extreme complexity is shown since the turning points are made evident by bright spots of light. If the mirror is slowly turned by hand, the production of the harmonic curve by the combination of vibratory and translatory motions is demonstrated. With a tuning fork the simplicity of the sine curve is exhibited; with two tuning forks the combination of sine curves is shown; the imperfect tuning of two forks is demonstrated by a slowly changing wave form; the relations of loudness to amplitude and of pitch to wave length may be illustrated.

The projection phonodeik is especially suitable for exhibiting the characteristics of sounds from various sources; as seen on the screen the sound waves are constantly in motion, changing shape and size with the slightest alteration in frequency, loudness, or quality of the source.

[*Editor's Note:* Material has been omitted at this point.]

Determination of Pitch with the Phonodeik

The photographs obtained with the phonodeik permit a very convenient and accurate determination of pitch; the time signals are given by a standard tuning fork, recording one hundred flashes per second; it is only necessary to compare the wave length and the time intervals to obtain the frequency. Various photographs, as Fig. 96, show the time signals.

A standard clock with a break-circuit attachment may be made to record signals simultaneously with the sound waves; by counting and measuring, the number of waves per second may be determined with precision. When two sounds are being compared by the method of beats, the exact number (including fractions) of beats per second may be determined by photographing the beats together with the time signals.

The phonodeik permits accurate tuning of all the harmonic ratios; if the spot of light is observed without the revolving mirror, its movements take place in a straight line; two tones sounding simultaneously give a composite wave form, the turning points of which are visible as circles of extra brightness on the line, like beads on a string. When the ratio of the component tones is inexact, there is a constant change of wave form which causes the beads to creep along the line; when the ratio is exact, the wave form is constant and the beads are stationary, signifying perfect tuning.

[*Editor's Note:* The preceding excerpt ends at the section "Photographs of Compression Waves," which is a commentary on work by Toepler, Sabine, and others. This subject is treated at length in Part III of this Benchmark volume. We continue with "Lecture V," which discusses the responses of horns and diaphragms and is itself a course in the design problems of those transducers that deliberately make use of multiresonances. (Wente circumvented the entire problem by placing the first resonance of his diaphragm-plus-cavity system far above his working frequency band. See Part II.)]

LECTURE V

INFLUENCE OF HORN AND DIAPHRAGM ON SOUND WAVES, CORRECTING AND INTERPRETING SOUND ANALYSES

Errors in Sound Records

The photographs and analyses of sound waves obtained by the complicated mechanical and numerical processes described in Lectures III and IV, are, unfortunately, not yet in suitable form for determining the tone characteristics of the sounds which they represent. Before these analyses can furnish accurate information they must be corrected for the effects of the horn and the diaphragm of the recording instrument, a correction involving fully as much labor as was expended on the original work of photography and analysis. For the sake of greater emphasis, it may be directly stated that the neglect of the corrections for horn and diaphragm often leads to wholly false conclusions regarding the characteristics of sounds, since horns and diaphragms of different types give widely differing curves for precisely the same sound.

For research upon complex sound waves, a recording instrument using a diaphragm should possess the following characteristics: (*a*) the diaphragm as actually mounted should respond to all the frequencies of tone being investigated; (*b*) it should respond to any combination of simple frequencies; (*c*) it should not introduce any fictitious frequencies; (*d*) the recording attachment should faithfully transmit the movements of the diaphragm; and (*e*) there must be a detérminate, though not necessarily simple, relation between the response to a sound of any pitch and the loudness of that sound.

It is well known that the response of a diaphragm to waves of various frequencies is not proportional to the amplitude of the wave; the diaphragm has its own natural periods of vibration, and its response to impressed waves of frequencies near its own is exaggerated in degrees depending upon the damping. The resonating horn also greatly modifies waves passing through it. Therefore, it follows that the resultant motion of the diaphragm is quite different from that of the original sound wave in the open air.

The theory of these disturbances for simple cases is complete, but what actually happens in a given practical apparatus is made indeterminate by conditions which are complicated and frequently unknown. There being no available solution of this problem, it was necessary to make an experimental study of these effects as they occur in the phonodeik.[66]

It has been proved that the phonodeik possesses several of the characteristics mentioned; (*a*) it has been shown by actual trial that it responds to all frequencies to 12,400; (*b*) various combinations of simple tones up to ten in number have been actually produced with tuning forks, and the photographic records have been analyzed; (*c*) in each case the analysis shows the presence of all tones used, and no others. We have then only to determine the accuracy with which the response represents the original tones, the qualifications (*d*) and (*e*) mentioned above; this requires the investigation of all the factors of resonance, interference, and damping.

[*Editor's Note:* Material has been omitted at this point.]

The irregular curve of Fig. 116 is the response obtained with one of the earliest forms of phonodeik; it shows an almost startling departure from the ideal response represented by the smooth curve. What produces the range of mountains, with sharp peaks and valleys? Why is there no response for the frequency 1460; why is it excessive for frequencies from 2000 to 3000? There were five suspected causes: (1) unequal loudness of the pipes; (2) the diaphragm effects; (3) the mounting and housing of the diaphragm; (4) the vibrator attached to the diaphragm; (5) the horn.

The investigation of these peculiarities was most tantalizing; the peaks acted like imps, jumping about from place to place with every attempt to catch them, and chasing and pushing one another in a very exasperating manner. Perhaps two months' continuous search was required to find the causes of "1460" and "2190" alone. The investigations led to many improvements in the phonodeik and to

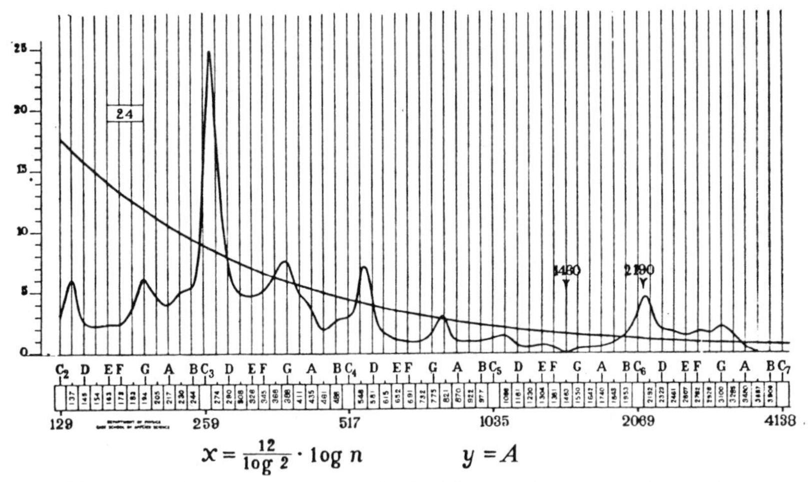

Fig. 116. A response curve obtained with an early form of phonodeik.

a practical method of correction for the departures from ideal response.

Response of the Diaphragm

Experiments have been made with diaphragms of several sizes and thicknesses, and of various materials, such as iron, copper, glass, mica, paper, and albumen. The experiments described in this section concern circular glass diaphragms having a thickness of 0.08 millimeter, and held around the circumference, either firmly clamped between hard cardboard gaskets and steel rings, or loosely clamped between soft rubber gaskets; the diaphragm is entirely free, there being no horn or housing of any kind. The silk fiber of the phonodeik vibrator is attached to the diaphragm to record its movements.

[*Editor's Note:* Material has been omitted at this point.]

Influence of the Horn

A horn as used with instruments for recording and reproducing sound is usually a conical or pyramidal tube, the smaller end of which is attached to the soundbox containing the diaphragm, while the larger end opens to the free air. The effect of the horn is to reinforce the vibrations which enter it due to the resonance properties of the body of air inclosed by the horn. The quantity and quality of resonance depends mainly upon the volume of the inclosed air and somewhat upon its shape. If the walls of the horn are smooth and rigid, they produce no appreciable effect upon the tone. But if the walls are rough or flexible, they may absorb or rapidly dissipate the energy of vibrations of

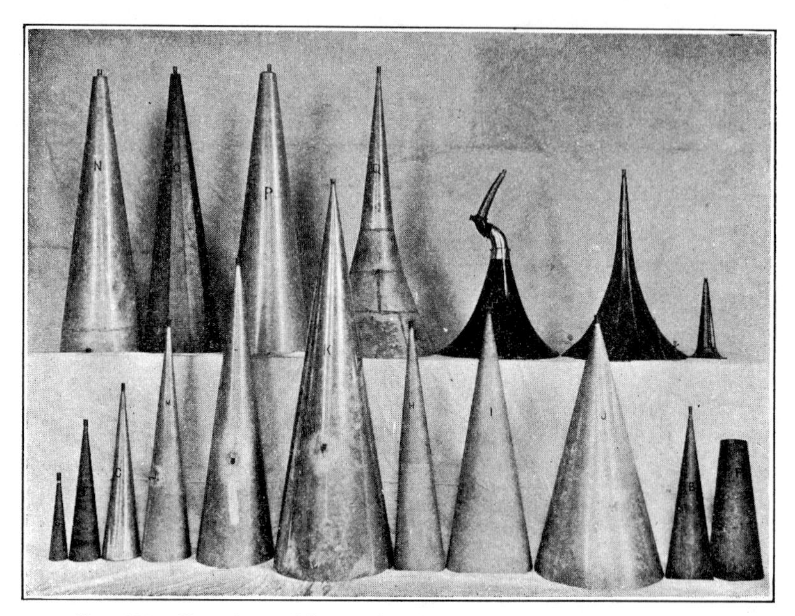

FIG. 122. Experimental horns of various materials, sizes, and shapes.

the air of certain frequencies and thus by subtraction have an influence upon tone quality. The horn of itself cannot originate any component tone, and hence cannot add anything to the composition of the sound. The horn is an air resonator and not a soundboard; any vibrations which the walls of the horn may have are relatively feeble and are received from the air which is already in vibration, while in the case of a soundboard, the air receives its vibration from the soundboard as a source. Because the horn operates through the inclosed air, it is a very sensitive resonator, and hence its usefulness when its action is understood and properly applied.

A horn used in connection with the diaphragm very greatly increases the response, but it also adds its own natural-period effects, which are quite complex. A variety of horns, shown in Fig. 122, were used in the experiments; these are of various materials, sheet zinc, copper, thick and thin wood, and artificial stone; one horn was made with double walls of thin metal, and the space between was filled with water. Probably the most rigid material, such as stone or thick metal, gives the best results. For convenience, however, sheet zinc is used with the phonodeik, and so long as the horn is supported under constant conditions, which are involved in the "correction curve" described later, this material is satisfactory.

A horn such as is used in these experiments has its own natural tones, which can be brought out by blowing with a mouthpiece as in a bugle; these tones are a fundamental with its complete series of overtones. The fundamental pitch can be heard by tapping the small opening with the palm of the hand.

When a horn is added to the diaphragm, the response is greatly altered; Fig. 123 shows the response curves for three horns of different lengths. In each curve the peaks corresponding to the fundamental of the horn, the octave, and the other overtones up to the seventh, are distinct. These peaks are indicated in the figure by H_1, H_2, etc., while the diaphragm peak is marked D.

In the upper curve, the peak due to the diaphragm comes between the peaks for the fundamental and the octave of the horn; the latter peaks are pushed apart, one being lowered in pitch and the other raised, so that the interval between them is two semitones more than an octave; the peak for the third partial is in its proper position. The horn reacts upon the diaphragm, causing it to have a period different from that which it had before the horn was applied.

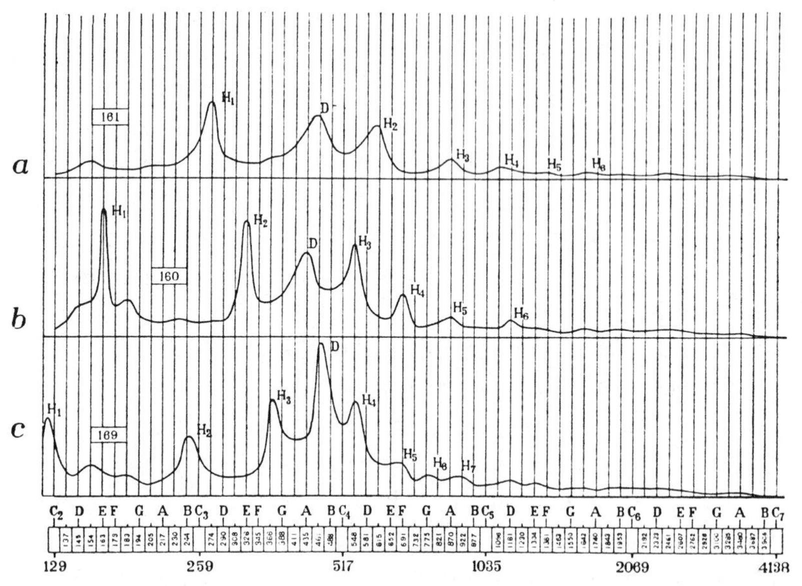

FIG. 123. Resonance peaks for horns of various lengths.

The middle curve shows the response with a longer horn; the diaphragm peak now comes between the peaks of the second and third partials, and both these and that of the fourth are displaced. The lower curve represents the response for a still longer horn.

A long horn seems to respond nearly as well to high tones as does a short one, while the response to low tones is much greater; the response below the fundamental of the horn is very feeble. The horn selected should be of such a length that its fundamental is lower than the lowest tone under investigation. For the study of vowel sounds, the horn employed has a length of about 48 inches, giving a fundamental frequency of about 125.

It is important that there are no holes, open joints, or leaks of any kind in the walls of the horn, because tones with a node in the position of a hole will be absent. A hole

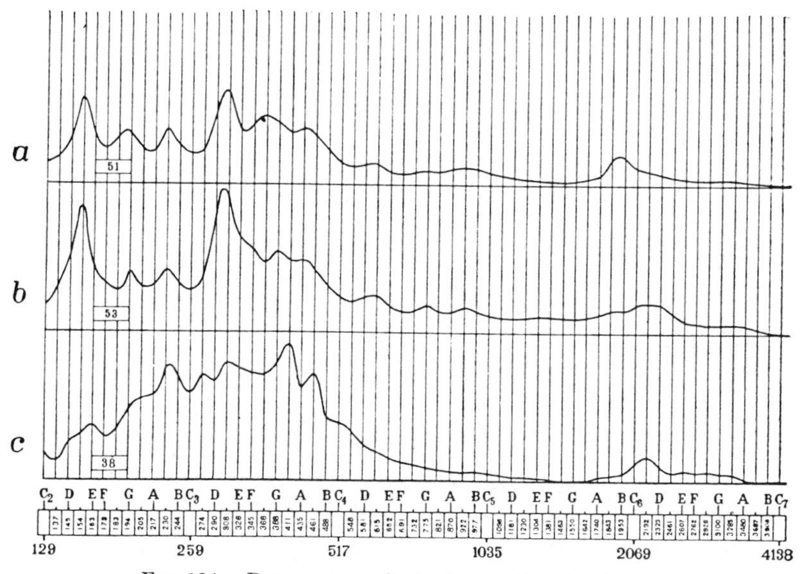

FIG. 124. Resonance peaks for horns of various flares.

one millimeter in diameter is sufficient to alter the response.

The flare of the horn has a great influence upon the response; Fig. 124 shows the responses obtained with three horns of the same length, but of different flares. The upper curve is the response for a narrow conical horn, the large end of which has a diameter equal to one fifth of the length; the middle curve is for a wide cone, the diameter of the open end being one half of the length; the lower curve is for a horn of flaring, bell shape. Widening the mouth increases the effect in a general way; the bell flare makes the natural periods indefinite, and heaps up the response near the fundamental, diminishing that for the higher tones.

The shape selected for use with the phonodeik is a cone

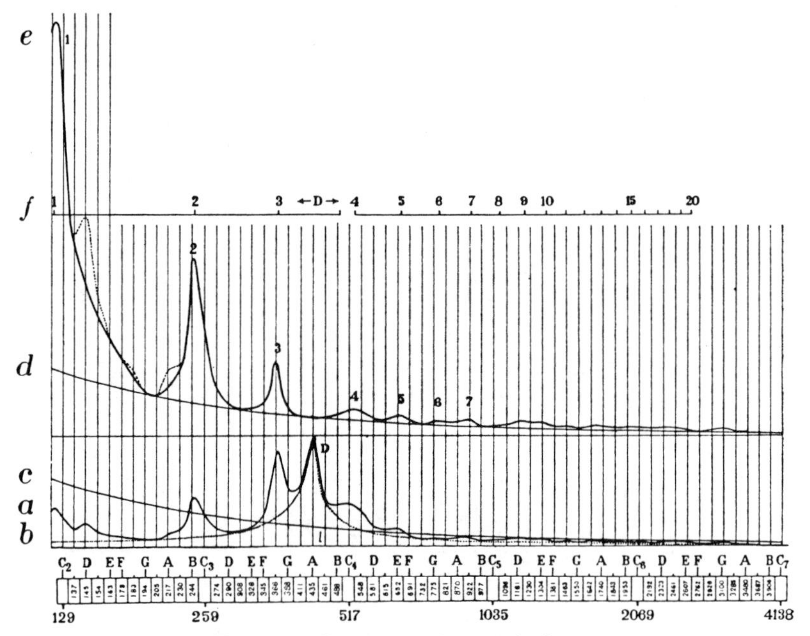

FIG. 125. Resonance effect of the horn.

of medium flare; this gives a good distribution of response, and the resonances are definite, but not too sharp to allow of correction.

The curve *a*, Fig. 125, is an actual response curve containing both horn and diaphragm effects, *b* is the diaphragm response, while *c* and *d* are the ideal curves previously explained. The effect of the natural period of the diaphragm is represented by the sharp peak *D*. If the diaphragm responded ideally, the curve *b* would coincide with *c* throughout its length, and *a* would then indicate the effects due to the horn alone. The ordinates of the curve *a* have been multiplied by such factors as will reduce the diaphragm curve to the ideal, and the results are plotted in *e*. The difference between the curve *e* and the ideal *d* is the effect due to the horn, corrected for the peculiarities of the diaphragm, in the manner explained later in this lecture.

The line *f* indicates the location of the natural harmonic overtones of the horn; the curve *e* shows by its peaks that the horn strongly reinforces the tones near its own fundamental, and, in a diminishing degree, those near all of its harmonics. The resonance of the horn increases the effects of all tones corresponding to the complete series of harmonics which the horn itself would give if it were blown as a bugle or hunting horn.

The diaphragm peak *D* comes between the peaks for the third and fourth partials of the horn, and it in effect divides the partials into two groups which are pushed apart in pitch; the amount of this displacement is shown by the gap in the line *f* near the fourth point.

Correcting Analyses of Sound Waves

The investigation of the effects of the horn and diaphragm of a sound-recording apparatus, a few details of which have been described, involved an unexpected amount of labor; it is estimated that the time required was equivalent to that of one investigator working eight hours for every working day in three and a half years. It has been shown that the horn and diaphragm introduce many distortions into the curves obtained with their aid, and that the distortions vary greatly with the conditions of the instrument. Unless these errors are eliminated and the true curves found, the records of the instrument will be without value because the incorrect and false curves can lead to no rational conclusions whatever. One may wonder, if the horn produces such disturbances, why it is not dispensed with in scientific research; the horn has been retained because the

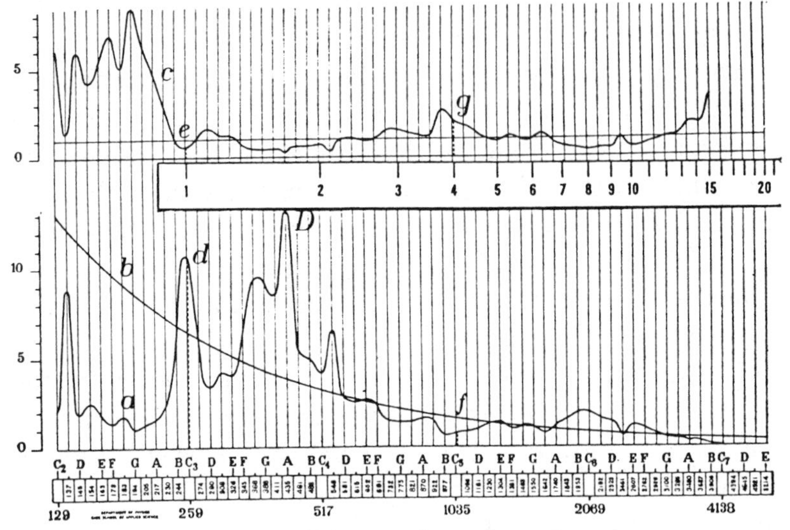

FIG. 126. Curves used in correcting analyses of sound waves.

sensitiveness of the recording apparatus is increased several thousand-fold by its use.

Whenever records are made for analytical purposes, the condition of the phonodeik is usually determined, as previously explained, by photographing its response to each of a set of sixty-one organ pipes of standard intensity, covering a range of frequencies from 129 to 4138. The actual response of the phonodeik in the form for research is shown by the irregular curve *a*, Fig. 126, while *b* is the desired or ideal response. The shape of the curve will vary with every change in the size or condition of the horn or diaphragm, with temperature changes, with the tightness of clamping of the diaphragm, with the nature of the room in which the apparatus is used, and with other conditions. The sharp resonance peak *D* is due to the free period of the diaphragm.

The ordinates of this curve give the amplitudes of the phonodeik records for sounds of various pitches, all being of the same intensity. The curve shows that the tone of any particular pitch, whether a single simple tone or a single component of a complex sound, in general produces a response either too large or too small as compared with the response due to other pitches. When any sound has been photographed with the phonodeik in the conditions here represented, and the amplitudes of all the components of the complex tone have been determined by analysis, it is necessary to correct each individual amplitude by multiplying it by a factor corresponding to its particular pitch. The factor for a tone of any pitch is the number by which the ordinate of the actual response curve for the given pitch must be multiplied to give the ordinate of the ideal curve.

These correction factors are obtained from a correction curve, *c*, Fig. 126, determined in the following manner. The sixty-one ordinates of the ideal curve *b* corresponding to semitones of the scale are divided by the corresponding ordinates of the response curve *a*; the quotients are the correction factors for these particular pitches. These factors are then plotted on the chart, and a correction curve is drawn through the points as shown at *c*. The correction curve is an inverse of the response curve, where one has a peak the other has a corresponding valley.

EDITOR'S NOTES

Page 79, line 9: Miller makes use of the additional amplifying ability of a resonant horn, like Webster, Stewart, and centuries of wind-instrument makers. His friend Slepian (see Paper 29, this volume) was one of the first to break away from this tradition by trying to build a nonresonant horn (which has quite different virtues from the resonant horn). This *may* be why Slepian refused to acknowledge any debt to Webster. Webster's ideal horn seems to have been the resonant, conical horn; Slepian's, the nonresonant, exponential horn. See C. R. Hanna and J. Slepian, The Function and Design of Horns for Loud Speakers, *Am. Inst. Electr. Eng. Trans.* **43**:393–404, 1924; Paper 17 in *Acoustic Transducers,* I. Groves, ed., Hutchinson Ross Publishing Company, Stroudsburg, Pa., 1981, pp. 155–166.

Page 148: In Figure 116 the abscissa is the log frequency of the sound source, while the ordinate is the (linear) particle displacement of the received sound. The solid curve is the ideal particle displacement response, a true hyperbola when both scales are linear. This hyperbola occurs when Miller's source is "a constant intensity source" whose diameter is $\ll\lambda$ because then the volume velocity $\dot{V}$ at the source drops 6 dB/oct (on a 20 log scale). (This comes about as follows: Radiated power = $\dot{V}^2 \times R_{rad}$. $R_{rad} = (\rho c/\text{Area})\,[k^2r^2/(1 + k^2r^2)]$. When $k^2r^2 \ll 1$, $R_{rad} \simeq (\rho c/\text{Area})\,k^2r^2$, which rises +6 dB/oct on a 10 log scale.) Now when $\dot{V}$ drops 6 dB/oct the pressure in the radiated wave is "flat" versus frequency. Then the particle velocity in front of the phonodeik would be flat and hence the particle displacement would drop 6 dB/oct which is really what the solid curve shows.

Page 158: On these last few pages Miller has clearly stated that he deliberately uses resonant horns for their amplifying capabilities. These are the basis of musical wind instruments but are not good for an acoustic phonograph (which is almost

Miller's phonodeik in reverse). The horn's amplification at steady state is very great at a few select frequencies, but not elsewhere. See also p. 163. A good discussion of this subject is given by G. W. Stewart (*Phys. Rev.* **16**:313, 1920), which is summarized in *Acoustics* by Stewart and Lindsay (Van Nostrand, 1930, Section 6.3.).

Page 161, lines 2 and 3: Miller here is almost describing the well-behaved exponential horn. Note his remark that "the natural periods are indefinite," which is just what is needed in a phonograph for a smooth broadband response; but of course the "gain" is then not nearly as large as with a resonant conical horn at the resonance peaks.

Page 164, last sentence: In Figure 126, the lower graph shows two response curves. Curve b, the smooth quasi-hyperbolic curve, is the theoretical sound particle displacement near the diaphragm, due to a set of 61 organ pipes of so-called standard intensity, covering the range 129–4138 Hz. Curve a is the experimental response curve (diaphragm displacement) of a phonodeik to this standard-intensity sound field. The upper graph is a correction curve, curve c. All ordinates are on a linear scale. If the ordinates had been plotted in decibels, the correction curve c would be essentially a mirror image of the response curve.

4

Reprinted from pages 45, 53, Plate II, and 54 of *Sound Waves: Their Shape and Speed*, Macmillan, New York, 1937, 164 p.

SOUND WAVES: THEIR SHAPE AND SPEED

D. C. Miller

[*Editor's Note:* These few pages from Miller's 1937 book are a natural continuation of Paper 3.]

Fig. 14 shows three records of the *same*

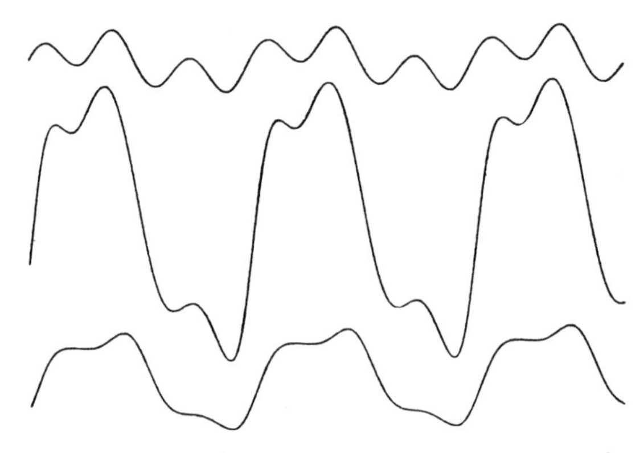

Fig. 14. Three curves for the same tone made under different conditions.

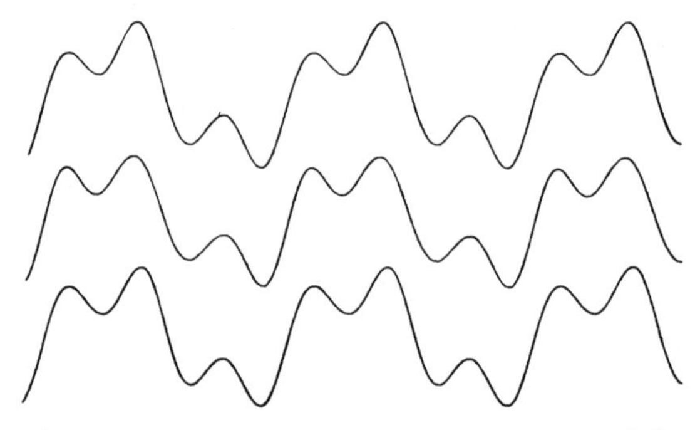

Fig. 15. The three curves of Fig. 14 corrected for resonance distortions.

tone from an organ pipe, made with three different horn-and-diaphragm combinations. The curves were each corrected by the method referred to, giving the three curves shown in Fig. 15. It is reasonable to infer that these curves represent the true shape of the sound wave in the free air, and are suitable for quantitative, comparative studies.

[*Editor's Note:* Material has been omitted at this point.]

Plate II shows photographs of the sound waves from various wind instruments. Curve *F* was made with a flute in G, sounding the tone $C_3 = 256$ vibrations per second. The fundamental constitutes the greater part of the tone; the larger kink in the top of the curve corresponds to the octave overtone, while the smaller kinks are due to the double octave or fourth partial tone. The general simplicity and smoothness of the wave indicates that there are no other overtones of appreciable intensity. A sound consisting of a fundamental and the octave and double octave only, wholly lacking in odd-numbered partials, produces the sensation of mellowness and simplicity characteristic of certain tones of the flute.

PLATE II. PHOTOGRAPHS OF SOUND WAVES.

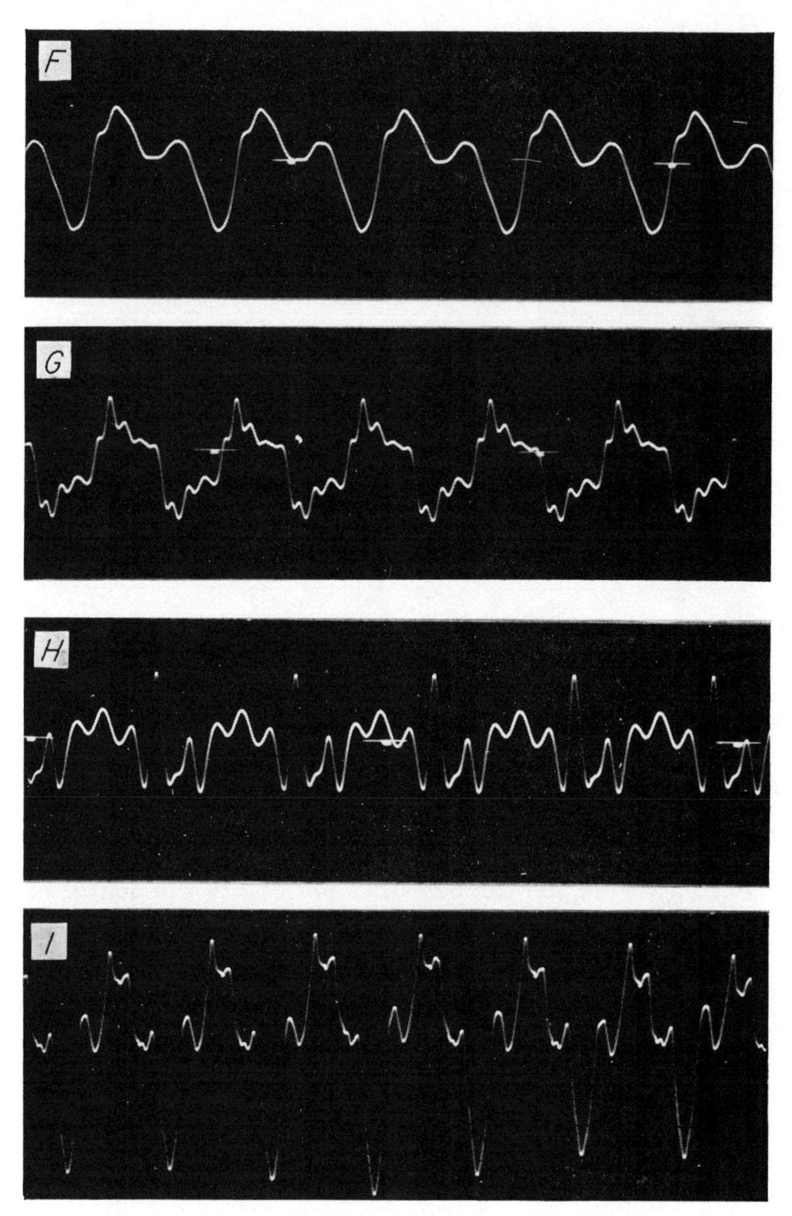

F. Flute. *G*. Clarinet. *H*. Oboe. *I*. Saxophone.

Curve *G*, Plate II, is the record from a clarinet sounding the tone $C_3 = 256$ vibrations per second. There is a strong fundamental together with at least two higher partials. The two partials beat with each other, once per period, hence they are consecutive in number and one is odd numbered and the other even numbered. Eight kinks are evident, and the stronger partials are either the seventh and eighth, or the eighth and ninth; which, it is difficult to determine from simple inspection; machine analysis, in this case, shows that the stronger partials are the eighth and ninth, while there are traces of the third, fifth, and seventh partials. It is the presence of these relatively loud higher partials that produces the reedy tone quality.

The record from an oboe sounding the tone $C_3 = 256$ vibrations per second is shown in curve *H*, Plate II. The fundamental component is almost wholly lacking; the fourth and fifth partials contain the greater part of the energy of the sound, while there are traces of a series of higher partials indicated by the long and narrow projections in the wave form.

The saxophone record, curve *I*, Plate II, shows a strong fundamental, with both even and odd harmonics, all starting in the same phase. The fourth partial, the double octave, is the strongest overtone, with the fifth and sixth of moderate intensity. Machine analysis shows the full series of overtones of diminishing strength to the fifteenth.

[*Editor's Note:* In the original, material precedes and follows these excerpts.]

5

Reprinted from *Nat. Res. Counc. Bull.* **4**(23):20 (1922)

DETECTION AND MEASUREMENT OF SOUND

D. C. Miller and A. G. Webster

[*Editor's Note:* We conclude the Miller papers with his one-page collaboration with Webster from the famous November 1922 *Bulletin of the National Research Council* entitled "Certain Problems in Acoustics." We note that the "refractometric" method is usually referred to as the schlieren method (see Part III of this volume). Note also that "need #4" is still unsolved.]

Probably the instruments available to the physicists for the detection and measurement of sound are less satisfactory than those for any other field of research. A wide variety of methods have been proposed and more or less developed, such as,—vibrating membranes and diaphragms, applied as in the phonautograph (phonograph), the phonometer, and the phonodeik; vibrating diaphragms in connection with electrical effects as in the carbon transmitter, the condenser transmitter, the thermophone; the Rayleigh disk; the hot-wire anemometer; sensitive flames; and the direct detection of waves in a medium by the "refractometric" method.

For the measurement of the energy of sound the methods are quite limited in number and in applicability. These are: the Rayleigh disk, the phonometer, the sound-pressure balance, and indirect methods by the use of electrical apparatus.

For the making of records of sound waves for quantitative investigation, various methods of detecting sound may be combined with some form of recording apparatus, usually photographic in nature. Among those available are: various miscrophones and transmitters in connection with the string-galvanometer or oscillograph, the "barreter," the phonodeik, and the refractometric method.

Under this section the needs involving research are many and urgent. In general these needs are:

(1) Sensitive instruments for detecting and measuring sounds, the instruments being free from all resonance effects which distort the record, and being capable of responding to the entire range of frequencies.

Specifically, some of these needs are:

(2) A direct-reading instrument capable of determining the absolute (and relative) intensity of a sound, equivalent to the photometer in optics.

(3) An instrument for recording sound waves in air, as in the phonodeik, but free from the disturbing effects of the horn and diaphragm.

(4) A method for directly photographing the sound waves in air, as in the refractometric method, but directly applicable to common sounds as to amplitude and frequency.

(5) Reference standards of sound intensity for various frequencies, and of various multiples of the unit.

BIBLIOGRAPHY

Koenig, R. Quelques Experiences D'Acoustique, Paris, 1882.
Nichols and Merritt. Photography of Manometric Flames. Phys. Rev., vii, pp. 93–101, 1908.
Winkelmann. Handbuch der Physik, Bd. II, Akustik, Leipzig, pp. 147–177, 228–254, 1909. This work contains extensive bibliographies.
Miller, Dayton C. The Science of Musical Sounds, New York, pp. 70–174, 1916.
Webster, A. G. Proc. Nat. Acad. of Sci., v, pp. 163–199, 275–282, 1919.
See also the lists of references on the *Measurement of Sound Intensity,* Section VII; on *Sound Generators,* Section VIII, and on *Photography of Waves,* Section XI, of this report.

6

Reprinted from *Nature* **110**:42–45 (1922)

Absolute Measurements of Sound.[1]

By Dr. ARTHUR GORDON WEBSTER, Professor of Physics, Clark University, Worcester, Mass., U.S.A.

IT is now more than thirty years since it occurred to me to devise an instrument that should be capable of measuring the intensity or loudness of any sound at any point in space, should be self-contained and portable, and should give its indications in absolute measure. By this is meant that the units should be such as do not depend on time, place, or the instrument, so that, though the instrument be destroyed and the observer dead, if his writings were preserved another instrument could be constructed from the specifications and the same sound reproduced a hundred or a thousand years later. The difficulty comes from the fact that the forces and amounts of energy involved in connexion even with very loud sounds are extremely small, as may be gathered from the statement that it would take approximately ten million cornets playing *fortissimo* to emit 1 horse-power of sound.

Before we can measure anything we must have a constant standard. In sound we must construct a standard which emits a sound of the simplest possible character, which we call a pure tone; it will be like that emitted under proper conditions by a tuning-fork, which is described by saying that the graph representing the change of pressure with the time shall be that simple curve known as the sinusoid or curve of sines. From this connexion we say that the pressure is a harmonic function of the time. Unfortunately, the pressure change is so small that at no point in a room, even when a person is speaking in a loud tone, does the pressure vary from the atmospheric pressure by more than a few millionths of an atmosphere. Thus we require a manometer millions of times as sensitive as an ordinary barometer, and, in addition, since the rhythmic changes occur, not once in an hour or day, but hundreds of times per second, if we wish the gauge to follow the rapid changes accurately, we have many mechanical difficulties.

The problem of a standard of emission has been solved by a number of persons, including Prof. Ernst Mach and Prof. Ludwig Boltzmann, and Dr. A. Zernov, of Petrograd, a pupil of the celebrated Peter Lebedeff. The problem of an absolute instrument for the reception and measurement of a pure tone has been also successfully dealt with by a number of investigators, among whom may be mentioned Prof. Max Wien, of wireless fame, the late Lord Rayleigh, and Lebedeff. But there remains a third step in the process, which is as important as the first and the second. Given the invention of the proper standard source of sound, which I have named the "phone," because it is *vox et praeterea nihil*, and of a proper measuring instrument, which should evidently be called a phonometer, there still remains the question of the distribution of the sound in space between the phone and the phonometer. Any measurements made in an enclosed space will be influenced by reflections from the walls, and, even if we had a room of perfectly simple geometrical form, say cubical, and were able to make the instruments of emission and reception work automatically without the disturbing presence of an observer, it would still be impossible to specify the reflecting power of the walls without a great amount of experimentation and complicated theory. Nevertheless, this is exactly what was done by the late Prof. Wallace C. Sabine, of Harvard University, who employed the human ear as the receiving instrument. Those who have made experiments upon the sensitiveness of the human ear for a standard sound will immediately doubt the possibility of making precise measurements by the same ear at different times, and particularly of comparing measurements made by one ear with those made by another. Nevertheless, Sabine attained wonderful success and was able to impart his method to pupils who carried on his work successfully, so that he was able to create the science of architectural acoustics and to introduce a new profession. Still, the skill that required three or four months to attain by Sabine's method may be replaced by a few minutes' work with the phonometer.

In order to avoid the influence of disturbing objects, the observer should take the phonometer to an infinite distance, which is manifestly impossible. The method employed was to get rid of all objects except a reflecting plane covered with a surface the coefficient of reflection of which could be measured. For this purpose the teeing ground of a suitable golf course was used. With the present instrument it can be determined in a few minutes, if there is no wind.

[1] From a Friday evening discourse delivered at the Royal Institution on June 10, 1921.

In 1890 I proposed to use a diaphragm made of paper, which should be placed, shielded on one side, at the point where the sound was to be measured. In order that the effect of the sound should not be distorted, the membrane, instead of having to do any work, as in the case of the diaphragm of the phonograph in digging up the wax, or in that of the microphone in compressing the carbon, was to be perfectly free, but was to carry a small plane mirror cemented on at its centre. In close juxtaposition and parallel with this was the plane side of a lens which, viewed in the light from a sodium flame, was to give Newton's rings, or interference fringes. Of course, when the sound falls upon the diaphragm the fringes vibrate rapidly and disappear from sight.

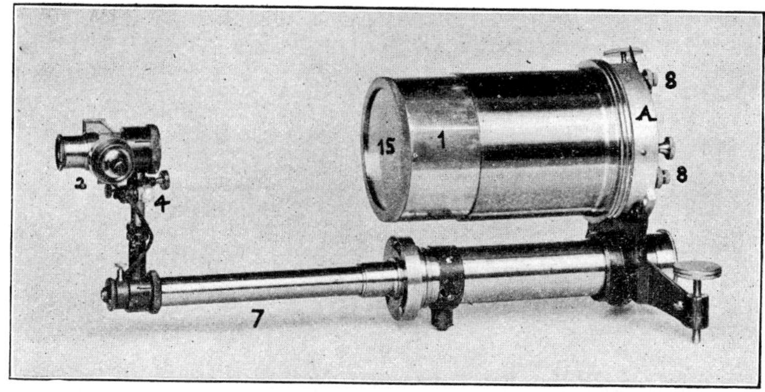

Fig. 1.—Phonometer. (Interferometer not shown.)

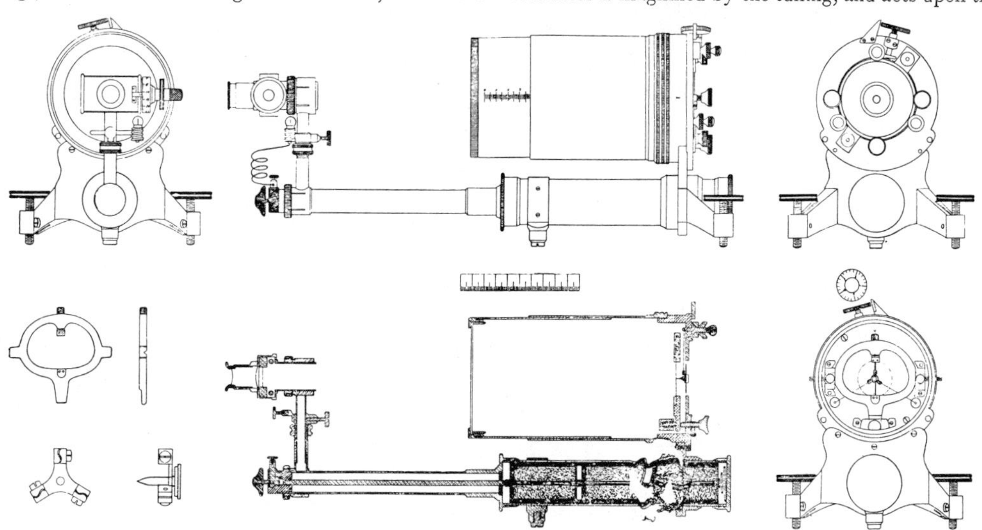

Fig. 2.—Parts of the phonometer.

By the introduction of a Michelson optical interferometer, two of the difficulties of this instrument were overcome, namely, (1) that of adjusting the lens so that it would not strike the vibrating mirror, since the mirrors in the interferometer could be as far apart as one pleased; and (2), more important still, it permitted the use of fringes in white light, so that it was possible to use gas, incandescent, or arc light with excellent effect. A further improvement was introduced by the use of a thin plate of mica for the diaphragm.

To obtain the sensitiveness necessary to measure sounds of ordinary intensity, the property of resonance is employed twice—*i.e.* a system of two degrees of freedom is used. First, the plate resounds to a sound more strongly as it is tuned more nearly to it; and second, a resonator that can also be tuned is put behind the plate. The sound entering by the hole in the resonator is magnified by the tuning, and acts upon the plate, which is also tuned. A graph can be plotted in which one co-ordinate represents the stiffness of the plate, or rather what may be called the mistuning, which is the stiffness lessened by the product of the mass by the square of the frequency. The other co-ordinate represents the corresponding quantity for the resonator, the stiffness of which depends simply on the volume into which the air is compressed, while the effective mass depends on the dimensions of the whole, and its damping on the sound radiated from the mouth. It is then found that the tuning should not be such as to make the representative point occur at the middle of the figure, making both mistunings zero, but that both mistunings should be of the same sign and a certain magnitude, depending on the coefficients of damping of the two degrees of freedom of the coupled system. The mathematical theory is precisely that of a wireless receiver. The ultimate sensitiveness depends on the smallness of the damping of the plate.

The apparatus as it was built several years ago was mounted upon a heavy bronze stand, covered at the back by a heavy bronze cover to keep out the sound, while the three shafts turning the screws of the interferometer adjustment protruded through sound-tight fittings. Upon the front of the instrument a properly tuned resonator was attached, and at the side was a

FIG. 3.—Front view of phonometer with annular opening.

small incandescent lamp with a straight, horizontal filament, an image of which was projected by a lens upon the first mirror of the interferometer. Upon this was focussed a telescope, giving in the reticule an image of the horizontal, straight filament, crossed by the vertical interference fringes seen with white light. In order to get these the plate must be in the proper position within a few hundred thousandths of an inch. The objective of the tuning-fork was carried by a tuning-fork which oscillated vertically, tuned to the pitch of the pure tone to be examined, and this, combined with the horizontal motion of the fringes, resulted in a figure of coloured fringes in the form of an ellipse. On slightly mistuning the fork, the ellipse could be made to go through all its phases, and when it was reduced to an inclined straight line its inclination was read off on a tangent scale. The amplitude of the compression of the air in the sound was then directly proportional to the scale-reading.

While the interferometer is still used for calibration, the movement of the diaphragm is recorded for actual measurements by a thin steel torsion strip carrying a concave mirror. A lamp with a vertical, straight filament is viewed through a telescope into which the small mirror focusses the image of the filament on the reticule, and a magnification of from 1200 to 1500 is used, so that the sensitiveness is about the same as with the interferometer.

At first the only method of tuning was the clumsy one of changing the mass of the diaphragm by adding small pieces of wax. This was not capable of continuous variation. Now the diaphragm has been discarded and replaced by a rigid disc supported by three steel wires in tension. The disc is made of mica or aluminium, and is carried by a little steel spider containing three clamps to hold the wire. The tension is regulated by three steel pegs, one of which is controlled by a micrometer screw. The disc is placed in the circular hole through which the sound enters the resonator. This has the advantage of reducing damping very largely, and thus of increasing the sensitiveness enormously. The instrument now competes with the human ear, and can be tuned over two octaves or more.

This sensitiveness can be demonstrated by projecting the coloured interference fringes on a screen and singing faintly in a remote part of the room, when the fringes will disappear. Using the telescope end of the apparatus, the instrument will indicate the sound of a tuning-fork when one can scarcely hear it. It is obvious that the disc may be made the diaphragm of a telephone and thus increase its sensitiveness. In fact, Prof. King has used with great success such a telephone to record wireless messages. He has also invented another sort of tunable diaphragm composed of a stretched steel membrane with compressed air behind it, which enables it to be tuned continuously, but over a smaller range.

I now come to the source of sound—the phone. This has been reduced to a reversed form of the phonometer. The disc is driven by an interrupted or alternating current by means of electromagnets, and tuned like the phonometer. Its excursion is measured by a powerful microscope, and the emission of sound is known in absolute measure. It is now driven by a triode valve tube, in the manner suggested by Prof. W. H. Eccles, of Finsbury Technical College, London, for a tuning-fork. This has been worked out for me by Dr. Eckhardt at the Bureau of Standards in Washington.

The third part of the investigation involves a determination of the coefficient of reflection of the ground. The phone is set at a convenient height, and the phonometer at a convenient distance. Either is then moved along at a constant height and the varying deflections of the phonometer are read while the sound remains the same. Interference sets in between the direct sound and its image reflected in the ground, and the existence of a minimum is obvious to the most naive observer by the ear alone. The reflection of either grass or gravel was found to be about 95 per cent., while, with a most carefully deadened room, the walls of which were covered with thick felt, there was perhaps 20 per cent. reflection. The whole measurement at both ends and the transmission checks up with an accuracy of about 2 per cent.

With this apparatus all sorts of acoustical experiments may be performed. By attaching to the phonometer a long glass tube or antenna, it has been possible to explore all sorts of places, such as the field within a horn or tube lined with an absorbent substance. The transmission of sound through fabrics, walls, and telephone booths may also be quickly examined. The instrument is used by psychologists and by telephone and acoustic engineers, and is of interest to navigators. An interesting by-product is an instrument for showing the direction of an acoustic signal in the fog. It has been called a phonotrope, on the analogy of heliotrope, which turns to the sun. It consists of two equal horns which bring the sound to the opposite sides of the disc. When the whistle blows, the band of light spreads out, and on turning the instrument it closes to zero when the sound is directly ahead. Thus at several miles the direction is given to within two or three degrees.

Finally, let us consider that mystery of sound, the violin, which has been studied by Prof. Barton of Nottingham, and by Prof. Raman at Calcutta. This may be described by the engineer as a box of curious shape, made of a curious substance, wood, of variable thickness, with two holes of strange figure to let the sound out of the resonating box. The latter is actuated by a curious substance, catgut, made of the intestines of a sheep, and set in vibration by another curious substance, the tail of a horse. Yet from this wonderful box we get the most ravishing sounds, which affect profoundly the emotions of the most civilised. Yet the physicist reduces all musical instruments to combinations of resonators with strings, membranes, bars, plates, and horns. The mathematical theory of strings was given by Euler two hundred years ago, of bars and plates less than a hundred years, of resonators by Helmholtz and Rayleigh, and I have recently added a theory of horns which, while only approximate, works well in practice, and investigations are now being carried out by such methods on vowels and the violin.

EDITOR'S NOTES

Page 42: It is probable that this paper, delivered in June 1921, was written down by a stenographer present at the meeting, rather than by Webster himself. Webster delivered dozens of papers orally and usually failed to write them up. This is stated by Webster himself in Paper 8 of this group. Moreover, the error "whole" for "hole" on p. 44, line 1, looks like an auditory error. This paper also appeared, a year later, in the American journal *Science* for August 23, 1923 (after Webster's death).

Page 42, right-hand column, line 7: *Phone* is the Greek word for either voice or sound. *Vox* is a Latin word for either voice or sound. *Vox et praeterea nihil* means (to Webster) "a sound and nothing else."

Page 43, right-hand column, 5th line from bottom: This factor $(K - \omega^2 M)$ is what Webster refers to elsewhere as acoustic impedance. Nowadays, following G. W. Stewart, we prefer to work with the factor $(K/\omega - \omega M)$, and we call this acoustic impedance.

Page 44, left-hand column, line 9: This disarmingly simple sentence masks a very sophisticated set of constraints on $Q_{primary}$, $Q_{secondary}$, the damping of the plate, and the volume of the Helmholtz resonator, or $V_{primary}$. See "A Note on the Coupled Circuits of Paris, Webster, and Radio Engineering" (Paper 11) at the end of Part I, but read Paris's paper first.

Paper 44, left-hand column, 16th line from bottom: "The objective of the telescope" was intended. Note that this part of the apparatus is pure Lissajous.

Page 44, right-hand column, 16th line: See Paper 11 at the end of Part I, explaining how the damping is reduced.

7A

Reprinted from *Am. Inst. Electr. Eng. Proc.* **38**(Part I):889–898 (1919)

THE ABSOLUTE MEASUREMENT OF THE INTENSITY OF SOUND

BY ARTHUR GORDON WEBSTER

ABSTRACT OF PAPER

This paper includes a description of a series of acoustical researches extending over a period of twenty-eight years. The properties of vibrating bodies and the subject of elastic hysteresis are discussed. Two fundamentally important instruments for the absolute measurement of sound have been developed and the theory given. The first is the standard of sound, called the phone, which is capable of reproducing at any time a sound of the simplest character and which permits the output of sound to be measured in watts of energy. The second is an instrument called the phonometer for measuring a sound in absolute measure. This instrument is now practically as sensitive as the human ear. Two essential features are the small damping of the vibrating system which results in extreme sensitiveness, but at the same time in great selectiveness, and the capacity for being tuned. Each of these instruments is fitted with a variable volume resonator and tuning over a range of about two octaves is accomplished by varying the volume of the resonator and by changing the tension of the wires to which the piston is attached. The determination of the space distribution of sound and of the effect of disturbing bodies, and the measurement of the reflecting coefficient of surfaces have been accomplished. The phonotrope is a third instrument designed and used to find the direction of a source of sound, for example a fog signal.

THE profession of acoustical engineering is a new one that has arisen during the last few years. The late Professor Wallace C. Sabine, of Harvard University, attracted the attention of the public to the very great importance of physical measurement and thoroughly scientific design in the construction of auditoriums which until his time had remained an entire mystery to architects.

The writer has been interested for many years in the problem of the measurement of sound in absolute units, involving the design of an instrument which is capable of determining, at any point in space, the pressure in the air wave as a function of the time. In order to carry out this object three things are necessary.

First, the construction of a standard of sound which will enable a given sound of the simplest possible character, namely, that in which the pressure varies as a simple harmonic function of the time, to be reproduced at any time, and which permits an output of sound to be measured in watts of energy. This problem has been solved by a number of persons including Professor Ernst Mach and Dr. A. Zernov of Petrograd.

Second, the invention of an instrument capable of measuring a sound of the simple harmonic character described, also in absolute measure; that is to say, in such a way that the amplitude of pressure variation may be measured in absolute measure, as dynes per square centimeter, or in millionths of an atmosphere. This problem has also been successfully traeted by a number of investigators, among whom may be mentioned Professor Max Wein, Lord Rayleigh and Lebedeff.

Third, a step in the process remains which is perhaps equally important as the first and second. Given the invention of the proper source of sound, which I have denoted a phone, because it is a sound and nothing else, and a proper measuring instrument, which I call a phonometer, there still remains the question of the propagation and the distribution of sound in space between the phone and the phonometer. Any measurements made in an enclosed space will be influenced by the reflection from the walls, and even if we had a room of perfectly simple geometrical form and were able to make the instruments of emission and reception work automatically without the disturbing presence of an observer, it would still be impossible to specify the reflecting power of the walls without a very great amount of experimentation and complicated theory. Nevertheless, this is exactly what Sabine did. He did it, however, by employing as a receiving instrument the human ear. Those who have made experiments upon the sensitiveness of the human ear for a standard sound will immediately doubt the possibility of making precise measurements by the same ear at different times, and particularly of comparing the measurements made by one ear with those made by another. Nevertheless, Sabine attained wonderful success and was able to impart his method to pupils and colleagues who carried on his methods successfully.

To proceed with the question of disturbing objects, one should take his phone and his phonometer to an infinite distance from all objects, which is manifestly impossible. The plan which I followed was to attempt to get rid of all objects except an infinite space covered with a surface the reflecting coefficient of which could be measured. I found the proper weather and the proper space on the golf links of a country club where conditions appeared to be favorable. Several days were spent in determining the reflection coefficient. This was done about ten years ago, whereas with the instrument available to-day it could be done in a very few minutes.

The history of my investigations is briefly as follows: In 1890 I proposed to use a diaphragm made of paper or some similar substance which should be placed, shielded on one side, at the spot where the sound was to be measured. Upon the center of the diaphragm was cemented a small plane mirror. In close juxtaposition and parallel with this was the plane side of a plano-convex lens which viewed in the light from a sodium flame was to give Newton's rings, or as we now call them, interference fringes. Of course, when the sound falls upon the diaphragm, the fringes vibrate rapidly and disappear from sight. I then determined to use the stroboscopic arrangment which would permit the viewing of the fringes as they slowly shifted their position. All sorts of difficulties occurred. The light from the sodium flame was not strong enough to permit the fringes to be seen through the stroboscopic device. It was not at that time easy or perhaps possible to make an electric motor that would drive a stroboscopic disk at a constant speed. One of the first difficulties was to devise a method of controlling a synchronous motor. This was accomplished by means of interrupting the current by a tuning fork.

Later I made the acquaintance of Professor A. A. Michelson and of his remarkable optical interferometer, and I immediately saw that one of the difficulties which I had met, namely, the difficulty of adjustment of the lens so that it should not strike the vibrating plate, could be overcome at once, as the two mirrors in the Michelson interferometer could be any distance apart. Also the trouble due to the faintness of the light disappeared, for in Michelson's instrument white light could be used, and it was possible to use gas, incandescent or arc light with good effect.

In 1896 I put the subject in the hands of a student who produced an instrument which was successful. The examination of the substance suitable for a membrane or a diaphragm consumed much time, and all organic substances such as parchment, paper or anything of that sort had to be discarded because it was impossible to keep the tension constant under the changes of moisture in the air. Metal was found to be quite out of the question on account of the difficulties due to thermal expansion. We next decided upon thin glass diaphragms which we afterward found were used by phonograph companies. A slight change of temperature, however, such as was made by breathing upon the diaphragm, would sometimes produce a change of pitch of several tones, so I next decided to clamp the glass diaphragm upon a ring sawed out of plate glass and cement it on with silicate of soda. I finally found that it was possible to get mica so flat that it was superior to the best possible glass, and this gave me the best results up to that time.

Let me now state briefly the qualities which the diaphragm must have in order to be successful. It is well known to physicists that a flat diaphragm can vibrate in an infinite number of modes which the physicist calls normal vibrations. In the simplest mode the diaphragm vibrates so that every point moves in the same direction. The natural period of this vibration is the lowest that the diaphragm can have. Next is a mode in which the central portions of the membrane move in one direction while all those outside of a certain circle, which we call a nodal line, move in the opposite direction. Other vibrations have radial, as well as circular, nodal lines on the opposite sides of which the diaphragm is moving in opposite directions. When a mass is screwed to any point of the diaphragm its motion is entirely changed. And any apparatus that may be attached to the diaphragm and required to do work, such as the carbon button of the telephone, or anything that interferes with the free response of the diaphragm to the variations of the air pressure will alter the action in such a way as to make it extremely difficult of interpretation. My idea, then, was to make a diaphragm which did not do any work at all, but merely carried the mirror which moved practically by itself.

Now, in spite of the great variety of forms and vibrations, it can easily be shown that under proper circumstances the diaphragm will move essentially as a whole, and may, therefore, be compared to a single body which has three characteristic constants: First, its mass; second, its stiffness, defined as the force required to produce a unit deflection of its nodal point; and third, and most difficult to control, the coefficient of damping, which is defined as the force of resistance that will be exerted from any cause when the velocity of the body or the nodal point of the diaphragm is unity. This damping must be due to a large number of causes. In the first place, the resistance of the air will cause damping. In the second place the energy of the sound waves that are radiated from the vibrating diaphragm will cause damping. In the third place and by far the most important, the bending of any body is resisted by elastic forces and we therefore have the phenomenon of elastic fatigue. In order to settle the question "what is the best material from which to make tuning forks", which has been variously answered as tool steel, bronze, bell metal, quartz, etc. I perceived that it was necessary to have an exact method of experimentation.

In striking a steel bar supported by a string at two points it will be observed that the high overtones emitted by the bar rapidly die away, and it occurred to me that if the bar was thrown into one normal vibration there would be a bar rate of damping for every normal vibration, and that by stating the rate of natural damping a natural hysteresis could be studied. This was demonstrated experimentally by my assistant Mr. James L. Porter with very successful results as far as the experiment went. The mathematical theory has been lacking up to the present time. Two theories are possible, first the theory of solid viscosity, and second the theory of heredity, or elastic hysteresis. The attempts to compute the results of our experimental measurements are now approaching completion, and we soon shall be able to give a theory of elastic hysteresis. My conclusion is that quartz is the best substance from which to make tuning forks.

This digression from the investigation is told in order to show the very great difficulties that this subject presented.

The illustrations herewith show the apparatus as it was built several years ago. It was mounted upon a heavy stand made of bronze, covered in at the back by a heavy bronze cover, through which protruded through air-tight fittings the shafts turning the screws of the interferometer adjustments, three in number. Upon the front of the instrument was attached a resonator properly tuned and at the side was a small incandescent lamp with a straight filament arranged horizontally, an image of which was projected by a lens upon the first miror of the interferometer. A telescope was focused upon this image, giving an image in the reticle of the horizontal straight filament. This was crossed by the vertical interference fringes. The objective of the telescope was carried by one prong of a tuning fork which oscillated vertically and this, combined with the horizontal oscillation of the fringes due to the vibration of the diaphragm, resulted in a figure in the form of an ellipse. On tuning the two sounds together this ellipse could be caused to go through its various phases as slowly as desired. At the moment when the ellipse degenerated into an inclined straight line the reticle, which was tilted by an observer, was brought into line with the interference lamp and the tangent of the angle was read off on a tangent scale. The pressure on the sound wave was then proportional to the number read off on the tangent scale. The apparatus was used in this form for a good many years, a precursor of it being shown at the Congress of Physics at Paris in 1900.

Professor D. C. Miller's phonodeik which is a very beautiful instrument, was designed for quite other purposes than the phonometer of the speaker. It was intended to photograph sound curves of any sort and by means of it Professor Miller has obtained most beautiful reproductions of speech, and sounds of musical instruments. Unfortunately, this instrument does not adapt itself in the least to absolute measurements. One objection to Professor Miller's instrument is that it employs a spring made of rubber material. This spring has to be calibrated whenever measurements are taken, and it is well known that the properties of soft rubber are far from constant.

The speaker next undertook to devise a better mode of reading the amplitude of vibration of his diaphragm. In the first place, the mirror which was carried by the diaphragm in Professor Miller's apparatus, instead of being carried on an axis in jewels was placed on a thin steel torsion strip which could be made at far less expense and could be rapidly adjusted. The straight filament lamp was now viewed through a telescope into which the mirror focused the filament in the reticle of the eyepiece. A magnification of about 1200 to 1500 is used. The instrument was also used photographically just as the old instrument had been used to photograph the motion of the interference fringes. Photographing moving interference fringes, however, was attended by many experimental difficulties.

A new instrument in the form described was made for Professor Louis Vessot King, of McGill University, who had a commission from the Canadian Government to make experiments upon fog signals and had secured permission to investigate the great siren at Father Point on the St. Lawrence. Professor King spent the whole summer at Father Point in making the first survey of an acoustical field ever made. The Canadian Government had furnished him a large steamer with which he went out every day, blew the "criard" as the natives call it, the very perfect siren which can be heard thirty miles away in good weather, and explored the gradient of sound under all meterological conditions, studying the wind, the temperature, and the temperature of the water, etc. This spurred the speaker on to undertaking the same thing in this country. It was impossible to get from the Lighthouse Board the loan of a steamer, but permission was given to go on any lighthouse steamer going along the coast of Maine. The first step was to tune the phonometer for the different pitches, and at that time the only method of tuning the diaphragm was to load upon it small pieces of wax. This, although it worked perfectly, seemed very clumsy, and not an engineering method. It was therefore decided to redesign the whole instrument so that it could be tuned continuously. It occurred to me that this could be accomplished by furnishing potential energy to the diaphragm by means of a spring parallel to it, and bearing against it by means of a bridge, and tunable like a violin string. The very first attempt showed this to be entirely successful. Then the idea immediately presented itself that if a string could be tuned continuously why use the energy of bending at all and why not get entirely rid of all difficulties of elastic fatigue. Since that time the diaphragm has been entirely discarded and replaced by a stiff disk supported upon three strings in tension. The disk is made of aluminum or mica and is carried by a little spider made of aluminum containing three clamps to hold the wire. The wire is made of steel and is under tension about steel pegs, two of which are turned roughly by means of a screw driver, the other by means of a lever actuated by a micrometer screw. We have now a very perfect instrument at least ten times as good as the one Professor King had.

A few figures will show the advantages in the reduction of the damping. In a system of one degree of freedom the amplitude of the vibration for the most perfect tuning is inversely proportional to the quotient of the damping coefficient k

divided by the mass m. When we began with the good glass diaphragm we had a value of k something like 150. In 1913 the mica diaphragm lent to King had reduced this to 30. With the phonometer as built now we have reduced this to 1.34. In all these cases the mass was about the same so that so far as the sensitiveness goes, the instrument is more than one hundred times as good as it was ten years ago. The instrument now competes in sensitiveness with the human ear and may be tuned to pitches varying by perhaps two octaves. If I strike a tuning fork and allow its amplitude to die away, a person looking into the eye-piece of the instrument cannot say with certainty whether he can see the sound or hear it the longest. Unfortunately, however, this very great sensitiveness obtained by a small damping is attended with a very great disadvantage. The instrument is extremely selective and if the tuning varies by a very small amount the amplitude falls off very greatly.

Within the last three months I have devised a plan of doing away with this, and making an artifcial ear. For, as you know, the human ear has a wonderful sensitiveness, not for all sounds, but to sounds of a frequency of perhaps 30 per second up to perhaps 30,000 per second, attaining a maximum sensitiveness for frequencies of from 1000 to 1500. We know how this affects the telephone and how the damping of the telephone and microphone disks enables the response to be carried over a large range, so that they answer very well for the frequencies involved in speech. Nevertheless, the sounds of *s*, *f*, *t*, and certain others, on account of the very high harmonics involved, are most difficult to transmit telephonically.

Having described the construction of the diaphragm, the new piston with wires, and the method of reading the vibration by means of the inclined mirror which is also reduced to absolute measure by means of the interferometer temporarily attached, or more simply by displacing the diaphragm by a micrometer screw, I come to the theory of the instrument. The resonator into which the air enters, and the hole which is now constituted by an angular opening around the disk, furnishes through the movement of the air in and out an additional degree of freedom. We are then in the possession of a system with two degrees of freedom, statically coupled by means of the increased pressure in the resonator when the piston is forced in. The theory of such a system under the action of the periodic force is well known and has been treated by Professor Max Wien in great detail. If the frequency in 2 π seconds be n and if the mass be m, I shall call the stiffness minus the product of the mass and square of the frequency the uncompensated stiffness or mistuning. If this were equal to zero for a system of one degree of freedom we should have resonance. In the case of two degrees of freedom I plot the mistunings horizontally and vertically, and the amplitude of response as a third co-ordinate. I thus get a surface that may be described as two mountain peaks with a pass between them, the summit of the pass being when both mistunings are zero; but this is by no means the highest point of the region. To attain either summit we must mistune both systems by a certain amount which is porportional to their two dampings. The addition of the resonator to this instrument may produce an increase of sensitiveness of fifty times. You now see, I hope, how we have obtained an engineering instrument, everything in which is measurable in absolute measure. It is obvious that the piston may be made the diaphragm of the telephone, that the instrument may be used by the psychologist, the engineer, the physiologist, for instance in a stethoscope, and in many other applications.

I turn now to the source of the sound, the phone. With the advent of the tuneable diaphragm came the new phone. It is very light. The amplitude of the diaphragm is measured by a microscope. I use a hot wire vacuum tube as a source of alternating current, tuneable at will, and tune the phone to it. We thus have in small compass a very perfect set of acoustical instruments.

The third portion of my investigations involved a determination of the coefficient of reflection of the ground. In order to accomplish this the phone is set at a convenient height and the phonometer at a convenient distance. The latter is then moved back and forth at the same height, when it is immediately found that interference between the phone and its image in the ground sets in, producing a variation of the intensity of the sound. Different curves have been plotted for different coefficients of reflection. When the reflection is zero or the ground is acoustically perfectly black we have a rectangular hyperbola. The existence of the minimum is obvious to the most unskilled observer. We found the coefficient of reflection of grass, or gravel surface, to be about 95 per cent. I may say that the whole measurement of the instruments of the two ends and the transmission between checks up with an accuracy of probably better than two per cent. With this apparatus all sorts of experiments have been performed. By attaching to the phonometer a long glass tube or antenna it has been made possible to explore all sorts of places, such as the field within a horn or a tube lined with absorbent substance. The theory has been always completely verified. In order to examine the transmission of sound a piece of substance is clamped between two heavy cast iron rings cemented against a hole in a brick wall which will exclude all sound except that that comes through the fabric. The transmission through doors, windows, walls, and telephone booths may be carried out very quickly, and the coefficients of absorption and reflection determined.

Finally, there has resulted an instrument for determining the direction of a fog signal blowing in the fog which I call a phonotrope; since the heliotrope turns toward the sun, the radiant of heat, so the phonotrope turns toward the radiant of sound. This instrument consists of two equal horns bringing the sound to the opposite side of the same disk. It is arranged to be rapidly tuned to the whistle, and when the whistle blows the band of light spreads out; the whole instrument is then revolved until the band reduces its width to zero when the whistle is directly ahead. This instrument was taken to Pensacola to see whether it would determine the direction of an aeroplane in the night. It was found to be as sensitive as the ear, but owing to Doppler's principle, the continual coming and going of the aeroplane changed the pitch so as to put it out of tune. A new modification that I have devised will obviate this I hope.

I have now given you briefly and without any mathematics an account of the principles which I think must always be involved in any measurements of sound. I have always been very anxious to join forces with Professor Miller and to calibrate his instruments so as to render his wonderful results of serious interest to the physicist. I am also glad to co-operate with all my colleagues whether engineers, physicists, physiologists or physicians. It will be a great pleasure to me to know that this apparatus may be of use in solving any of the multitude of questions that confront us.

[*Editor's Note:* In the original, material follows this excerpt. The figures have been omitted.]

EDITOR'S NOTES

Page 893, 5th line from bottom: According to the description here and on p. 891, we can visualize a mason jar or canning jar of bronze, with a thin diaphragm as the top cover, inherently "shielded on one side." Then a Helmholtz resonator with an opening at the top and at the bottom (like the glass chimney of an oil lamp) is placed on top of the diaphragm, giving a second "mesh." This diaphragm is a true measurer of the pressure of the sound wave.

Page 894, line 2: Compare Lissajous's setup.

Page 895, 14th line from bottom: Note that the disc held by three wires (mentioned also in the *Nature* article) is surrounded by an annulus of air and hence acts like a very loose plug or stopper in a bottle. This plug is now sensitive, not to the pressure of the sound wave but rather to the pressure gradient (Webster never tells us this). So he now has what we today call a pressure-gradient microphone rather than a pressure microphone. Therefore he should, and does, place the disc in the *mouth* of the Helmholtz resonator where the kinetic energy is greatest (as is correct also for a velocity microphone). When earlier he used the shielded diaphragm, he arranged for the diaphragm to act as the *closed end* of the Helmholtz resonator where the potential energy is greatest.

Page 897, line 8: The two mountain peaks with a pass are described in some detail by E. L. Chaffee, *Proc. IRE* **12**:299–359, 1924. Plate I is the most pertinent (but still does not shed enough light on Webster's achievement).

7B

Reprinted from pages 721–723 of *Am. Inst. Electr. Eng. Proc.* **38**(Part I):713–723 (1919)

DISCUSSION ON "THE ABSOLUTE MEASUREMENT OF THE INTENSITY OF SOUND" (WEBSTER)

H. S. Osborne, A. G. Webster, and A. E. Kennelly

[*Editor's Note:* These remarks are part of the discussion period that took place when this paper and others were presented in Boston. This discussant was H. S. Osborne, a telephone engineer.]

The instruments and methods which Prof. Webster has shown to us are certainly of the very greatest interest to any one who has to deal with acoustic problems. However, electrical engineers, from the prejudice of their training, very much prefer to read their results on electric instruments when it is possible, rather than to observe them through a microscope as mechanical displacements.

Mr. Jones has spoken of the fact that, in many cases, we are very much interested in measuring sound with an instrument which is not highly selective, but which, on the contrary, responds so as to give as nearly as possible a faithful reproduction of a complicated wave shape, that is, an equal response to the different harmonic components of the wave. A good solution of this problem has been discussed in two papers which were printed in the *Physical Review* in July, 1917, which I would like to call to your attention. These are the "Thermophone as a Precision Source of Sound," by H. D. Arnold and I. B. Crandall, and "A Condenser Transmitter as a Uniformly Sensitive Instrument for the Absolute Measurement of Sound Intensity," by E. C. Wente. By means of the instruments described in these papers it is possible to set up a system for transforming acoustic into electric waves and vice versa with an efficiency of conversion which can be made sensibly uniform throughout the range which is important in telephonic work. Of course, these instruments give very small responses, but in these days of vacuum tube amplifiers that is not a difficulty.

Mr. Taylor has spoken of the possibility of using Prof. Webster's instruments to study the characteristics of interfering currents in telephone circuits. However, for such complicated wave shapes as those or for the study of the characteristics of telephone speech, instruments giving a uniform response, such as are described in the papers just mentioned are very much more valuable than instruments which are sharply tuned to particular frequencies.

A. G. Webster: I was very much interested in these papers when I heard them read. I went to New York with some qualms as to whether I had been anticipated. When I saw the amplifying tube I said, "That let's me out." Who knows whether the amplification of that tube is the same for all these frequencies. I said to Major Whitehead last year in his laboratory—"I have seen this wonderful description of the amplifier. Does it do it?" I will not tell you what he said, but those of you who have heard of experiments of that sort, will, I am sure, say that the electrical engineer is mistaken. We know that this is an exact thing, no hysteresis or eddy current. This is a perfection of simplicity, and therefore I recommend it. As to its being selective, it is selective. You can tune it to any one of these sounds and make a photograph and analyze it.

I believe I can give more satisfactory answers to all of these telephone engineer's queries than can be got by the instruments he gets up himself. They are handy, no doubt, and all that. I remember Lord Kelvin seeing one of my instruments several years ago, and he said "It was important that sound could be measured by electrical reading apparatus." I do not do it that way.

A. E. Kennelly: In reference to the set of equations which Dr. Jones has so clearly explained, it is interesting to observe how his algebraical and our graphical presentations of the behavior of the telephone receiver converge towards similar conclusions.

We are all indebted to Dr. Webster for his exhibition here of this ingenious acoustic apparatus. The fact that he has been able to reduce the damping coefficient to so low a value as has been mentioned is a very valuable achievement in itself.

[*Editor's Note:* The handwriting was now on the wall, but Webster would not look. See paper by H. B. Miller, *Acoust. Soc. Am. J.* **61**:174–181 (1977).]

8

Reprinted from *Nat. Acad. Sci. Proc.* **6**:316, 317, 318 (1920)

EXPERIMENTS ON THE VIBRATION OF AIR IN CONICAL HORNS

BY ARTHUR GORDON WEBSTER

At the meeting of the Committee on Sound of the National Research Council recently held at Geneva, Illinois, it was represented to me that my habit of not publishing results had been a serious detriment to the progress of certain investigators in sound and I was strenuously urged to reform my habits. I shall, therefore, take the liberty of bringing out of my drawers a certain number of papers some of which I have had for many years but which for one reason or another have not been published.

The curves presented herewith were shown at a meeting of the American Association for the Advancement of Science in Columbus in 1915,

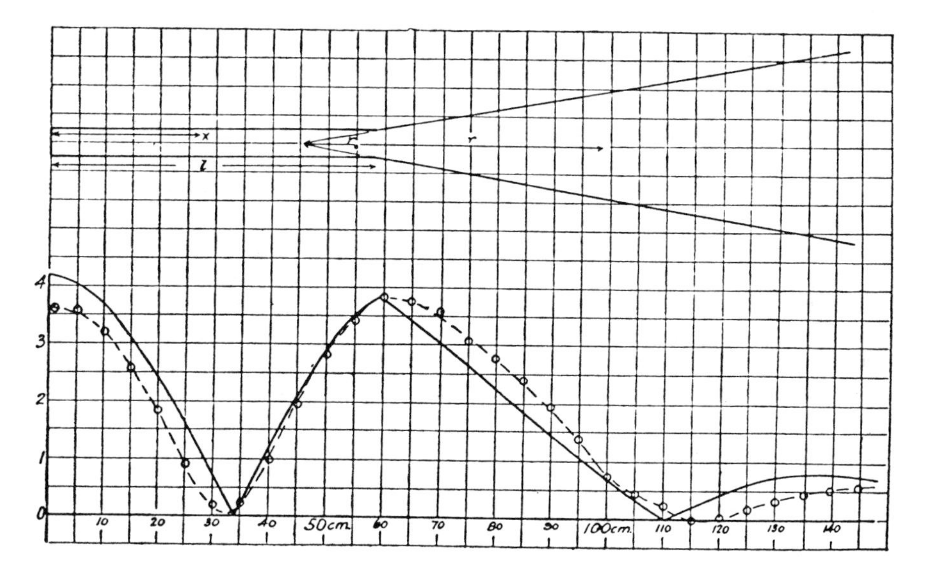

but I believe have not been published. They were made by means of the phonometer and phone described in my paper in these PROCEEDINGS. **5**, May, 1919 (163–166), and they are intended to verify the theory of horns given by me in these PROCEEDINGS, **5**, July, 1919 (275–283).

The horn is always a portion of a circular cone mounted on one end of a cylindrical tube, the other end of which is closed. The sound was made by the phone emitting a constant sound of a pitch of 256 per second. The standing waves inside the horn were explored by an antenna consisting of a glass tube of three millimeters internal diameter and several feet long attached to a disk closing the end of a cylindrical tube screwed into the opening of the phonometer. This could be put into any part of the space to be explored. To be sure, the pressure measured by the phonometer is not the same as that of the antenna, but inasmuch as standing waves are formed in the antenna the pressure measured by the phonometer is proportional to that at the end of the antenna and, therefore, to that in the place to be observed.

The figures printed herewith show on the dotted lines the actual observations and on the full lines the theory. The shape and dimensions of the horns are also shown on the figures. The experiments were made by my then assistant, Dr. H. F. Stimson, to whom I am indebted for his careful work.

It will be seen that the theory is very substantially verified, the defect in it being due to the very rough estimate made about the correction of the open end which, as I stated in my paper on horns, is far from being accurate. But inasmuch as it is the only theory that has ever been given I feel that the results constitute a distinct advance. Prof. G. E. Stewart has done a large amount of work on horns and he also confirms my theory . . . .

[*Editor's Note:* In the original, material follows this excerpt. Figures containing graphs have been omitted.]

EDITOR'S NOTES

Page 316, paragraph 2: The correct page numbers are not 163–166 but 173–179. The July NAS paper is much like the AIEE 1919 paper, containing much qualitative theory. The final sentence reads: "The theory of the phonometer and the phone is given in another paper." This devoutly wished consummation has yet to show itself.

Page 317, line 6: Some related measurements are described by P. B. Flanders in Paper 38 in this volume. See, for example, Figure 7.

9

Reprinted from pages 16–18 of *Nat. Res. Counc. Bull.* **4**(23):16–19 (1922)

THE MEASUREMENT OF SOUND INTENSITY IN ABSOLUTE UNITS

P. Sabine and A. G. Webster

The complete theoretical and experimental solution of this problem calls for an extended investigation. It involves the production of a constant and reproducible source of sound, of determinable acoustic output, a measuring instrument whose readings may be reduced to absolute dynamical units, and means for varying sound intensity in a known way. Up to the present, these three aspects of the question have never been thoroughly investigated in a single experimental research. A brief summary of the various modes of attack heretofore used on these different parts of the general problem follows:

I. Methods of Measuring Acoustical Output

(1) The density variation in a vibrating air column has been determined by measuring the shift of the interference fringes formed by two beams of light, one of which traverses the vibrating column. The fringes were viewed stroboscopically and the sound output computed from the theory of open pipes.

(2) The pressure variation in a sounding organ pipe has been measured by means of a delicate manometer and a synchronously operating valve. Obviously the method is limited to sounds of low pitch and great intensity.

(3) The acoustical output from telephonic sources has been determined. The method calls for an analysis of the diaphragm motion as a function of the electrical input, by actual study of this motion, using optical methods under varying conditions of current amplitude and frequency.

(4) The Phone, developed by A. G. Webster, is an instrument whose output may be computed from easily made measurements. It consists of a cylindrical resonator capable of tuning over a considerable range of frequency, with a tuned diaphragm mounted in the resonator opening. The latter is supported by three wires, and moves as a whole when actuated by alternating current of the same frequency as that of the resonator and the diaphragm. The amplitude of vibration is measured microscopically.

(5) The Thermophone has been suggested and used as a standard source of sound of easily calculable output. The passage of an alternating current through a very thin strip of metal foil produces rapidly damped temperature waves in the air near the metal strip. The

expansion and contraction of the air results in the generation of a sound wave whose intensity may be computed from the electrical and thermal coefficients of the strip, its dimensions, and the value of the alternating current.

II. Methods of Absolute Measurement

Various methods have been used for the comparison of sound intensities. Some which are susceptible of giving results in absolute units are as follows:

(1) The determination of the static pressure due to a train of waves gives immediately a value of the energy density of the train. The method has been employed in the analogous case of light. Such a method would be applicable only to sounds of very high pitch and short wave length.

(2) Wien and others have employed a form of vibration manometer. The corrugated diaphragm of an aneroid barometer is made part of the inner wall of a resonator tuned to the frequency of the sound to be measured. The motion of the diaphragm under the varying pressure of the sound was measured by the width of a band of light reflected from a small mirror, one point of which was in contact with the diaphragm and which rotated with the motion of the latter.

(3) The Condenser Transmitter has been proposed as a means of measuring in absolute dynamical units the pressure variations of sound. It is essentially an electrostatic telephone transmitter, the diaphragm constituting one plate of a variable air condenser. The fixed plate is very close to the diaphragm, the air film intervening being so small as to be an important factor in the diaphragm damping. The motion of the diaphragm varies the capacity of the condenser and produces a minute alternating current in the circuit. The latter is amplified by means of vacuum tubes to give measurable currents with a vacuum thermocouple and microammeter. The instrument must be calibrated in absolute units. The thermophone has been used for this purpose. Both instruments are in a closed space, the pressure changes being computed from the alternating currents in the thermophone that produce the sound.

(4) The Rayleigh Disc and Resonator has been used successfully for the determination of relative values of sound intensities. If the modulus of torsion of the fiber supporting the disc be known, the absolute value of the alternating air stream velocity may be determined from the deflection produced by the sound. This gives the intensity at the disc and within the resonator. For the determination of the absolute value of the sound as unaffected by the presence of the resonator,

correction must be applied for the effect of the suspended disc upon the motion of the air within the resonator, and for the ratio of the intensities inside and outside such a resonator.

(5) The Phonometer is in all respects similar in construction to the Phone already described. The diaphragm is tuned to unison with the sound to be measured, by varying the tension in the three supporting wires. Instead of using the microscope for measuring the diaphragm motion, however, a small concave mirror is caused to rotate by this motion. The width of the band of light caused by this motion measures the diaphragm motion. Theoretical treatment shows that the pressure amplitude of the air motion is proportional to the amplitude of the diphragm. The proportionality factor involves the mechanical properties of the diaphragm, its dimensions and the dimensions of the resonator, the frequency of the sound and constants of the medium. The construction of the instrument is such that all the necessary data for the reduction of the instrumental readings to absolute units of pressure on the diaphragm are determinable.

(6) If a very fine wire, heated to a dull red incandescence, be exposed to sound waves, there will result a periodic change of resistance. The wire may thus act as a receiver and the measurement of the alternating currents produced in the secondary of a transformer may be used as a basis for the comparison of sound intensities. The method has not been employed for absolute measurements, but it possesses possibilities of being adapted to this purpose.

[*Editor's Note:* In the original, material follows this excerpt. The Sabine writing here with Webster was not Wallace but Paul. This paper is part of the famous November 1922 *Bulletin of the National Research Council* on "Certain Problems in Acoustics," published by the National Academy of Sciences.]

10

Reprinted from *Sci. Progress* **20**:68–82 (1925)

THE MAGNIFICATION OF ACOUSTIC VIBRATIONS BY SINGLE AND DOUBLE RESONANCE

By E. T. PARIS, D.Sc., F.Inst.P.
Experimental Officer in the Signals Experimental Establishment, Woolwich

The problems which engage the attentions of investigators in acoustics may generally be placed under one or other of three headings, according as they refer to the production, transmission, or reception of sound. The first of the groups so formed is concerned with the mechanism of sound-sources, the second with the transmission of sound-waves through the atmosphere and other media, while the third has reference to the means by which we are able to detect and record sound-waves and to measure their amplitudes and wave-lengths. It is with one of the problems of the third group, relating to the employment of resonance with certain types of sound-receiving apparatus, that it is proposed to deal in the present article.

Existing sound-receiving instruments may be divided into two classes according as they are intended for the reproduction or recording of speech and musical compositions, or for purely scientific purposes such as the measurement (or comparison) of the amplitudes of sound-waves. Examples of the first class are the microphones used in telephony and " wireless," and the apparatus employed in the manufacture of gramophone records. In this class, also, we have instruments intended for the examination of wave-form, such as the phonodeik of D. C. Miller [1].[1] One of the principal requirements of such instruments is that they shall be capable of giving a uniform response over a wide range of frequencies. Resonance is to be avoided, for its existence implies that some sounds will be recorded more vigorously than others, so that the reproduced sound (or wave-form) will not have the same character as the original, and may, in fact, be very different. As the removal of all traces of resonance is generally impossible, it is usual to attempt to overcome this difficulty by making the resonance frequencies of the instruments either much higher or much lower than any that are likely to occur in the received sound.

[1] The numbers in brackets refer to the notes and references at the end of the article.

In instruments of the second class, on the contrary, tuning, or resonance, is generally an advantage. One reason for this is that, in experiments involving the measurement of sound-amplitudes, it is usually desired to concentrate attention on some particular wave-length, and the most convenient way of doing this is by employing a receiving instrument tuned to the frequency of the tone on which observations are to be made. A tuned instrument, moreover, has the advantage that it is less susceptible to disturbances by stray sounds—an important consideration when it is remembered that many acoustical experiments are necessarily conducted in the open air or in buildings the walls of which provide but an indifferent protection against unwanted noises.

The special advantage, however, of tuned reception lies in the very large " magnifications " which are obtainable by the employment of resonance. One of the principal difficulties in the way of making accurate quantitative measurements of acoustic phenomena is due to the minuteness of the motions and pressure-variations occurring in sound-waves, but the magnification of these by means of a resonator renders it possible to construct instruments in which the effects to be observed are comparatively large and easily measurable. The type of instrument here contemplated consists of two essential parts—a tuning device or resonator, and a sound-detecting device or detector. It is the function of the resonator to select sounds of a particular wave-length and to magnify the motions or pressure-variations to such an extent that they may be easily recorded by the detector. The last named is exposed to the magnified acoustical effects and translates them into terms of some easily observed quantity, such as the deflection of a light moving part, a change in ohmic resistance, and so on. Typical detectors are the hot-wire microphone grid and the Rayleigh disc. In the former it is a change in ohmic resistance which is observed, while in the latter it is the rotation of a small circular mirror suspended by a fine quartz fibre.

It is evident that the magnification of acoustic vibrations (whether motions of the air particles or pressure-variations) obtainable by the use of a resonator is an important element in determining the effectiveness of tuned receiving instruments. We will therefore consider the principal factors affecting magnification in some well-known instruments, beginning, as an introduction, with the simple case of the resonator as used aurally by Helmholtz in his classical work on the quality of musical notes.

The type of resonator used by Helmholtz—generally known as the " Helmholtz resonator "—consists of a spherical or cylindrical vessel (the exact shape does not matter) in which there is a single orifice providing communication between the interior of the vessel and the outside air. One of the forms used by Helmholtz—spherical in shape and having an orifice with a short neck—is shown diagrammatically in Fig. 1. The air in such a vessel has a natural frequency of its own, and the natural tone is easily elicited by tapping the sides of the vessel or blowing across the edge of the orifice. When the resonator is exposed to sound-waves having the same frequency as that of its own natural tone, the air in the orifice is set into sympathetic vibration, and moves with an amplitude which is many times greater than that of the original sound-wave. Similarly, the fluctuations in pressure inside the resonator, corresponding with the movements of the air in the orifice, are many times greater than those in the sound-wave, so that if the ear is placed in communication with the interior of the vessel, the particular tone to which the resonator responds is heard with increased loudness. It was in this way that Helmholtz used the resonator, the communications with the ear being by way of a small nipple shown at N, Fig. 1. The resonator, together with the ear applied to the nipple, constitutes a tuned receiving instrument in which the ear plays the part of the detector.

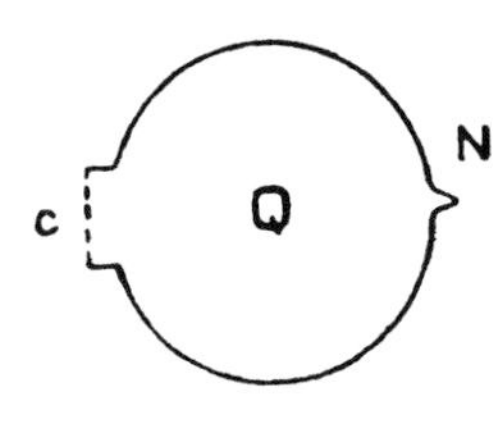

FIG. I.
Resonator as used by Helmholtz.

In order to form some estimate of the assistance which such a device gives to the ear, it is necessary to refer to the simple theory of the Helmholtz resonator. This theory is well known and is given in most textbooks on Sound, and a very brief recapitulation will suffice for our present purpose.

Let dq/dt be the instantaneous current of air (c.c. per second) in the orifice—directed inwards—and let L be the length of the neck. Then the rate of change of momentum in the orifice is $\frac{d}{dt}\left(\rho L \frac{dq}{dt}\right)$, ρ being the density of air. In order to obtain the equation of motion, this quantity must be equated to the force acting on the air in the orifice due to the rise of pressure dp inside the resonator caused by the introduction of the extra volume of air q. If σ represents the cross-sectional area of the neck, this force is $-\sigma.dp$. The changes of pressure corresponding with the movements of the air in the orifice are assumed to take place adiabatically, and accordingly $dp = a^2 d\rho = a^2\rho(q/Q)$, where a is the velocity of sound in air and Q is the internal volume of the resonator. The equation of motion of the air in the orifice is thus:

$$\frac{d^2q}{dt^2} + \frac{a^2}{Q}\left(\frac{\sigma}{L}\right)q = 0 \quad . \quad . \quad . \quad . \quad . \quad . \quad . \quad . \quad . \quad . \quad (1)$$

Actually, the motion of the air is not confined entirely to the orifice, but extends for a little way both inside and outside the resonator. The momentum of the moving air as given above is therefore underestimated, but a correction on this account can be made by adding to the length L, another length α, called the "end-correction." For a cylindrical orifice of radius R, Rayleigh has shown that the end-correction (for each end) is 0·8 R if there is a flange, and 0·6 R if there is no flange. Generally one end is flanged and the other unflanged, so that the total correction is 1·4 R. Let L^1 be the corrected value of L, then the linear quantity (σ/L^1) is constant for a given orifice. It is called the "conductance" of the orifice and is usually denoted by c and expressed in centimetres. Substituting c for (σ/L) in equation (1), we see that the frequency of the resonator is given by:

$$n = \frac{a}{2\pi}\sqrt{\frac{c}{Q}} \quad . \quad . \quad . \quad . \quad . \quad . \quad . \quad . \quad . \quad . \quad . \quad (2)$$

which is the formula given by Rayleigh. Frequencies calculated from (2) are found to agree very satisfactorily with experimental values. In order to determine the motion of the air in the orifice due to a sound-wave of given pressure-amplitude, say F, allowance must be made for the dissipation of sound-energy from the resonator. This may arise from two causes: (1) From the escape of sound-energy by way of the orifice into the surrounding air, and (2) from the operation of viscous forces in the orifice giving rise to the degradation of sound-energy into heat. The first of these is analogous to the escape of energy by radiation from an electrical oscillating circuit, while the second is analogous to the losses due to the ohmic resistance of the circuit. Allowance for the losses from both causes can be made in the equation of motion by the insertion of a term proportional to the current in the orifice, and the final equation from which we can determine the forced vibration due to a pressure-variation F cos wt at the mouth is [2]:

$$\frac{d^2q}{dt^2} + 2h\frac{dq}{dt} + w_0^2 q = \frac{cF}{\rho}\cos wt \quad . \quad . \quad . \quad . \quad . \quad . \quad . \quad (3)$$

where w_0 is written for $2\pi \times$ the natural frequency of the resonator, that is, for $a\sqrt{c/Q}$. The forced vibration is thus:

$$q = \frac{cF/\rho}{\sqrt{\{(w^2 - w_0^2)^2 + 4w^2h^2\}}} \text{ os } (wt - \theta) \quad . \quad . \quad . \quad . \quad . \quad (4)$$

It may be assumed that the resonator is exactly tuned to the frequency of the sound which it is desired to hear, so that $w = w_0$, and the amplitude of q as given by equation (4) is then simply $cF/2\rho w_0 h$.

The loudness of the sound heard by the ear depends on the amplitude of the pressure-variation occurring in the resonator. In order to obtain some idea of the advantage to be gained by the use of the resonator, we will define the "pressure-magnification" as the ratio of the maximum pressure-variation inside the resonator to the maximum pressure-variation in the incident sound-wave. Since $dp = a^2\rho(q/Q)$, it follows that the maximum pressure-change inside the resonator is $\frac{a^2\rho}{Q}\cdot\frac{cF}{2\rho w_0 h}$, or $\frac{w_0 F}{2h}$. The pressure-magnification is the ratio of this quantity to F, that is, $\frac{w_0}{2h}$, or $\frac{\pi n_0}{h}$, where n_0 is the number of vibrations per second in the resonant tone. It is thus proportional to the frequency of the resonator and inversely proportional to the damping factor.

It may be noted here that there is another kind of magnification shown by resonators, namely, the velocity-magnification, to which we shall have occasion to refer later. As it is important to distinguish between the two kinds of magnification, we will denote the pressure-magnification by M_p and the velocity-magnification by M_v. Thus $M_p = \frac{\pi n_0}{h}$.

To find a numerical value for M_p in any particular case it is necessary to discover the value of h. It has previously been mentioned that the magnitude of h is dependent on two factors, namely, the rate at which sound escapes from the resonator by radiation, and the rate at which it is converted into heat by the operation of viscous forces in the orifice. In the case of resonators with wide mouths the losses due to viscosity are small compared with those due to radiation, and in these circumstances the value of h may be calculated by means of a well-known formula, viz. $h = (\pi n_0^2 c/2a)$ seconds^{-1} [3], the modulus of decay being the reciprocal of this quantity, *i.e.* $2a/\pi n_0^2 c$ seconds. Substituting this value of h in the expression for M_p we obtain $2a/n_0 c$, by means of which we can calculate how many times the pressure-variation inside the resonator is greater than the pressure-variation in the original sound-wave.

As an example, consider the case of a resonator tuned to respond to a tone of 256 vibrations per second, and having a short-necked circular orifice of 3 cm. in diameter. The conductance of such an orifice is approximately equal to its diameter [4]. The pressure-magnification is therefore $\frac{2a}{n_0 c} = 88$. This shows how great an increase in sensitiveness to a particular tone can be obtained with the help of such a simple piece of apparatus as a Helmholtz resonator. In practice there will inevitably be some loss of acoustical energy in the communication of the pressure-changes to the ear through the nipple (N, Fig. 1), so that an estimate of the increase in sensitiveness based on the above figure would err somewhat on the side of being too great.

The above expression for the pressure-magnification gives some indication of the lines along which we should proceed in order to discover the best dimensions for the construction of a resonator of given frequency. If n_0 is fixed, the value of M_p increases as c is made smaller, but it must be remembered that in obtaining the expression for M_p allowance was made only for the radiation losses from the resonator. If c is diminished by reducing the size of the orifice, a point is soon reached at which (as will be seen later) viscosity losses must also be taken into account. There is, in fact, a limit to the extent to which the orifice may be reduced, the advantage gained by any further diminution in the value of c being outweighed by the increase in damping due to viscosity. Hence, for every frequency, there exists an optimum value of c which, if employed for the construction of a resonator, will give the maximum value to the pressure-magnification. In the present imperfect state of our knowledge regarding the viscosity losses in orifices of various shapes and sizes, however, it is not possible to give any precise information concerning these optimum values of c.

A modern application of the Helmholtz resonator to sound-reception is to be found in the selective hot-wire microphone. The resonator as used aurally after the manner of Helmholtz is a qualitative instrument, useful only for identifying and establishing the existence of tones which would otherwise be faint or inaudible to the unaided ear. In order to obtain quantitative observations it is necessary to replace the ear by a detector, the response of which is some function of the amplitude of the vibration induced in the resonator. This is, in effect, what is done in the hot-wire microphone invented by Major W. S. Tucker [5]. In this instrument a short length of fine platinum wire, $6\,\mu$ in diameter, is mounted in the orifice of a resonator. The wire is arranged in the form of three loops attached to a small porcelain bridge, the arrangement being known technically as a "grid." The grid is heated to just below red heat by means of an electric current of about 25 milliamperes, and when hot has a resistance of 200 to 300 ohms. When the microphone is exposed to sound-waves in tune with the resonator, the vibratory motion of the air in the orifice cools the grid and there is a consequent drop in its ohmic resistance. This can be measured by one of the usual standard methods (say, by a Wheatstone bridge), and provides a ready means of estimating the strength of the sound affecting the microphone.

It will be seen that there is an essential difference between the way in which the resonator was used by Helmholtz and the way in which it is used in the hot-wire microphone. In the former case it was employed to obtain enhanced pressure-variations, which were then transmitted to the ear by way of the nipple. In the latter case the resonator is used to obtain increased movements of the air particles, which movements are recorded by the hot-wire grid. As will be seen later, it is the velocity-magnification (M_v) which is the important factor in determining the sensitiveness of hot-wire instruments.

The resonators used in the construction of hot-wire microphones are usually cylindrical in shape and have the form shown in section in Fig. 2. The orifice (*c*) in which the heated grid (M) is mounted is also cylindrical, and on account of the necessity for shielding the grid from accidental currents of air it is usually made to quite small dimensions, a common size being 10 mm. long and 5 mm. internal diameter. The instrument is made in brass with ebonite insulation where necessary, and tuning is accomplished by adjusting the internal volume Q of the cylindrical body of the microphone.

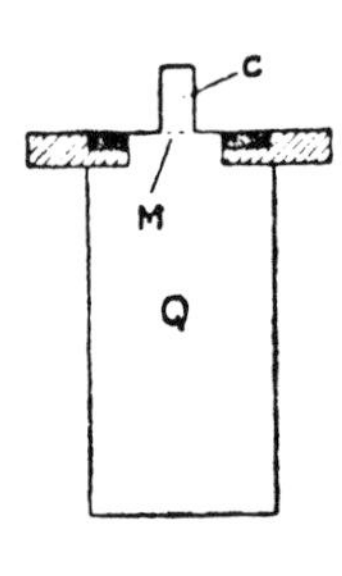

Fig. 2.
Form of Helmholtz resonator used in hot-wire microphone.

In order to understand fully the working of this type of microphone it would be necessary to establish two relations, one connecting the pressure-amplitude in the sound-wave with the amplitude of the vibration in the orifice, and another connecting the amplitude in the orifice with the ohmic change in the grid. If both these relations were known it would be possible to employ the hot-wire microphone for measuring the amplitude of sound-waves in C.G.S. units. The first part of the problem concerns the performance of the resonator and can be solved by means of the equations already given; but the second part of the problem, relating to the behaviour of the grid, still awaits a complete and satisfactory solution. It is known, however, that for small amplitudes at a given frequency the ohmic change in the grid is proportional to the square of the maximum velocity in the orifice [6], while for larger amplitudes experiments by R. C. Richards [7] indicate that the ohmic change is a function of the maximum velocity only and is independent of the frequency. We see, therefore, that in examining the performance of the resonator in relation to the sensitiveness of the instrument, it is the velocity-magnification (M_v) to which attention must be directed. This velocity-magnification may be defined as the ratio of the maximum velocity occurring in the orifice of the resonator to the maximum (particle) velocity in a *plane* wave which is stimulating the microphone [8]. It is necessary to specify the kind of wave because it is the pressure-amplitude in the wave which determines the motion in the orifice, and the relation between pressure-amplitude and particle-velocity is dependent on the shape of the wave fronts.

The maximum velocity in the orifice of the resonator is $\frac{1}{\sigma}\left|\frac{dq}{dt}\right|$, and, considering only the case of exact tuning $(w = w_0)$, we see from equation (4) that $\frac{1}{\sigma}\left|\frac{dq}{dt}\right| = \frac{cF}{2ph\sigma}$. The theory of plane waves of sound shows that the maximum particle-velocity associated with a pressure-amplitude F is F/ap. The velocity-magnification obtainable from the resonator is therefore $M_v = \left(\frac{cF}{2ph\sigma}\right)\Big/\left(\frac{F}{ap}\right) = \frac{ac}{2h\sigma}$. This expression is, of course, applicable not only to hot-wire microphones, but to other instruments in which a detector is placed in the orifice of a resonator. Such an instrument, for example, is the late Prof. A. G. Webster's " phonometer " [9], in which a very light disc carrying a small mirror is mounted in the orifice. The disc is set in forced vibration by the oscillatory air-currents, and the amplitude of its motion is observed by means of a beam of light reflected from the mirror, after the manner of a vibration-galvanometer.

Comparing the value of M_p with that of M_v, we see that $\frac{M_p}{M_v} = \frac{2\pi n_0 \sigma}{ac} = \frac{2\pi\sigma}{\lambda c}$, λ being the wave-length of the sound. Since the conductance c is equal to σ/L^1, it follows that, for any particular resonator, the pressure-magnification is to the velocity-magnification in the proportion $2\pi L^1/\lambda$. Now, all the dimensions of a Helmholtz resonator are small compared with the wave-length of the natural tone, so that L^1/λ is a very small fraction (generally between ·01 and ·05), and hence it follows that, for the same resonator, M_p is very much less than M_v.

It is found by experiment that the values of the damping factor h for the rather narrow orifices used in the construction of hot-wire microphones are very much greater than would be required to account for radiation losses alone. To illustrate this point the case may be quoted of a resonator which responded to a tone of 256 vibrations per second, and which had a cylindrical orifice 10 mm. long and 4·5 mm. in diameter. The damping factor of this resonator, determined experimentally by a method which need not be described here, was found to be 49 seconds^{-1}, whereas the damping factor calculated from the radiation formula $\left(h = \frac{\pi n_0^2 c}{2a}\right)$ is only 0·7 seconds^{-1}. Thus in the case of so small an orifice, the radiation losses are almost negligible, and practically the whole of the damping must be attributed to the effects of viscosity.

The expression for M_v given above indicates that the magnification of velocities obtainable by the use of a resonator can be increased by reducing the cross-sectional area σ of the orifice. This reduction in cross-section, however, is accompanied by a rise in the damping factor owing to the greater effects of viscosity. Hence, as in the case of pressure-magnification, there is a limit beyond which it is no longer advantageous to reduce the size of the orifice. In spite of the adverse effects of viscosity the magnifications obtainable from resonators with quite small orifices are nevertheless surprisingly large. Thus, in the case of the resonator mentioned previously with a frequency of 256 vibrations per second, we have $\sigma = \frac{\pi(0 \cdot 45)^2}{4}$ cm.^{2}, and $h = 49$ sec.$^{-1}$. The value of the conductance c calculated from equation (2) by inserting the appropriate values of n_0 and Q, was 0·115 cm., so that putting $a = 33{,}760$ cm. per second, we find that $M_v = \frac{ac}{2h\sigma} = 249$. That is, the motions in the sound-wave are magnified nearly two hundred and fifty times in the orifice of the resonator.

Although this result shows that the velocity-magnifications obtainable by the use of Helmholtz resonators are very great, it is found in practice that the sensitiveness of singly resonated instruments is insufficient for the requirements of some quite ordinary experiments. In order to procure still greater sensitiveness, tuned microphones are made with double instead of single resonators. This application of double resonance to sound-receiving instruments for the purpose of enhancing sensitiveness is due to Prof. C. V. Boys, who used the principle many years ago for making a very delicate Rayleigh disc instrument now known as the " Tonomicrometer " [10].

A simple form of double resonator, called the Helmholtz double resonator, is shown in Fig. 3. It can be used with either Rayleigh disc or hot-wire grid, and consists of two Helmholtz resonators in series. The sound enters through the orifice c_1 of the large outer resonator Q_1. The small inner resonator Q_2 communicates with the interior of the outer resonator through the narrow orifice c_2. It is in this orifice that the detector (Rayleigh disc or hot-wire grid) is placed.

Very little experience with singly and doubly resonated

instruments is sufficient to show that the use of a properly tuned double resonator in place of a single one is accompanied by a very marked increase in sensitiveness. The construction of a double resonator which will be very sensitive at a particular frequency is, however, by no means a simple matter. Suppose, for example, that we have a small resonator tuned to a frequency n_0 and that it is desired to enhance its sensitiveness by the use of double resonance. If we combine this small resonator with another larger one, also tuned to n_0, in the manner shown in Fig. 3, it will not in general be found that the sensitiveness at n_0 is greatly increased, although this may happen in special cases. It will, however, be found that the instrument is very sensitive at *two* frequencies, one of which is greater and the other less than n_0. The two resonators, in fact, when combined together, react one on the other, forming a coupled system with two degrees of freedom, and exhibiting the characteristics usually associated with such systems. The extent to which the two affect each other is determined by the relative size of the orifices c_1 and c_2. If these are of comparable dimensions, the reaction between the two resonators will be great, for it is obvious that the presence of the inner resonator will then have considerable affect on the pressure-changes in the outer one. The coupling in this case is " tight." If, on the other hand, the inner orifice is very small compared with the outer one, the presence of the inner resonator has but little effect on the pressure-changes in the outer one, and the coupling is said to be " loose."

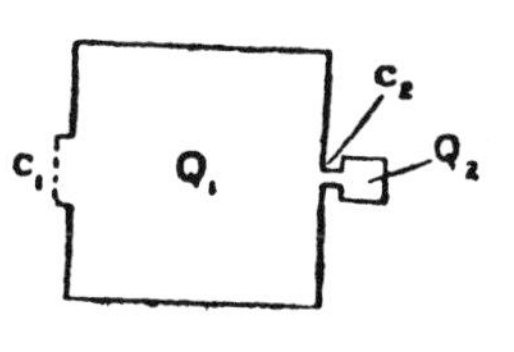

FIG. 3.
Helmholtz double resonator.

The case of extremely loose coupling, when the reaction between the component resonators may be neglected, is susceptible to easy theoretical treatment, and the result, although of limited application, is of some interest as it illustrates the idea underlying the use of double resonance for the purpose of obtaining increased magnification. Suppose, then, that the inner orifice is so small that the motions and pressure-changes in the outer resonator may be regarded as approximately unaffected by the presence of the inner one. If both the component resonators are tuned to the same frequency the sound may be supposed to be magnified first of all by the outer resonator and then again by the inner one. The magnification due to the outer resonator is that of the *pressure*-variations in the incident wave, the magnified pressures being produced inside the outer resonator and thence operating on the inner one where the second magnification takes place. Thus, let $M_p{}^1$ be the pressure-magnification ordinarily associated with the outer resonator, so that if F is the pressure-amplitude in the incident wave, the magnified pressure-amplitude inside Q_1 will be $M_p^1 . F$. If the orifice c_2 is a narrow one, as is supposed to be the case, the radiation losses from it may be disregarded, since they will be small compared with the viscosity losses. Hence the pressure-variation in Q_1 will produce an oscillation in the orifice c_2 which will have approximately the same amplitude as if the inner resonator alone were exposed to a plane sound-wave of pressure-amplitude $M_p^1 . F$. Let M_v^{11} be the velocity-magnification belonging to the inner resonator, then we see that the overall velocity-magnification due to the double resonator is the product $M_p^1 . M_v^{11}$, and the advantage of the double over the single resonator is expressed by the factor M_p^1.

If the above theory were strictly applicable to resonators used in practice, the magnifications obtainable would be very large indeed, since the fullest use would be made of the magnifying powers of both components. It cannot, however, be used for any but extreme cases of loose coupling, and then only as a means of obtaining an approximation which should be regarded as an upper limit to the magnification.

In general, account must be taken of the reaction between the resonators, and for this purpose it is necessary to develop a more complete theory, the principal results of which may be briefly stated. In common with all coupled systems of this kind, the double resonator exhibits two resonance frequencies, one of which is always higher and the other lower, than the frequency of either component. By an extension of the method outlined above for obtaining the frequency of a single Helmholtz resonator (or by other means, such as an application of Lagrange's equations), an equation can be derived by means of which the resonance frequencies of the double resonator can be calculated. This equation, which was first given by Rayleigh in 1870 [11], may be written in the form:

$$n^4 - n^2 \{(\mu + 1)n_1^2 + n_2^2\} + n_1^2 n_2^2 = 0 \quad . \quad . \quad . \quad . \quad (5)$$

in which n_1 is the frequency of the outer resonator, n_2 the frequency of the inner one, and μ is the ratio $\frac{c_1}{c^2}$. The roots of this equation, treated as a quadratic in n^2, give the squares of the resonance frequencies of the double resonator. It can also be shown [12] that the amplitude of the oscillations in the orifice c_2 corresponding to a pressure-variation F cos $2\pi nt$ at the mouth of the resonator is:

$$\left|\frac{dq_2}{dt}\right| = \frac{n n_1^2 c_2 F / 2\pi\rho}{[\{(n_1^2 - n^2)(n^2_\mu - n^2) - \mu n_1^4 - h_1 h_2 (n/\pi)^2\}^2 + (n/\pi)^2 \{h_1(n^2_\mu - n^2) + h_2(n_1^2 - n_2)\}_2]^{\frac{1}{2}}} \quad . \quad . \quad (6)$$

when h_1 and h_2 are the damping factors associated with the orifices c_1 and c_2, and for brevity n^2_μ is written for $n_2^2 + \mu n_1^2$. The velocity-magnification obtainable from the double resonator is the ratio of the maximum velocity in the inner orifice to the maximum velocity in the incident plane wave. If σ_2 is the cross-sectional area of the inner orifice, this ratio is:

$$\left(\frac{1}{\sigma_2}\left|\frac{dq_2}{dt}\right|\right)\Big/(\mathrm{F}/a\rho).$$

The above equations allow us to calculate the velocity-magnification, say $\mathrm{M}_v^{\mathrm{III}}$, obtainable from a double resonator in any particular case. As an example let us consider the case of a double resonator made from the two resonators (tuned to 256 vibrations per second) which have been previously mentioned. In this case we have $n_1 = n_2 = 256$. The orifice of one of these resonators was supposed to be wide and to have a conductance of 3 cm., while the orifice of the other was narrow with a conductance of 0·115 cm. The damping factor h_1 for the large orifice is calculated by means of the formula already given. The damping factor of the smaller orifice is 49 sec.$^{-1}$. From these values we find that the resonance frequencies of the double resonator are 232 and 282 vibrations per second, and that the way in which $\mathrm{M}_v^{\mathrm{III}}$ changes with the frequency of the incident wave is that shown in Fig. 4. The magnification for a tone of 256 is not much greater than with the small single resonator, being only about 389 against the previous value of 250. At the resonance frequencies, however, the magnification is much greater, being about 1,180 at the lower tone of 232 vibrations per second, and 1,020 at the upper tone of 282 vibrations per second.

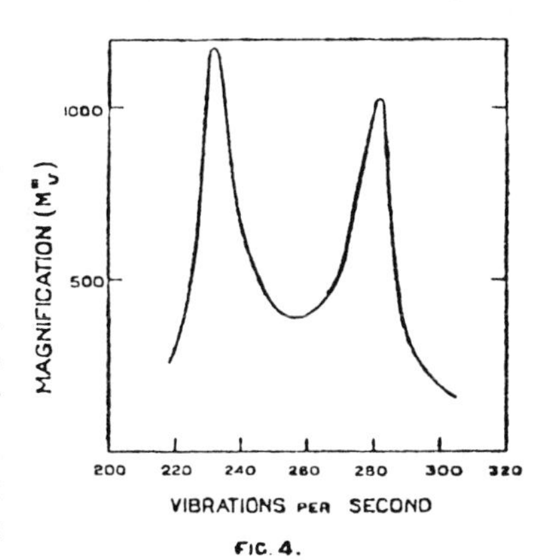

FIG. 4.

Variation of Magnification with frequency of sound.

It is obvious that to obtain good magnification for a tone of 256 vibrations per second it would be necessary to design a resonator having components with natural frequencies either both above or both below 256.

The greatest magnifications are obtained with this type of resonator when the coupling is loose, that is, when a condition is approached such that the simple theory is applicable and $\mathrm{M}_v^{\mathrm{III}} = \mathrm{M}_v^{\mathrm{I}} . \mathrm{M}_v^{\mathrm{II}}$. In the example quoted above, the coupling is fairly tight, so that the resonance frequencies are well separated. In order to reduce the coupling, it is necessary to make the orifice c_2 very small, and this, as explained in the case of single resonance, tends to increase the damping of the resonator on account of viscosity losses, so that there is no advantage to be gained by making the orifice less than a certain size. Moreover, the fact that the detector must be mounted in the orifice c_2 sets a limit to the amount by which it can be reduced. Much looser coupling can, however, be obtained than is indicated by the curve in Fig. 4, without seriously impairing the sensitiveness of the instrument.

There is another type of double resonator which it is convenient to use for certain purposes, in which the coupling can be varied in a different manner. This is the Boys pattern double resonator, first used by Prof. C. V. Boys for the construction of very sensitive Rayleigh disc instruments. The Boys resonator consists of a Helmholtz resonator combined with a resonator of the "stopped-pipe" variety, after the manner shown diagrammatically in Fig. 5. The detector is placed in the orifice of the Helmholtz resonator at *c*. The general behaviour of a double resonator of this kind is very similar to that of the Helmholtz pattern, although the theory of its action (which need not be given here) is a good deal more complicated. In order to obtain sensitiveness, the stopped pipe must have the same resonant tone—which may either be the fundamental or one of the overtones of the pipe—as the Helmholtz resonator.

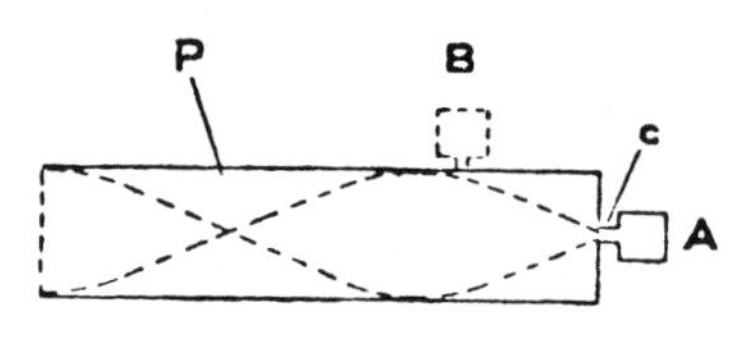

FIG. 5.
Boys Double Resonator.

The way in which the coupling can be varied in this type of resonator is as follows: It is found that the coupling is closest when the Helmholtz resonator is placed with its orifice through the wall of the pipe at a position which would normally be a node. Such a position, for example, is the closed end of the pipe, as at A, Fig. 5. If, however, the Helmholtz resonator is moved away from the closed end and towards a position where a loop is to be expected in the stopped pipe, the coupling is found to be loosened, as is evidenced by the coming together of the resonance frequencies, while at the same time the double resonator becomes more sensitive. After a certain position is passed (approximately three-quarters of the distance from node to loop), the sensitiveness again falls off, and in the vicinity of the loop there is practically no response, although the coupling is very loose. These properties of the Boys resonator are illustrated by the curves in Fig. 6, which are resonance curves taken by means of a hot-wire microphone [13]. The ordinates are proportional to the ohmic change in the resistance of the grid, and hence approximately to the energy of the forced vibration, while the abscissæ are the frequencies of the incident sound. The curve (*a*) is that which was observed when the Helmholtz resonator was at the closed end of the pipe, as at A, Fig. 5. The curve (*b*) corresponds to a position similar to B, Fig. 5, about three-quarters of the way to the first loop. Comparing (*b*) with (*a*) we see that the resonance peaks have come much closer together and the sensitiveness is much greater, although the response is over a shorter range of frequencies. The curve (*c*) shows the very small effect which is observed when the Helmholtz resonator is placed at a loop.

It is seen, therefore, that if a Boys double resonator is used, there is this alternative method of varying the coupling so as to obtain increased sensitiveness. In this method there is no necessity for the reduction of the orifice of the inner resonator to very small dimensions, and thus excessive losses due to viscosity may be avoided.

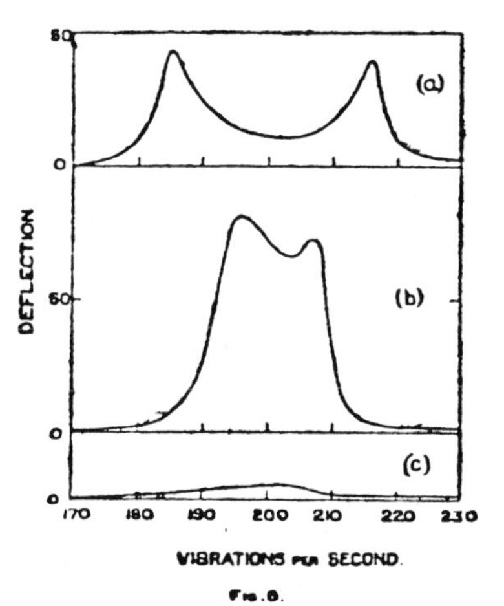

Fig. 6.
Resonance curves for Boys double Resonator.

In addition to increased sensitiveness, there is another advantage which a doubly resonated instrument possesses which should be mentioned here. In sensitive singly-resonated instruments the tuning is necessarily sharp, so that any slight mistuning between the sound to be observed and the instrument itself may result in a considerable change in the numerical value of the magnification, and this may give rise to errors in the measurements which are being made. Doubly resonated instruments, however, may be designed in such a way that the possibility of error on this account is to a great extent removed. This can be done by arranging for loose coupling so that the resonance frequencies are sufficiently close together for the trough between them, as shown in the resonance curve, not to fall very far below the peak values. The magnification for such a resonator is then approximately the same over a short range of frequencies between the peak positions, and if the frequency of the sound to be measured comes within this range, the errors due to inexact tuning are greatly minimised. The curve shown at (*b*), Fig. 6, is an example—though not a particularly good one—of fairly uniform magnification over a short range of frequencies, in this case from about 195 to 205 vibrations per second.

Enough has now been said to indicate the general nature of the problems which are being investigated in connection with the application of single and double resonance to the construction of sound-receiving instruments. The theory and use of such resonators is of great importance in the design of sensitive acoustical receivers, and the determination of the numerical value of the magnification is a necessary preliminary to the employment of such instruments for the measurement of sound-amplitudes in absolute units. For the measurement or comparison of sound-amplitudes, a doubly resonated instrument is generally to be preferred, both on account of its greater sensitiveness and broader tuning.

In conclusion, it may be noted that the interest attaching to the problems discussed above is not confined entirely to acoustics. Most coupled acoustical systems have their counterparts in electrical circuits and the same mathematical treatment is applicable to both sets of phenomena. The pursuit of these electrical and acoustical analogies is a matter of considerable interest, and may be attended with results beneficial to both sciences.

NOTES AND REFERENCES

1. *The Science of Musical Sounds*, p. 79 (1916).
2. Rayleigh, *Theory of Sound*, vol. ii (2nd ed.), pp. 194–5.
3. Rayleigh, *loc. cit.*, pp. 194–5; Lamb, *Dynamical Theory of Sound*, p. 265.
4. Rayleigh, *loc. cit.*, p. 178.
5. Tucker, W. S., and Paris, E. T., " A Selective Hot-wire Microphone," *Phil. Trans.*, A, vol. 221, pp. 389–430 (1921).
6. Tucker, W. S., and Paris, E. T., *loc. cit.*, p. 410.
7. *Phil. Mag.*, vol. xlv, pp. 926–34 (1923).
8. Cf. Tucker and Paris, *loc. cit.*, p. 401, where the term " magnification " was inadvertently applied to the ratio of the *current* in the neck to the *velocity* in the sound-wave. The meaning intended was that now given.
9. Described in *Nature*, vol. cx, pp. 42–5 (1922).
10. *Dictionary of Applied Physics*, vol. iii, p. 723.
11. *Theory of Sound*, vol. ii, p. 191; *Sci. Papers*, vol. i, p. 43.
12. *Proc. Roy. Soc.*, A, vol. ci, p. 408 (1922).
13. For details see *Phil. Mag.*, vol. xlviii, p. 783 (1924).

EDITOR'S NOTES

Page 70, Equation 1: We could rewrite Equation 1 as:

$$\frac{d^2q}{dt^2} + \frac{c^2}{V}\,[\rho/(\rho L/\sigma)]q = 0.$$

Using modern terminology, c is the velocity of sound; V is the volume of the resonator; $(\rho L/\sigma)$ is the acoustic mass

m_{ac} of the air in the orifice; $\rho c^2/V$ is the acoustic stiffness K_{ac}. Then we have $m_{ac} d^2q/dt^2 + K_{ac} q = 0$. And of course $\omega_0 = \sqrt{K_{ac}/m_{ac}}$. Note that conductance C or (σ/L) can always be replaced by ρ/m_{ac}.

Page 73, lines 16–19: This is where Rayleigh went astray in the conclusion of his 1918 paper, "Note on the Theory of the Double Resonator," *Phil. Mag.* **36**:231–234. This paper could be considered a collector's item—the rare occasion when a Rayleigh solution missed the mark.

Page 75, center of page: The equation for M_v is basically u_1/u_0, as defined in the first line of page 75. Today, remembering that F is pressure, we would write u_1 or $CF/2\rho h\sigma$ as $(\rho p/m_{ac}) \cdot (2\rho h\sigma)^{-1} = p/2m_{ac} h\sigma = p/R_{ac}\sigma$. That is: in the usual damping term $\exp[(-R_{ac}/2m_{ac})t]$, $R_{ac}/2m_{ac} = h$; and $2m_{ac}h = R_{ac}$. So $u_1 = p/R_{ac}\sigma$. And $u_0 = p/\rho c$. Therefore, $M_v = \rho c/R_{ac}\sigma$. And indeed Paris's $aC/2h\sigma$ can go directly to $\rho c/R_{ac}\sigma$. Note that R_{ac} is in acoustic ohms and that $R_{ac}\sigma$ is in "specific ohms," like ρc. The beauty of working with the specific resistance $R_{ac}\sigma$ (which is really the resistance, acoustical or mechanical, normalized to an orifice area of 1 cm^2) is that it allows us to work directly with pressure and linear velocity. If we worked with volume velocities and with various orifices having different areas, we would obscure the point that normalizing brings out—namely, just how the linear velocity varies.

Page 76, lines 6–8: See note for page 73. Also note that $R_{ac}\sigma = [R_{radiation} + \text{Viscosity Coeff.}\ (L'/\sigma^2)]\sigma$. The second term is the viscous flow resistance. Hence, as σ becomes small, $R_{ac}\sigma$ becomes controlled by the viscosity coefficient times L'/σ, where L' is the effective length of the orifice.

Page 77, center of page: The two cases, tight coupling and loose coupling, are discussed in Paper 11 at the end of Part I.

Page 78, line 14: Recall that a quantitative example of M_p' was worked out on page 72—namely, $M_p' = 88/1$, or +39 dB.

Page 79, Figure 4: The reader would do well to draw onto Figure 4 a single-mesh circuit's resonance curve, whose peak value is 250, as calculated on page 76. This value is approximately 4 dB lower than the dip value of 389 shown in Figure 4.

11

A NOTE ON THE COUPLED CIRCUITS OF PARIS, WEBSTER, AND RADIO ENGINEERING

H. B. Miller

This article was prepared expressly for this Benchmark volume.

We will start with Figure 4 of Paris's paper and the Editor's Note pertaining to that figure, which suggests superposing a single-mesh resonance curve and finding that this peak is then 4 dB below the two-mesh dip. This result is hard for a modern communications engineer to accept because it is contrary to his experience. Referring to Morecroft's[1] Figures 119, 120, and 121, if the two-mesh circuit has a coupling coefficient $k \simeq 36\%$ or even 18%, and $Q_1 = Q_2 \simeq 15$, the two-mesh response curve is everywhere lower than the one-mesh curve's peak value. This is shown qualitatively in our Figure 1, which is a sketch of linear current versus frequency and is shown more quantitatively in Morecroft and in Terman,[2] where a flat-topped bandpass filter is seen to be obtainable.

How do we get from our Figure 1 to Paris's (modified) Figure 4—or, equally well, to Webster's double-peaked output curve, which we shall treat as being identical with Paris's Figure 4? We start with a single mesh, which we will call mesh #2, as exemplified, for example, by the small inner Helmholtz resonator of Paris's Figure 3, having volume V_2. Call the stiffness reactance at resonance $|X_2| = 1000$ ohms, and the resistance $R_2 = 100$ ohms. We will assume that we have already done everything possible to minimize the loss resistance R_2 (= 100 ohms) and to maximize the linear velocity i'_2 through the orifice by making conductivity or diameter c_2 as small as possible (since c_2 acts as a velocity transformer). Nevertheless, we still have a single-mesh circuit with a $Q_2 = 10$. At resonance the velocity, or current in the electric analog, is $i'_2 = E_0/R_2$, where E_0 is the generator voltage or the pressure of an incoming sound wave, and the single prime in i'_2 refers to the single-mesh case. Since $R_2 = 100$ ohms, $i'_2 = E_0/100$.

How can we reduce the effective value of R_2, at least in the neighborhood of resonance? Professor C. V. Boys came up with the solution. He realized that you can reduce the effective value of R_2 around resonance by coupling of a higher-Q added mesh (#1) to the given mesh (#2), if you follow certain constraints. Then it can be shown that in the region of resonance $i''_2 \simeq E_0/(|X_1| + R_2/Q_1)$. Here the double prime in i''_2 refers to the secondary current in the two-mesh case; Q_1 is the Q of the primary or added mesh (#1) when isolated; and $|X_1|$ is the stiffness reactance of mesh #1, as seen, for example, in Paris's Figure 3. If $Q_1 \simeq 100$, then $R_2/Q_1 = 1$, which is now negligible.

So we have reduced the effective R by 100 times. Unfortunately, we have introduced an additional impedance $|X_1|$, in series with E_0 and R_2/Q_1.[3] So the next task is to make $|X_1|$ as small as possible. This can be done by making the outer Helmholtz resonator, or mesh #1, as large in volume V_1 as possible, say, $V_1/V_2 = 100/1$. (This requires diameter ratios of only 4.6/1.) Then $|X_1|$ is 1/100 of $|X_2|$, or 10 ohms. As a bonus, as V_1 increases, its mouth area also increases (cf. Fig. 3) and so viscous losses in mesh #1 are negligible and therefore it is easy to obtain a Q_1 ten times higher than Q_2. Then $i''_2 = E_0/(10 + 1)$.

So the increase in gain is $i''_2/i'_2 = 100/11 = 9/1$ or +19 dB. This is true at the dip region of the two-mesh response curve. At the two peaks, things are even better. We can pick up an extra 1 to 4 dB at the peaks, which is shown in our Figure 2.

The brilliant daring of Webster was to let a high-Q acoustical resonator react on his low-Q mechanical resonator in order to drastically reduce the effective value of his irreducible mechanical losses. This kind of coupling was done about thirty years ahead of Hanna and Slepian, who coupled a metal diaphragm to an air chamber with an exit hole, which then fed a horn.[4]

One further point: $|X_1|$ in Paris refers to a shunt acoustic stiffness as the mutual element, but $|X_1|$ in Webster[5] refers to a shunt acoustic mass as the mutual element. However, the circuit shown in our Figure 2 handles both cases if we merely interchange the locations of L_1 and C_1 when dealing with Webster. To say it differently, in Paris (and in early Webster) the Helmholtz-resonator ***stiffness*** is bypassed by the given resonator; whereas in later Webster, the Helmholtz-resonator ***mass*** is bypassed by the given resonator. This allows Webster's moving plate to take advantage of the transformer action of the orifice.

NOTES AND REFERENCES

1. J. H. Morecroft, 1928, *Principles of Radio Communication,* 2nd ed., Wiley, New York.
2. F. E. Terman, 1937, *Radio Engineering,* 2nd ed., McGraw-Hill, New York. See especially Figure 38.
3. An intuitively appealing derivation comes by way of Thevenin's theorem. Thus, at resonance, i''_2 is:

$$\text{(a) } Q_1E_0/(Q_1{}^2R_1 + R_2),$$
$$\text{or (b) } Q_1E_0/(Q_1|X_1| + R_2),$$
$$\text{or (c) } E_0/(|X_1| + R_2/Q_1).$$

Form (b) implies that if $Q_1|X_1|$ is made $\ll R_2$, then $i''_2 \approx Q_1E_0/R_2 = Q_1i'_2$.

This is the limit, never remotely approached, that Paris was referring to in his $M_p' \cdot M_v''$ relation shown on his page 78. A qualitative explanation for why the system works at all is as follows: if $X_2 \gg X_1$ (say 100:1), the resonance curve for mesh #1 alone is not "punctured" into a dip surrounded by two peaks, when the high-impedance mesh #2 goes into resonance. The second mesh's response then tends to ride on the shoulders of the first mesh's response, as is shown in Paris's Figure 4 and Figure 6b. This puncturing, however, which is implied in our Figure 1, is precisely what happens when X_2 is of the order of X_1, or even ten times higher. Thus note that when $X_2 \approx 10X_1$, this means

Current (Linear)
k ≅ 36 %
ω′
ω
ω″
Frequency

Fig. 1

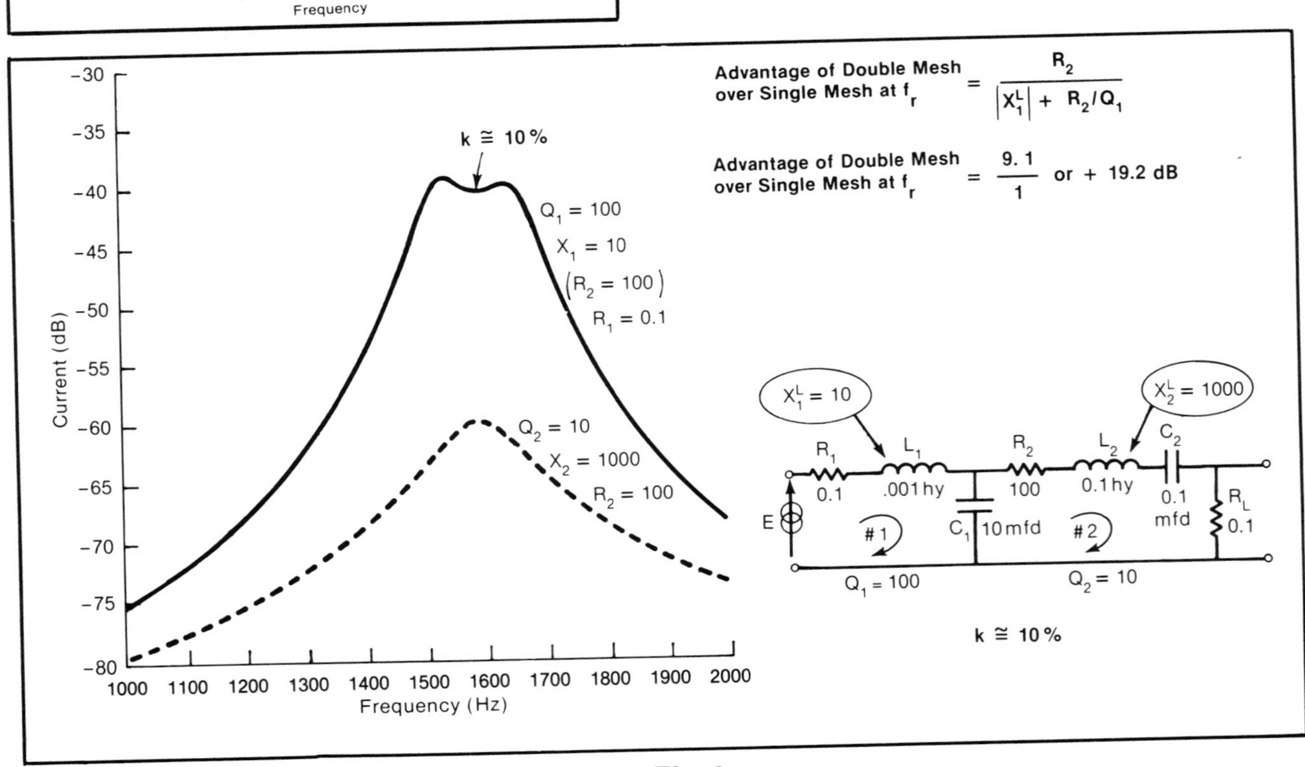

Current (dB)
-30
-35
-40
-45
-50
-55
-60
-65
-70
-75
-80
1000 1100 1200 1300 1400 1500 1600 1700 1800 1900 2000
Frequency (Hz)
k ≅ 10 %
Q1 = 100
X1 = 10
(R2 = 100)
R1 = 0.1
Q2 = 10
X2 = 1000
R2 = 100
Advantage of Double Mesh over Single Mesh at fr = R2 / (|X1L| + R2/Q1)
Advantage of Double Mesh over Single Mesh at fr = 9.1 / 1 or + 19.2 dB
X1L = 10
X2L = 1000
R1
0.1
L1
.001 hy
C1
10 mfd
R2
100
L2
0.1 hy
C2
0.1 mfd
RL
0.1
E
#1
#2
Q1 = 100
Q2 = 10
k ≅ 10 %

Fig. 2

a fairly tight coupling coefficient, of the order of 0.30. But note that when $X_2 \simeq 100X_1$, this means a loose coupling coefficient, of the order of 0.10. Hence the repeated recommendation of Webster and Paris is to use a very loose coupling since, tracing our steps backward, this will insure the high ratio of Volume 1/Volume 2, or C_1/C_2, which is essential to success.

Form (c) shows how the effective damping of R_2 is reduced by the resonating of the high -Q mesh #1. It also presents the undesired added resistance Q_1R_1 in its most reduced form $|X_1|$, or $(1/\omega C_1)$. The gain of a double mesh over a single mesh is thus:

$$[E_0/(|X_1| + R_2/Q_1)]/[E_0/R_2] = R_2/(|X_1| + R_2/Q_1).$$

4. "The Function and Design of Horns for Loud Speakers," C. R. Hanna and J. Slepian, *Trans. AIEE* 43:393 (1924).
   This paper will be found in I. Groves, ed., *Acoustic Transducers*, Hutchinson Ross Publishing Company, Stroudsburg, Pa., (1981).
5. Refer to the first comment on Webster (Paper 7) describing a glass "chimney" placed over a shielded stiff diaphragm. This is very close to Slepian's air-chamber transformer, coupling a stiff diaphragm to the throat of a horn. Each man's air chamber was a true transformer, transforming force as well as velocity, but Webster's second aim was to let his air chamber encourage a single high-Q resonance, while Slepian's second aim was to damp out this resonance (and many additional resonances).

BIBLIOGRAPHY

Aiken, C. B., 1937, Two-Mesh Tuned Coupled Circuit Filters, *Inst. Radio Eng. Proc.* 25:230.

Chaffee, E. L., 1924, Regeneration in Coupled Circuits, *Inst. Radio Eng. Proc.* 12:299-359.

Purington, E. S., 1930, Single- and Coupled-Circuit Systems, *Inst. Radio Eng. Proc.* 18:983.

Part II

ELECTROACOUSTIC MICROPHONES AND CLOSED-CAVITY CALIBRATIONS

Editor's Comments on Papers 12 Through 15

While Webster, Miller, and Paris were building and using their acoustic oscillographs and wave analyzers, Wente, Arnold and Crandall, and Fletcher were attacking the acoustical-measurements problem from a different approach. Their job was to improve the telephone transmitter (microphone) that was of course an *electro*acoustic transducer. Arnold and his group tried to calibrate a carbon transmitter using their "best receiver" (earphone) as the standard source of sound, but the data were not repeatable (Fletcher, 1976, private communication). So Wente decided to build an electroacoustic microphone that would have a flat frequency response and that could be calibrated (at low frequencies) merely by measuring capacitances C_0 and ΔC (or C_1) and polarizing voltage E_0, as described in his paper, and hence be indifferent to the sound source.

Paper 12 is Wente's first paper on his condenser microphone. It is a classic. It demonstrates Wente's gift for solving a problem by himself from two approaches: design and build an experimental model via intuition; do a theoretical analysis to optimize the intuitive design.

This paper is followed by one page of Wente's second paper, Paper 13, on the same subject wherein are discussed the quality-control problems of producing a condenser microphone to act as a secondary standard for the calibration of carbon microphones, the microphone used in large quantities by the American Telephone and Telegraph Company (AT&T). (The complete Wente [1922] paper has been reprinted in Groves, 1981.)

However, an absolute calibration of the condenser microphone by measurement of C_0 and C_1, which is possible in the quasi-static frequency region, fails near resonance—say, at frequencies greater than an octave below resonance. For this region Wente needed a precision source of sound, operable at any frequency. Paper 14, written by H. D. Arnold and I. B. Crandall in 1917, presents the theory and design of a thermophone that served the purpose. These two men were colleagues of Wente at AT&T (Arnold was laboratory director at the time). The thermophone had been known in Germany and used in England prior to 1917 but only as an imprecise source of sound. It was the contribution of Arnold and Crandall to turn the thermophone into a precision source of sound. The volume editor has provided notes that should facilitate an electrical engineer's understanding of the paper.

The Arnold and Crandall paper revealed some unexplained discrepancies in the calibration of the thermophone as, for example, when gold foil was substituted for platinum foil. In 1922 Wente wrote his own analysis of the thermophone, in which he cleared up the discrepancies. Space constraints force us to bypass Wente's direct words. However, a fine summary of Wente's thermophone paper was written by Harvey Fletcher as Appendix A of his book *Speech and Hearing* (1929) and is presented here as Paper 15.

REFERENCE

Wente, E. C., 1922, The Sensitivity and Precision of the Electrostatic Transmitter for Measuring Sound Intensities, *Phys. Rev.* **19**:498–503. (Paper 13 in *Acoustic Transducers*, I. D. Groves, ed., Hutchinson Ross Publishing Company, Stroudsburg, Pa., 1981, pp. 128–133.)

ANNOTATED BIBLIOGRAPHY

Ballantine, S., 1932, Technique of Microphone Calibration, *Acoust. Soc. Am. J.* **3**:319–360.

Ballantine's Equation 4 (p. 329) is similar to Fletcher's first equation.

Clemens, J., 1978, Capacitive Pickup, *RCA Rev.* **39**:33–59.

See especially Section 1 on the capacitive signal pick-up for video phonograph records.

Everitt, W., 1936, *Communication Engineering,* 2nd ed., McGraw-Hill, New York.
See Chapter III, "Network Theorems," especially p. 44. Also see the 3rd ed., 1958, Chapter III, Sections 3–24 and 3–25. The compensation theorem is discussed here.

Firestone, F., 1933, A New Analogy Between Mechanical and Electrical Systems, *Acoust. Soc. Am. J.* **4**:249–267.
See especially p. 250 on the arbitrariness of analogies.

Fleming, J., 1927, *The Propagation of Electric Currents,* Constable & Co., London.
See Chapter V on propagation of currents in a submarine cable, especially p. 164.

Heaviside, O., 1950, *Electromagnetic Theory,* Dover Publications, New York.
See volume II, Chapter 6, especially Sections 248, 249, and 253 on the diffusion of electric current along an unloaded line.

Morse, P., 1948, *Vibration and Sound,* 2nd ed., McGraw-Hill, New York.
See Chapter V. Section 20, especially p. 202 and Equation 20.10, on membrane theory.

Pipes, L., 1946, *Applied Mathematics for Engineers and Physicisits,* McGraw-Hill, New York.
See Chapter 18, "Heat Conduction and Diffusion," especially Equations 4.9, 4.11, and 8.6.

Ramo, S., and J. Whinnery, 1944, *Fields and Waves in Modern Radio,* Wiley, New York.
See especially Equations 6.04-10 and 6.5-4, on the skin effect.

Riegger, H., 1924, *Wiss Veröff. Siemens-werken* **3**:67.
This paper discusses subsituting an a.c. carrier for Wente's d.c. carrier.

Sears, F., and M. Zemansky, 1949, *University Physics,* Addison-Wesley, Reading, Mass.
See Section 24-6 on units used in electrostatics and Section 27-5 on the dimensions of capacitance. See Chapters 16 and 17 on heat, especially Sections 16-5 and 17-1.

Skilling, H., 1957, *Electrical Engineering Circuits,* Wiley, New York.
See Chapter 11, "Network Theorems," especially pp. 342–348. These pages give a fine discussion of the compensation theorem.

Slater, J., and N. Frank, 1933, *Introduction to Theoretical Physics,* McGraw-Hill, New York.
See Chapter 18, "Heat Flow," especially Section 131 and Figure 34.

Weber, E., 1956, *Linear Transient Analysis,* vol. 2, Wiley, New York.
See especially Section 7.2 on the noninductive cable.

Wente, E. C., 1922, The Thermophone, *Phys. Rev.* **19**:333–345.

Reprinted from *Phys. Rev.* **10**:39–63 (1917)

A CONDENSER TRANSMITTER AS A UNIFORMLY SENSITIVE INSTRUMENT FOR THE ABSOLUTE MEASUREMENT OF SOUND INTENSITY.

By E. C. Wente.

THE various methods that have been used with more or less success for measuring the intensity of sound may be divided into five general classes: observation of the variation in index of refraction of the air by an optical interference method; measurement of the static pressure exerted on a reflecting wall; the use of a Rayleigh disc with a resonator; methods in which the motion of a diaphragm is observed by an optical method; the use of some type of telephone transmitter in connection with auxiliary electrical apparatus. The apparatus of either of the first two methods is non-resonant and hence the sensitiveness is fairly uniform over a wide range of frequencies. These methods are not sufficiently sensitive, however, to be of use in general acoustic measurements. On the other hand, instruments of the last three classes possess a natural frequency and are consequently very efficient in the resonance region. However, in the neighborhood of the resonant frequency the efficiency varies greatly with the pitch of the tone. It is possible to use a Rayleigh disc without a resonator, but its sensitiveness in that case is so low that it is of little practical value.

Because of the recent advances in the development of distortionless current amplifiers, the last class, in which use is made of some form of telephone transmitter, seems to offer the greatest possibilities. In the following pages a transmitter is described which has been calibrated in absolute terms for frequencies from 0 up to 10,000 periods per second and which has a nearly uniform sensibility over this range. The apparatus is easily portable, and possesses no delicate parts, so that, when once adjusted, it will remain so for a long period of time.

Except in cases where measurements are made with a single, continuous tone, it is desirable that the instrument for measuring the intensity of sound should have approximately the same sensibility over the entire range of frequencies used. This is especially important if the sound under investigation has a complex wave form. To avoid any great variation with frequency in the sensibility of a phonometer employing a vibrating system, it is necessary that the natural frequency lie outside the range of frequencies of the tones to be measured. Even if the natural frequency be compensated for in other ways, small variations in the constants of the instruments, which are always likely to occur, may change conditions appreciably at this frequency. It is pretty well recognized that for several reasons the natural frequency should lie above rather than below the acoustic range. If the instrument is to be used in studying speech, the natural frequency must indeed be very high. The upper limit of the frequencies occurring in speech is not definitely known, but it probably does not come below 8,000 periods a second. Titchener[1] found that if a Galton whistle was set so as to give a frequency of 8,500, the tone emitted could not be distinguished from an ordinary hiss.

An instrument that is to be used in studying speech should have high damping as well as a high natural frequency in order to reduce distortion due to transients. This is not so important if the natural frequency lies beyond the acoustic range, but nevertheless is desirable even in this case. Aperiodic damping is the best condition, but it is in general hard to obtain when the natural frequency is very high.

It seems best in this paper to give a rather complete treatment of the condenser instrument; for the sake of clearness, however, breaking up the matter into a number of sections as follows:

1. Theory of the Operation of an Electrostatic Transmitter.
2. General Features of the Design of the Instrument.
3. Deflection of the Diaphragm under a Static Force. Measurement of Tension and Airgap.
4. Sensitiveness of the Transmitter at Low Frequencies.
5. Sensitiveness at Higher Frequencies Determined by the Use of a Thermophone.
6. Natural Frequency and Damping of the Diaphragm.
7. Possibilities of Tuning.
8. Characteristic Features of the Instrument.
9. The Electrostatic Instrument used as a Standard Source of Sound.
10. Summary.

Some of these sections deal with theory and some with experimental work as need arises, the general aim being to put in proper order the material necessary for a full account of the condenser instrument.

1. Theory of the Operation of an Electro-static Transmitter.

The device to be described is a condenser transmitter, the capacity of which follows very closely the pressure variations in the sound waves. The use of such a device as a transmitter is not a new idea; in fact it

[1] Proc. Am. Phil. Soc., 53, p. 323.

was suggested almost as early as that of the corresponding electromagnetic instrument.[1] However, before good current amplifiers were available little or no use was made of electrostatic transmitters because of their comparatively low efficiency.

A simple circuit that may be used with such a transmitter is shown in Fig. 1. When the capacity of the transmitter is varied, there will be a corresponding drop of potential across R, which may be measured with an A.C. voltmeter or some other suitable device.

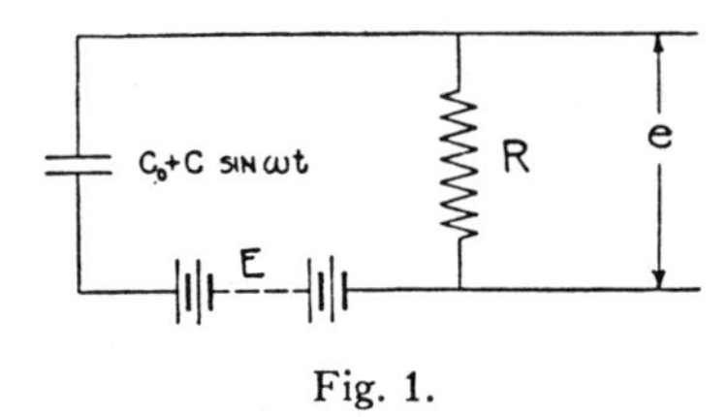

Fig. 1.

In order to get a quantitative expression for the magnitude of this voltage let us assume that the capacity at any instant is given by

$$C = C_0 + C_1 \sin \omega t,$$

in which $\omega = 2\pi \times$ frequency. For the circuit shown in Fig. 1

$$E - Ri = \frac{1}{C}\int i\,dt. \tag{1}$$

By differentiation and substitution we obtain

$$(C_0 + C_1 \sin \omega t)R\frac{di}{dt} + (1 + RC_1\omega \cos \omega t)i - EC_1\omega \cos \omega t = 0. \tag{2}$$

In order to evaluate this equation let us assume as a solution

$$i = \Sigma I_n \sin (n\omega t + \phi_n).$$

Substituting this value of i in (2) and determining the coefficients, we have

$$\left.\begin{aligned} i = {} & \frac{EC_1}{C_0\sqrt{\left(\frac{1}{C_0\omega}\right)^2 + R^2}}(\sin \omega t + \varphi_1) \\ & - \frac{EC_1^2 R}{C_0^2\sqrt{\left[\left(\frac{1}{C_0\omega}\right)^2 + 4R^2\right]\left[\left(\frac{1}{C_0\omega}\right)^2 + R^2\right]}}\sin (2\omega t + \varphi_1 - \varphi_2) \\ & + \text{terms of higher order in } C_1/C_0, \end{aligned}\right\} \tag{3}$$

in which

$$\phi_1 = \tan^{-1}\frac{1}{C_0\omega R} \quad \text{and} \quad \varphi_2 = \tan^{-1}\frac{1}{2C_0\omega R}, \quad \text{etc.}$$

[1] La Lumiere Electrique, Vol. 3, p. 286, 1881.

For the best efficiency R should be made large in comparison with $1/C_0\omega$. In this case, the expression for the voltage e becomes

$$e = Ri = \frac{EC_1}{C_0} \sin(\omega t + \varphi_1) - \frac{EC_1^2}{2C_0^2} \sin(2\omega t + \varphi_1 - \varphi_2) + \cdots.$$

From this equation we see that in order to get a voltage of pure sine wave form for a harmonic variation of capacity, C_1 must be small in comparison with $2C_0$. This condition is satisfied as long as the A.C. voltage is small compared with E.

Retaining only the first term in (3) we have

$$e = Ri = \frac{EC_1R}{C_0\sqrt{\frac{1}{C_0^2\omega^2} + R^2}} \sin(\omega t + \varphi_1). \tag{4}$$

This equation shows that, so far as its operation in the circuit is concerned, the transmitter may be considered an alternating current generator giving an open circuit voltage $E(C_1/C_0)\sin(\omega t + \varphi_1)$ and having an internal impedance $1/C_0\omega$. It can also be shown that the transmitter can be regarded from this point of view if R is replaced by a leaky condenser or an inductance, so that this result may be said to be true in general.

2. General Features in the Design of the Instrument.

The general construction of the transmitter is shown in Fig. 2, from which the principal features are evident. The diaphragm is made of steel, 0.007 cm. in thickness, and is stretched nearly to its elastic limit. The condenser is formed by the plate B and the diaphragm. Since the diaphragm motion is greatest near the center, the voltage generated, which is proportional to C_1/C_0, will be greatest if the plate is small. On the other hand, since C_0 is proportional to the size of the plate, it cannot be made too small or the internal impedance of the transmitter will be too great. Therefore from the standpoint of efficiency, a compromise has to be made in determining the area of the plate. However, if it is made much smaller than the diaphragm, the natural frequency of the vibrating system will be decreased, as is explained below. On the basis of these factors the size of the plate indicated was judged to be about the best for the transmitter.

After some experiments with various dielectrics between the plate and the diaphragm it was concluded that air was most suitable. The dielectric constant of air is not so high as that of some other materials, but its insulating properties are better. However, the principal advantage of using air is, that it has a high minimum value of sparking potential which lies in the neighborhood of 400 volts, below which there is no appreciable conduction. When E is less than this voltage, the air gap may be decreased without decreasing E, so that the efficiency of the instrument is limited practically only by the fact that when the gap is decreased below a certain value, the electrostatic force between the plate and diaphragm deflects the latter sufficiently to short circuit the condenser. When a potential difference of 320 volts was applied to the transmitter shown in Fig. 4, no appreciable current flowed across the air gap, certainly not more than 10^{-8} amperes. The fact that the air has such a high minimum sparking potential is one of the principal reasons why it is possible to design a successful condenser transmitter of the type shown in Fig. 2.

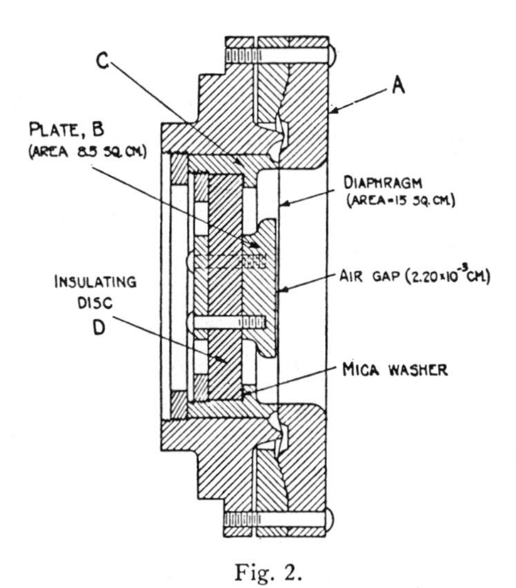

Fig. 2.

Sectional drawing of transmitter.

A word may be said in regard to the method of adjusting the transmitter so as to obtain a small uniform air gap. The surface of part A, next to the diaphragm, was ground plane before assembling. Small irregularities in the surface of the diaphragm facing the plate were removed by grinding with fine carborundum. Parts B, C and D were first assembled without the mica washer. The face of the plate and the ends of part C were then ground to the same level. Finally the mica washer was inserted between C and D and the whole apparatus assembled as shown. The mica may be split into washers of very even thickness, and a uniform air gap so obtained. The diaphragm is clamped between parts A and C, and is thus held in a true plane. In assembling the parts, the greatest care must be taken that no dust is caught between the plate and the diaphragm, for the insulation may be considerably reduced by the presence of any small particles in the gap.

Part C does not fit so perfectly against the diaphragm that the space surrounding the plate is shut off completely from the outside air. Changes in temperature and atmospheric pressure will therefore not affect the equilibrium position of the diaphragm.

The instrument used in these experiments was constructed just as shown in Fig. 2. It is evident from this figure that the diaphragm may be brought into contact with the plate if a mechanical pressure is accidentally exerted on the diaphragm. This will cause a spark to pass, if the transmitter was previously charged. In order to avoid damaging the metal surfaces in this way it may be advisable to glue to the face of the plate, *A*, a very thin layer of mica of uniform thickness, while still retaining an air-gap sufficient to allow free motion of the diaphragm.

3. Deflection of the diaphragm under a Static Force; Measurement of Tension and Air Gap.

It is not difficult to calculate the sensitiveness of the transmitter for low frequencies from the dimensions of its various parts, provided the magnitude of the deflection of every point of the diaphragm produced by a given static force is known. Since the diaphragm is made of very thin material and the tension is high, we may expect the diaphragm to behave very much as an ideal membrane, at least for frequencies near zero. In order to determine how closely this condition is approximated the following experiment was carried out.

When a static potential is applied between the plate and the diaphragm, the latter is deflected by the electrostatic force. The deflection produced

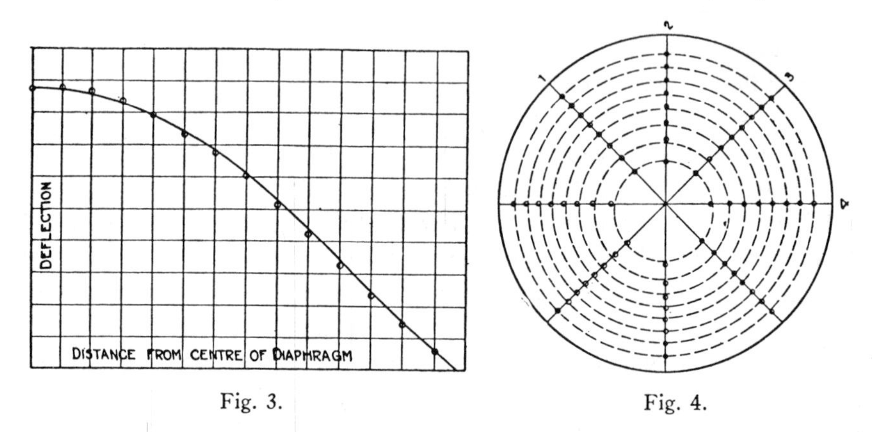

Fig. 3. Fig. 4.

in this way by a known potential was measured by a device very similar to that used by Prof. D. C. Miller in his phonodeik.[1] By this arrangement the deflection of the diaphragm was magnified 30,000 times. The mean values of the deflections produced at various points along eight evenly spaced radii when a potential of 320 volts was applied are shown in Fig. 3. Points of equal displacement of the diaphragm are plotted in Fig. 4. The fact that the curves drawn through these points are

[1] D. C. Miller, Science of Musical Sounds, p. 79.

practically circles shows that the tension of the diaphragm was very nearly the same in all directions.

The distance between the plate and diaphragm was also measured with this apparatus by applying a mechanical force until the diaphragm touched the plate. The value obtained in this way was 2.20×10^{-3} cm. The capacity of the transmitter was measured on a capacity bridge and found to be 335×10^{-12} farads, from which the computed value of the air gap is 2.25×10^{-3} cm. The mean of the values obtained in these two ways is 2.22×10^{-3} cms.

In order to determine how closely the diaphragm approximates an ideal membrane, we may calculate the form that the latter would have assumed under the conditions of the preceding experiment.

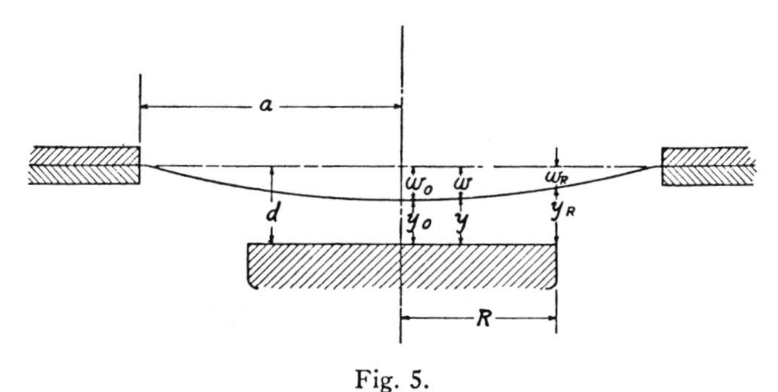

Fig. 5.

Referring to Fig. 5, if V is the potential between the plate and the diaphragm, and T, the tension of the membrane, we have

$$T\left[\frac{d^2w}{dr^2} + \frac{1}{r}\frac{dw}{dr}\right] + \frac{V^2}{8\pi y^2} = 0.$$[1]

This relation holds from $r = 0$ to $r = R$. Let

$$A = \frac{V^2}{8\pi T}$$

and $x = \log r$, then since $(w - w_R) = (Y_R - y)$

$$\frac{d^2y}{dx^2} = A\frac{\epsilon^{2x}}{y},$$

or, since $(w - w_R)/(w_0 - w_R)$ is very nearly equal to $(R^2 - r^2)/r^2$ and $(w_0 - w_R)$ is small compared with y_0,

$$\frac{d^2y}{dx^2} = \frac{A}{y_0}\left(1 - 2k\frac{r^2}{R^2}\right)\epsilon^{2x},$$

in which

$$k = (w_0 - w_R)/y_0.$$

[1] Rayleigh, Theory of Sound, II., p. 318.

From this we get

$$w = \frac{A}{4y_0^2}(R^2 - r^2)\left[1 - \frac{k}{2} - \frac{k}{2}\frac{r^2}{R^2}\right] + w_R. \tag{5}$$

The total force on the diaphragm is

$$F = \frac{\pi R^2 V^2}{8\pi d_1^2}.$$

where d_1 is defined by the equation

$$\frac{\pi R^2}{d_1} = \int_0^R \frac{2\pi r dr}{y^2} = \frac{2\pi}{y_0}\int_0^R \left(1 - 2k\frac{r^2}{R^2}\right) r dr = \frac{\pi R^2}{y_0}(1 - k).$$

so that

$$F = \frac{\pi R^2 V^2}{8\pi y_0^2}(1 - k).$$

In the region extending from $r = R$ to $r = a$,

$$F = -2\pi r T\frac{dw}{dr}.$$

From this

$$w_R = \frac{AR^2(1 - k)}{2y_0}\log\frac{a}{R}. \tag{6}$$

From (5) and (6)

$$w = \frac{A}{4y_0^2}\left\{(R^2 - r^2)\left[1 - \frac{k}{2}\left(1 + \frac{r^2}{R^2}\right)\right] + 2R^2(1 - k)\log\frac{a}{R}\right\}. \tag{7}$$

This equation gives the form into which the diaphragm will be bent if it behaves like an ideal membrane. The curve representing this equation is shown in Fig. 3. The observed points do not lie very far from this curve. We therefore conclude that the diaphragm behaves sufficiently like an ideal membrane, so that no great error will be incurred if this assumption is made in calculating the sensitiveness of the transmitter for low frequencies.

From equation (7)

$$w_0 = \frac{V^2}{32\pi T y_0^2}\left\{R^2\left(1 - \frac{k}{2}\right) + 2R^2(1 - k)\log\frac{a}{R}\right\}$$

or

$$T = \frac{V^2 R^2}{32\pi y_0^2 w_0}\left\{\left(1 - \frac{k}{2}\right) + 2(1 - k)\log\frac{a}{R}\right\}. \tag{8}$$

Hence, if the deflection at the center of the diaphragm produced by a known voltage is measured, the tension may be calculated from (8). Results obtained in this way for the diaphragm used in these experiments are tabulated below.

Volts.	Deflection (w_0) (cm.).	Tension (T) (dynes) / (cm.).
200	6.0×10^{-5}	6.59×10^{7}
240	6.8	6.58
280	12.4	6.55
320	16.9	6.55
Mean.		6.57×10^{7}

4. Sensitiveness of the Transmitter at Low Frequencies.

Having satisfied ourselves that the diaphragm behaves sufficiently like a perfect membrane, and having determined the tension and air gap, we can now proceed to calculate the efficiency of the transmitter for low frequencies. To do this it is necessary to find the change in capacity produced by a given pressure on the diaphragm, since by equation (4) the voltage generated is proportional to C_1/C_0.

Referring to Fig. 5, we see that the capacity is $R^2/4d$ if the diaphragm is not deflected. From the curve of deformation when a potential is applied (Fig. 3), it is evident that w_R is very nearly equal to 0.45 w_0. Hence the air gap at any point is given by

$$d - w_0 + \frac{.55}{R^2} w_0 r^2,$$

and since the surface of the diaphragm deviates but little from a plane area, the normal capacity to the first approximation is

$$C_0 = \int_0^R \frac{2\pi r dr}{4\pi \left(d - w_0 + \frac{0.55 w_0}{R^2} r^2\right)} = \frac{R^2}{4d^2}\left[1 + \frac{w_0}{d} - \frac{.55 w_0}{2d}\right] = \frac{R^2}{4d'}, \quad (9)$$

in which d' may be called the effective air gap.

If a pressure, P, uniform all over the diaphragm produces a deflection, u, the capacity of the condenser will have been changed by the amount

$$C_1 = \int_0^R \frac{u \cdot 2\pi r dr}{4\pi y^2}. \quad (10)$$

The quantity in brackets of equation (9) does not differ greatly from unity in any practical case, so that no great error will be incurred if we set y in (10) equal to the constant value d'. Since

$$u = \frac{P}{4T}(a^2 - r^2),[1]$$

equation (16) may be written

$$C_1 = \frac{P}{8Td'^2}\int_0^R (a^2 - r^2) r dr = \frac{PR^2}{32Td'^2}[2a^2 - R^2]. \quad (11)$$

[1] Lamb, Dynamical Theory of Sound, p. 150.

C_1 is the change in capacity produced by a static pressure, P; but this differs very little from the maximum value of the alternating capacity resulting from a pressure, $P \sin \omega t$, provided $\omega/2\pi$ is small compared with the natural frequency of the diaphragm.

Having determined C_1 per unit value of P from equation (11), and C_0 from (9), we may calculate C_1/C_0 and hence the sensitiveness, *i. e.*, the volts per unit pressure. In practically all the experiments that have been made with the electrostatic transmitter, the D.C. voltage was 321. Under this condition we obtain 315 E.S.U. for C_0 from (9) and 1.96×10^{-3} E.S.U. per dyne per sq. cm. for C_1/P from (11). Hence we have for the sensitiveness

$$\frac{EC_1}{PC_0} = \frac{1.96 \times 10^{-3} \times 321}{315} = \frac{2.00 \times 10^{-3} \text{ volts}}{\text{dynes per sq. cm.}}.$$

In order to check this value directly by experiment, the apparatus diagrammatically shown in Fig. 6 was constructed. A receptacle was

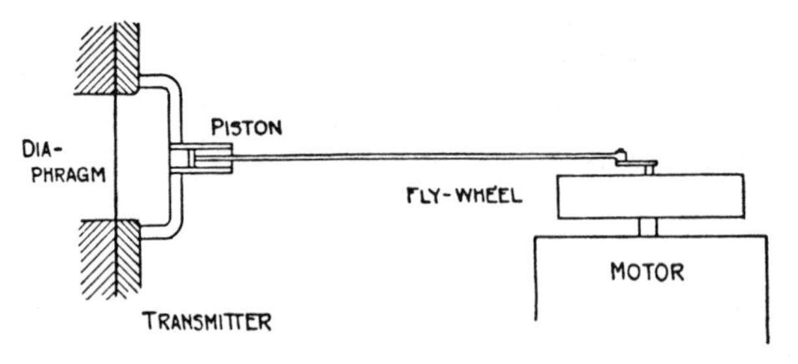

Fig. 6.

placed over the diaphragm as shown in the figure, thus forming an air-tight enclosure. Connected to this was a cylinder containing a piston. The connecting rod was long compared with the stroke of the piston so that with the motor running, the piston was given practically a simple harmonic motion. The fly wheel was fairly heavy and the connecting rod was made of stiff tubing, so that but little vibration was noticeable even when the motor ran at the highest speed.

The pressure variation is given by

$$\delta P = 1.4P\frac{\delta V}{V},$$

in which δV is one half the total piston displacement and P is the maximum value of the alternating pressure.

$$V = 45.2 \text{ c.c. (volume of chamber)}$$

$$\delta V = \frac{0.68 \times 0.418}{2} = .142 \text{ c.c.}$$

Hence

$$\delta P = 1.4 \times 10^6 \times \frac{1.42}{45.2} = 4{,}400 \text{ dynes per cm.}^2$$

The root mean square value of the pressure is

$$\frac{4{,}400}{\sqrt{2}} = 3{,}120 \text{ dynes per cm.}^2.$$

The circuit used in this test is shown in Fig. 7. The electrostatic voltmeter had a very small capacity, giving it at low frequency an impedance large compared with the 80 megohm resistance in shunt.

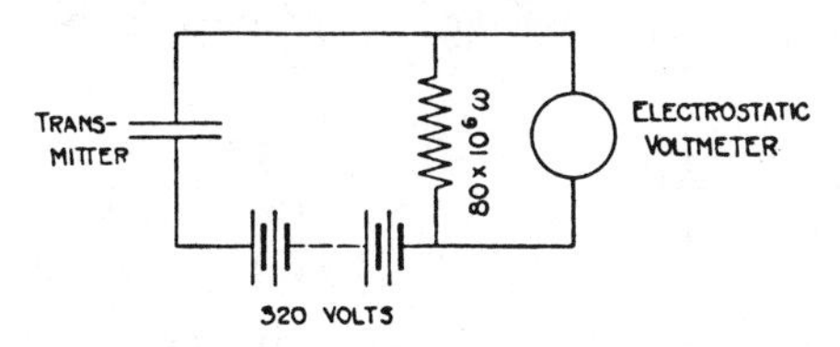

Fig. 7.

We may then calculate the open circuit voltage given by the transmitter from the voltmeter reading and the constants of the circuit, remembering that the transmitter may be regarded as a generator having an internal impedance $1/C_0\omega$. The following values were obtained in this way.

Motor Speed, R.P.M.	Frequency (P.P.S.).	Voltmeter Reading (Volts).	Open Circuit Volts.
1239	20.7	5.31	6.22
1074	17.9	5.31	6.27
950	15.8	5.31	6.33
824	13.75	5.20	6.29
584	9.75	4.92	6.27
Mean..........	....	....	........6.28

We therefore have for the sensitiveness,

$$\frac{6.28}{3120} = 2.02 \times 10^{-3} \frac{\text{Volts}}{\text{dynes per sq. cm.}}.$$

This value is in very close agreement with that given before, so that we may consider 2.00×10^{-3} volts per dyne as a reasonably correct value for the sensitiveness at low frequencies.

5. Sensitiveness at Higher Frequencies as Determined by the Use of a Thermophone.

By the methods just described the values of sensitiveness may be determined for very low frequencies only. In order to measure the sensitiveness at higher frequencies and also to get an idea of the natural frequency and damping of the vibrating system, use was made of the principle involved in the action of the thermophone as described by Arnold and Crandall.[1] A block of lead about 1.5 inches thick was placed against the face of the transmitter so as to form a cylindrical enclosure in front of the diaphragm, $1\frac{3}{4}$ inches in diameter and $\frac{3}{8}$ inch long. The general arrangement is shown in Fig. 8. All crevices were sealed up so that the only openings to the cavity were two capillary tubes several inches long and of about 0.01 cm. bore. Two strips of gold foil were mounted symmetrically inside of this enclosure, the ends being clamped between small brass blocks. The supports were arranged in such a way that a current could be passed through the two strips in series. The connection between them was in electrical connection with the diaphragm.

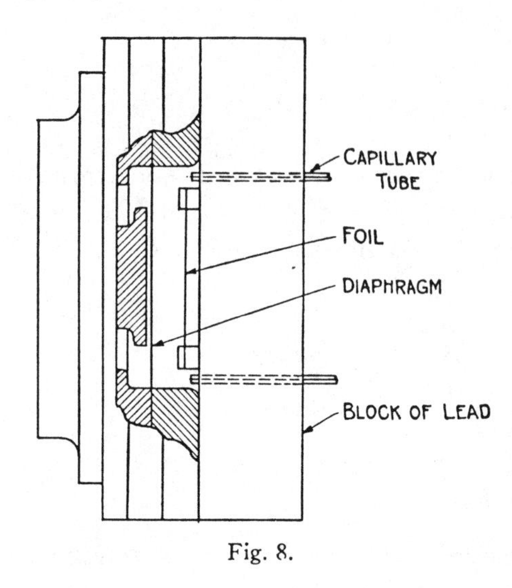

Fig. 8.

In the paper just cited it is shown that within an air-tight enclosure

$$\delta P = \frac{.0106 R i^2 P \sqrt{K_0} \left(\frac{\theta_1}{273} \right)^{1/4}}{\gamma V_0 \theta_2 f^{3/2}}, \tag{12}$$

in which

δP = maximum value of the alternating pressure within the enclosure.
P = normal pressure within the enclosure.
R = resistance of the foil.
i = r.m.s. value of the alternating current passing through the foil.
K_0 = diffusivity at 0° C. of the gas within the enclosure.
θ_1 = mean absolute temperature of the foil.
θ_2 = mean absolute temperature of the gas.
γ = heat capacity per unit area of the foil.
V_0 = volume of the enclosure.
f = frequency of the alternating current.

Equation (12) may be used for calculating the pressure variation provided the wave-length of sound is large compared with the dimensions of the enclosure. The velocity of sound in hydrogen is about four times

[1] Physical Review (Preceding Paper).

as great as in air; hence formula (12) holds for frequencies almost four times as high, when the enclosure is filled with hydrogen instead of air. Also, the diffusivity, K_0, is about six times as large for hydrogen as for air, so that greater pressure variation is obtained with the former. For these reasons, hydrogen was passed in a continuous stream through the enclosure by way of the capillary tubes, at a rate sufficiently slow to prevent any appreciable increase of the steady pressure above that of the atmosphere. The hydrogen was obtained from a Kipp generator and then passed through a solution of potassium permanganate and a drying tube containing phosphorus pentoxide.

In order to get the open circuit electromotive force of the transmitter, the circuit was arranged as in Fig. 9. The two resonant circuits were so

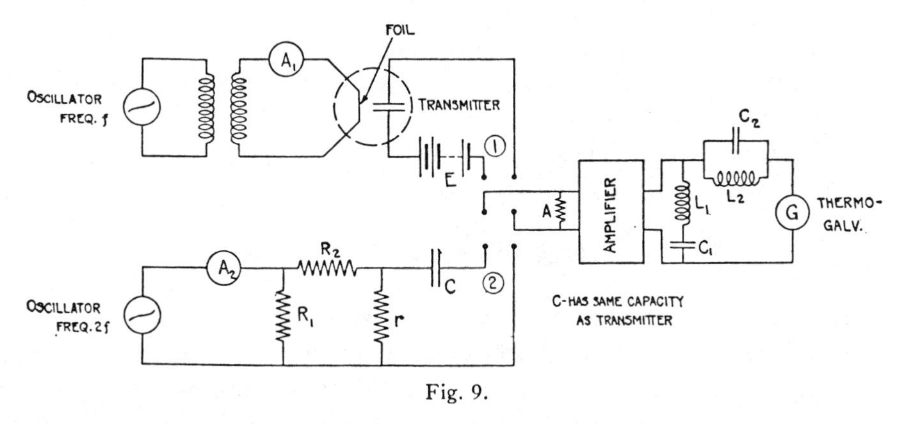

Fig. 9.

adjusted as to prevent current of the same frequency as that given by the oscillator from passing through the galvanometer. If a pure sine wave current passes through the foil, the pressure variation in the enclosure is of pure sine wave form and of double frequency. However, if there is any second harmonic present in the current, there will also be a component of the pressure variation of single frequency.[1] Putting in the resonant circuits eliminates this component from the measurements. The general procedure in making a measurement was as follows. The double-throw switch was first put in position 1, and the foil current and galvanometer current read. The switch was then thrown in position 2; R and r were then adjusted until the galvanometer read approximately the same as before. From the readings of A_2 and the values of R_1, R_2 and r, the voltage drop across r may be calculated. The open circuit voltage of the transmitter is then obtained by multiplying this voltage drop by the ratio of the galvanometer readings. That this gives us the open circuit voltage, follows from the fact that the transmitter

[1] Arnold & Crandall, *loc. cit.*

behaves as a generator having an internal impedance $1/C_0\omega$. Oscillator No. 2, of course, is set at double the frequency of Oscillator No. 1.

The current passed through the gold foil was about 0.5 ampere at all frequencies. Resistance measurements showed that with this current density, the foil was not heated more than 10° C. above the room temperature. The values of the quantities entering into the formula (12) for this experiment were as follows:

R = 4.18 ohms.
P = 10^6 dynes/cm^2.
K_0 = 1.48 C.G.S. units.
θ_2 = 295°.
θ_1 = 305°.
$\gamma = 4.15 \times 10^{-6}$ calories per sq. cm.
Thickness of gold foil = 7×10^{-6} cm.
Width of each strip = 1 cm.
Length of each strip = 2.6 cm.

Substituting these values in equation (12), we have for the root mean square value of the alternating pressure

$$2.59 \times 10^6 \frac{i^2}{f^{3/2}} \text{ dynes per sq. cm.}$$

Dividing the measured open circuit voltage by this value should give us the volts per unit pressure for all frequencies within certain limits. Measurements were made in this manner for frequencies from 160 to 18,000 cycles per second. The general shape of the curve obtained by plotting these values is shown in Fig. 10.

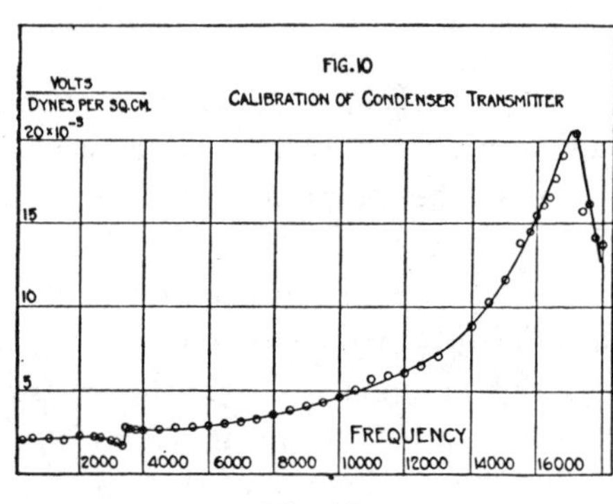

Fig. 10.

The absolute value of the sensitiveness at low frequencies as determined by this method was 0.121×10^{-3} volts per dyne, which is only about one sixteenth of that previously obtained by the piston method and by calculation from the dimensions of the instrument. In order to make further tests within the range of frequencies from 20 to 160 cycles, the gold was replaced by platinum foil, 4.42×10^{-4} cm. thick, and measurements were made as before. However, the size of the enclosure was increased in order to meet the conditions assumed in the derivation of formula (12), and for the same reason air was used instead of hydrogen.

Calculations made in a manner similar to that when gold foil was used gave a value of 1.93×10^{-3} volts per dyne per sq. cm. for the sensitiveness at low frequencies. This is in fair agreement with the value 2.00×10^{-3} obtained theoretically and with the piston apparatus.

Apparently when gold foil is immersed in hydrogen something takes place which is not taken account of in equation (12). The gold foil used was extremely thin (7×10^{-6} cm.) and when placed in hydrogen its specific heat per unit volume was apparently much greater than that of pure gold assumed in the calculations. On account of this discrepancy the gold leaf could not be relied upon for an absolute calibration, but it seemed reasonable to assume that the ratio between the true pressure and that calculated was independent of the frequency, so that a true relative calibration for different frequencies could be obtained. To get the absolute value of the efficiency at all frequencies, the values calculated from the readings on the gold foil were multiplied by the factor $2.0/.121 = 16.6$. The results so obtained are those plotted in Fig. 10.

6. Natural Frequency and Damping of the Diaphragm.

It is thought that this curve (Fig. 10) may be relied upon to give the sensitiveness in absolute value for frequencies up to 10,000 cycles. Above this frequency the wave-length of sound approaches the diameter of the cylindrical enclosure. The wave-length in hydrogen at 10,000 cycles is 13 cm. whereas the greatest distance from boundary to boundary of the enclosure is 4.4 cm. Although the absolute values of the sensitiveness above 10,000 cycles are probably not given by the points plotted in Fig. 10, nevertheless, this curve indicates in a general way the behavior of the transmitter at high frequencies. The principal peak in this curve comes at 17,000 cycles, which undoubtedly corresponds to the natural frequency of the diaphragm. The damping cannot be determined with any great assurance of accuracy, although the curve as drawn would indicate a damping factor of the vibrating system of about six or seven thousand.[1]

These high values of natural frequency and damping are in a large measure due to the cushion effect of the air between the plate and the diaphragm. Free lateral motion of the air is prevented by its viscosity. This increases the rate of dissipation of energy when the diaphragm is vibrating and also adds to its elasticity.

To see whether 17,000 cycles is a reasonable value for the natural frequency we may make an approximate theoretical calculation. When

[1] The term damping factor as here used may be defined as the reciprocal of the time required for the amplitude to fall to $1/2.718$ of its initial value.

the frequency is as high as 17,000 cycles it seems reasonable to assume that there is practically no lateral motion of the film of air. Let us further assume that the film of air is compressed and rarified adiabatically by the motion of the diaphragm and also that the plate is of the same size as the diaphragm. This latter condition is not quite satisfied in the case of the electro-static transmitter but no great error is introduced by this assumption, since the motion near the edge of the diaphragm is small. Under these conditions if d is the length of the air gap, P, the atmospheric pressure, ρ, the mass per unit area of the diaphragm, and T, the tension, the equation of motion of the diaphragm becomes:

$$\rho \frac{d^2w}{dt^2} = T\left(\frac{d^2w}{dr^2} + \frac{1}{r}\frac{dw}{dr}\right) - \frac{1.4Pw}{d},$$ [1]

or since w varies as $\epsilon^{j\omega t}$,

$$\frac{d^2w}{dr^2} + \frac{1}{r}\frac{dw}{dr} + \left(\frac{\rho\omega}{T} - \frac{1.4P}{Td}\right) w = 0. \qquad (13)$$

The solution of (13), consistent with the boundary conditions, is

$$w = J_0(lr),$$

in which

$$l = \sqrt{\frac{\rho\omega^2}{T} - \frac{1.4P}{Td}}.$$

The boundary conditions require that $J_0(la) = 0$. The lowest root of this equation is 2.4 so that

$$\left[\rho\omega^2 - \frac{1.4P}{d}\right]\frac{a^2}{T} = (2.4)^2$$

or

$$f_0 = \frac{\omega}{2\pi} = \frac{1}{2\pi}\sqrt{\frac{T}{\rho}}\left(\frac{2.4}{a}\right)\sqrt{1 + \frac{1.4Pa^2}{Td(2.4)^2}}.$$

This equation gives the natural frequency of the diaphragm when vibrating in its fundamental mode.

For the transmitter used in the preceding tests —

$$T = 6.57 \times 10^7 \text{ dynes per cm.}$$

$$\rho = .05 \text{ gm. per sq. cm.}$$

$$a = 2.18 \text{ cm.}$$

$$P = 10^6 \text{ dynes per sq. cm.}$$

$$d = 2.22 \times 10^{-3} \text{ cm.}$$

Hence

$$f_0 = 6{,}350\sqrt{8.9} = 19{,}000 \text{ P.P.S.}$$

[1] Rayleigh, Theory of Sound, L., 318.

which is slightly higher than the observed value. With the plate removed, the diaphragm would have a natural frequency of 6,350. This shows that the film of air between the plate and the diaphragm increases the elastic factor many times. It is due entirely to this fact that it has been possible to obtain natural frequencies above 10,000 without making the diaphragm exceptionally small.

We may satisfy ourselves that the maximum point in the efficiency curve is not due to resonance in the cylindrical enclosure by calculating its resonant frequencies. These frequencies are determined by the equation

$$J_n'(\sqrt{K^2 - p^2\pi^2 l^{-2}}\, R) = 0 \text{ [1]}$$

in which

$$K = \frac{2}{a}\pi f_0.$$

a = velocity of sound.

l = length of cylinder.

R = radius of cylinder.

p = an integer.

Since the foil was placed symmetrically in the enclosure, only the symmetrical modes of vibration need be considered, in which case $n = 0$.

The first root of the equation $J_0'(Z) = 0$ is 3.83. For the lowest resonant frequency, $p = 0$, so that we have

$$f_0 = \frac{a}{2} \cdot \frac{3.83}{R}.$$

In this problem

a = 127,000 cm./sec. (velocity of sound in hydrogen).

R = 2.18 cm.

hence

$$f_0 = 35{,}500 \text{ cycles per second.}$$

which is very much above the frequencies covered in the calibration.

If the enclosure is filled with air instead of hydrogen, the first resonant frequency comes at about one fourth of 35,500 or 9,000 p.p.s. A series of measurements were made with the circuit arranged as in Fig. 9, and air instead of hydrogen surrounding the gold foil. Points were calculated and plotted; the curve so obtained showed a sharp resonant point at 9,600 but none below. This may be taken as further evidence that the maximum point in Fig. 10 is not due to any resonance in the enclosure and so corresponds to the natural frequency of the diaphragm.

[1] Rayleigh, Theory of Sound, II, 300.

There is an irregularity in the calibration at about 3,500 periods per second. This is undoubtedly due to the natural frequency of the back-piece. At any rate, vibration of the plate would have an effect of this general character, *i. e.*, the efficiency would be decreased below, and increased above, resonance. In a later design, the plate and support have been made more rigid so as to form practically one solid piece. It is believed that with the newer model, the irregularity in the curve will have been eliminated.

This completes the account of the experimental work done in calibrating the instrument.

In order to obtain some idea of the sensitiveness of the electrostatic transmitter just described as compared with an electromagnetic instrument, the sensitiveness of the former was compared directly with an ordinary telephone receiver used as a transmitter, over a considerable range of frequencies. Except within a hundred cycles of the resonant frequency of the diaphragm of the receiver the electrostatic transmitter was found to generate a greater voltage for a given sound intensity.

7. Possibilities of Tuning.

Since an electrostatic transmitter is equivalent to an alternating current generator having an internal impedance $1/C_0\omega$, it is evident that, if in the circuit shown in Fig. 1, the resistance R is replaced by an inductance L, the voltage e will be a maximum for a frequency of

$$f = \frac{1}{2\pi\sqrt{LC_0}}.$$

The sharpness of tuning will of course depend upon the possibility of getting an inductance with a small resistance. In many problems in acoustics it is desirable to have a tuned system and in that case it is also better to have a diaphragm of low natural frequency and damping.

In order to get an expression for the sensitiveness as a function of the frequency, let us assume that we have a parallel plate condenser, one of the plates of which is fixed and the other moved perpendicularly to its own plane by a simple harmonic force. Practically this condition is approximated by a diaphragm, the center of which is separated a short distance from a plane plate as is shown in Fig. 2.

Let x = displacement of the diaphragm from its equilibrium position.

d = air gap, assumed large compared with x.

Then

$$\frac{1}{C} = \frac{1}{C_0}\left(1 + \frac{x}{d}\right).$$

The mechanical impedance of the diaphragm is

$$Z_1 = r + j\left(m\omega - \frac{s}{\omega}\right),$$

where r = resistance factor,

m = mass factor,

s = elasticity factor,

$\omega = 2\pi \times$ frequency.

If T = kinetic energy of the entire system,

W = potential energy of the entire system,

$2F$ = rate of dissipation of energy,

then

$$2T = m\dot{x}^2 + L\dot{y}^2,$$

$$2W = sx^2 + \frac{1}{C_0}\left(1 + \frac{x}{d}\right)(Y + y)^2,$$

$$2F = R\dot{y}^2 + r\dot{x}^2,$$

where Y is the permanent, and y the variable electric charge on the condenser. The equations of motion for the system are

$$\frac{d}{dt}\left(\frac{\partial T}{\partial \dot{x}}\right) - \frac{\partial T}{\partial x} + \frac{\partial F}{\partial \dot{x}} + \frac{\partial W}{dx} = P\epsilon^{j\omega t},$$

$$\frac{d}{dt}\left(\frac{\partial T}{\partial \dot{y}}\right) - \frac{\partial T}{\partial y} + \frac{\partial F}{\partial \dot{y}} + \frac{\partial W}{\partial y} = 0.$$

If second order quantities are neglected, and also the constant terms, which affect only the equilibrium position, these equations become

$$\left.\begin{aligned} pm\dot{x} + r\dot{x} + \frac{s}{p}\dot{x} + \frac{Y\dot{y}}{pC_0 d} &= P \\ pL\dot{y} + R\dot{y} + \frac{1}{pC_0}\dot{y} + \frac{Y}{pC_0}\frac{\dot{x}}{d} &= 0 \end{aligned}\right\} \qquad (14)$$

in which p is written for

$$j\omega = \frac{d}{dt},$$

solving equations (14) for $\dot{y}$ and substituting the values,

$$E = Y/C_0, \quad Z_1 = (r + s/p + pm), \quad Z = (R + 1/pC_0 + RL),$$

we have

$$\dot{y} = \frac{PE}{pd[(E/pd)^2 - Z_1 Z]},$$

or

$$e = \dot{y}(R + pL) = \frac{PE(R + pL)}{pd[(E/pd)^2 - Z_1Z]}.$$

In any practical case $(E/pd)^2$ is small compared with Z_1Z so that we may write without much error

$$e = \frac{EP(R + pL)}{pdZ_1Z}.$$

In order to obtain a large value of e/P, which is a measure of sensitiveness, Z_1Z should be made small, *i. e.*, the diaphragm should have a natural frequency equal to the frequency, $\omega/2\pi$, and the electrical circuit should be in resonance at the same point.

No extensive measurements have been made with the circuit arranged in this way, although enough has been done to show that it is feasible in some cases. The chief difficulty lies in the fact that the transmitter capacity is so small that the inductance has to be very large to get resonance for ordinary sound frequencies. This difficulty may be overcome by shunting the transmitter with a condenser, which of course reduces the generated voltage.

8. Characteristic Features of the Instrument.

Because of the high internal impedance of the electrostatic transmitter, it is possible to use the instrument efficiently only with high impedance apparatus, such as an electrostatic voltmeter or a vacuum tube amplifier. However, this is no special disadvantage if an amplifier is to be used, because it is not desirable to use a transformer in connection with an instrument for measuring sound intensities, since the ratio of transformation of a transformer is not independent of either frequency or load.

The method for calibration of this instrument as explained in the preceding pages is rather elaborate and requires considerable care. But since the efficiency depends primarily on the air gap and tension, it should not be difficult to make duplicate transmitters to which the same calibration applies, since the desired values of air gap and tension may be obtained without great difficulty, the former being tested by measuring the capacity, and the latter by determining the deflection produced by a known potential between the plate and diaphragm.

The fact that the sensitiveness of this instrument is independent of any properties of material, such as magnetization or electrical resistance, is of considerable advantage. For this not only allows us to make instruments which are almost exact duplicates, and so let the calibration for one instrument serve for all the rest, but, the calibration is also constant with the time. The metal parts are of machine steel throughout; from the construction as shown in Fig. 2, it is therefore evident that temperature can affect the sensitiveness but little. The tension of the diaphragm is, of course, not absolutely independent of temperature, nor is the action of the cushion of air between the plate and the diaphragm independent of the barometric pressure: but these effects are hardly worth considering. Being made of heavy material, the transmitter satisfies the requirement in the way of ruggedness; having once been adjusted, it should remain so, even if subjected to considerable rough usage.

The sensitiveness of the transmitter is not absolutely uniform, but varies only about a hundred per cent. between zero and 10,000 cycles, as the curve in Fig. 10 shows. This variation is much less than would be the case with an electromagnetic instrument with a diaphragm having the same natural frequency and damping. Except for eddy current and iron losses, the voltage generated by an electromagnetic transmitter is proportional to the *velocity* of the diaphragm, whereas that given by the electrostatic transmitter is proportional to the *amplitude*. Below the natural frequency, the variation of velocity with frequency is much greater than the variation of amplitude since the velocity is proportional to the product of the frequency and amplitude.

In most problems the transmitter would be used with an amplifier. Now, the sensitiveness of the transmitter increases, whereas the efficiency of an amplifier sometimes decreases with the frequency; at any rate, it is possible to design a circuit for the amplifier, so that the combination of the two has a constant sensitiveness over a wide range of frequencies.

Since the natural frequency of the transmitter is very high, instantaneous records of sound waves obtained in combination with a distortionless oscillograph would not only give the relative amplitudes of the different frequencies into which the sound may be analyzed, but also the phase relations should be practically unchanged for frequencies up to 10,000 p.p.s.

As yet no instrument is available which will record without distortion currents of frequencies as high as 10,000 cycles. Only after such an instrument has been developed will it be possible to get a true record of consonant sounds. The same is true in regard to the quantitative study of the quality of musical instruments. However, by using an ordinary high frequency oscillograph in connection with a condenser transmitter and amplifier, it should be possible to get curves equal to or better than any obtained heretofore.

9. The Electrostatic Instrument Used as a Standard Source of Sound.

There is of course no theoretical reason why the instrument described in the preceding pages cannot be used in a reversible manner: that is, as a source of sound when an alternating voltage is applied between the plate and the diaphragm. If the instrument is to be used in this way, it is better to have the plate the same size as the diaphragm, in order to get the maximum electrostatic force for a given voltage and air gap. The resulting increase in capacity is in general no disadvantage in this case. Also for convenience in using the instrument it may be desirable to have the face of the plate covered with a thin layer of mica.

Because of the simplicity of this type of instrument it is not difficult to calculate the output of sound energy for a given voltage *after its efficiency as a transmitter has been determined.* It is evident that the instrument can be excited in two different ways; (*a*) the alternating voltage can be applied alone, and (*b*) it may be superimposed on a static potential maintained by a battery in exactly the same way as when the instrument is used as a transmitter. The main principles underlying the two kinds of excitation in this case are quite similar to those discussed by Arnold and Crandall in connection with the excitation of the thermophone by pure A.C. and by A.C. with D.C. superimposed. For this reason neither type of excitation need be discussed at length; but a brief treatment of the condenser instrument excited by pure alternating current will be given.

When a pure alternating voltage is applied, the mean deflection of the diaphragm will depend on the magnitude of this voltage and the efficiency may vary somewhat because of the change in mean air gap, and the consequent change in the cushion effect of the air sheet on the motion of the diaphragm. It is therefore necessary to have curves corresponding to the curve in Fig. 10 but for a series of applied static potentials. These are most easily obtained by determining for a number of frequencies the generated voltage as a function of the static potential when sound of a fixed intensity falls on the transmitter. It will be found that the alternating voltage generated is so nearly proportional to the static potential that for most acoustic work this may be assumed to be the case.

When an alternating potential $\sqrt{2}v \sin \omega t$ is applied to the plates, the electrostatic force per unit area acting on the diaphragm is

$$\frac{v^2}{8\pi d^2}(1 - \cos 2\omega t). \tag{15}$$

Now refer to Fig. 10, assuming that the curve there shown gives the efficiency of the instrument, *used as a transmitter*, for an applied static potential v. If we multiply the ordinate (*i. e.*, the voltage per unit pressure) at frequency $\omega/\pi = f$ by the quantity

$$\frac{C_0}{v} \equiv \frac{C_0}{E}$$

we can obtain (cf. (4)) $\overline{C}_f$, the change in capacity per unit pressure. The total change in capacity due to the electrostatic force is then, (if $\overline{C}$ is the change per unit pressure at zero frequency)

$$C_1 = \frac{v^2}{8\pi d^2}(\overline{C} - \overline{C}_f \cos 2\omega t) \tag{16}$$

from which we can proceed to calculate the amplitude of motion of the diaphragm.

It is necessary of course to have a mean value of d, the air gap, but it is sufficiently accurate to take an arithmetic mean of the values at the center and at the edge of the diaphragm. The motion at the center is greater, but the motion near the edges extends over a greater area.

In computing the mean amplitude of the diaphragm we shall introduce very little error if we take the form of the diaphragm as that of a paraboloid. u, the amplitude at any radius, r, is given by the relation already quoted

$$u = \frac{P}{4T}(a^2 - r^2), \tag{17}$$

in which a = the radius of the diaphragm and plate. Equation (11) gives the total change in capacity in terms of P and T, that is (since $a = R$)

$$C_1 = \frac{a^4}{32d^2}\frac{P}{T}, \tag{11'}$$

or, eliminating P/T between (11′) and (17) we have, for displacement at any radial distance r, in terms of maximum capacity change

$$u = \frac{8d^2}{a^4}(a^2 - r^2)C_1.$$

Substituting for C_1 the value given in (16) we have

$$u = \frac{v^2}{\pi a^4}(a^2 - r^2)(\overline{C} - \overline{C}_f \cos 2\omega t), \tag{18}$$

in which v is the r.m.s. value of the applied alternating voltage, and $\overline{C}_f$

is the change in capacity per unit pressure, determined in the manner described from the calibration curve of the instrument used as a transmitter. Equation (18) is rigorously true for all frequencies within the range of calibration, because the quantity $\overline{C}_f$ is taken from the calibration curve.

If, however, T is known, we can obtain an approximate value of u good at low frequencies, without any knowledge of C_1. (This is merely "equilibrium theory" and makes use only of the elastic factor, leaving the inertia and mechanical resistance of the moving system out of account). Substituting the value of electrostatic force (15) for P in (17) we have

$$u = \frac{v^2(a^2 - r^2)}{32\pi d^2 T}(1 - \cos 2\omega t). \qquad (19)$$

The actual acoustic effect may be determined by the usual methods. If the diaphragm forms a wall of a small enclosure, the intensity is determined by the ratio of the volume displaced by the diaphragm as it vibrates to the volume of the enclosure. In other cases the intensity at a given point is calculated by determining the velocity potential due to the motion of the diaphragm.

It has been tacitly assumed that the amplitude of motion of the diaphragm is small compared with the air gap. This is necessary in order to get a pure tone when a sine wave E.M.F. is applied. While the instrument will not take care of a very large amount of energy, sound of the same order of intensity may be obtained as from an ordinary telephone receiver without appreciable distortion.

Summary.

1. A description is given of a transmitter of the electrostatic type which is especially adapted for measurement of sound intensities over a wide range of frequencies. The instrument is portable and is sufficiently rugged to retain its calibration.

2. A discussion is given of the necessary auxiliary apparatus and the precautions necessary for proper use.

3. A theory of the transmitter has been developed by which its operation can be predicted from a few simple measurements.

4. A description is given of the calibration of such an instrument in absolute terms over a wide range of frequencies. It is found that its efficiency may be made practically uniform for frequencies up to 10,000 cycles per second, and the results of the calibration are in agreement with the theory.

5. The apparatus when once adjusted may be used for the measurement of the intensities of sound at any frequencies throughout this wide range without further special adjustment.

6. Due to the uniform response through this wide frequency range it will be possible to secure correct indications of complex wave forms and to determine not only the relative intensities of the components but also their phase differences.

7. When properly calibrated this apparatus can be used as a precision source of sound.

RESEARCH LABORATORY OF THE AMERICAN
TELEPHONE & TELEGRAPH CO. AND WESTERN ELECTRIC CO., INC.

EDITOR'S NOTES

In the following notes, the approach sometimes is to start with the individual components that comprise Wente's exact equation and then to combine these components as if they were lumped impedances, thereby attempting a rough approximation to Wente's exact equation. The purpose is to give the reader additional insight into what is going on.

Page 39: Of the five general classes referred to, the first is the schlieren method discussed in Part III of this volume in Papers 16, 17, and 19 by Toepler and Dvorak. The second is the radiation-pressure method discussed by R. B. Lindsay in *Physical Acoustics* (Benchmark Papers in Acoustics, volume 4, Dowden, Hutchinson & Ross, Stroudsburg, Pa., 1974) and particularly by F. Borgnis in his paper "Acoustic Radiation Pressure of Plane Compressional Waves" (*Rev. Mod. Phys.* **25**:653–664, 1953; Paper 13 in Lindsay's *Physical Acoustics*). The third is the Rayleigh disc method, discussed in Part IV of this volume. The fourth embodies the optical methods used by D. C. Miller and by A. G. Webster, discussed in Part I of this volume. The fifth, using a telephone transmitter, is the subject of Wente's paper.

Note that "transmitter" is nowadays called "microphone" (except by Wente's employer, AT&T) and that "receiver" is nowadays called "earphone" or "headphone." Also note that Wente's "efficiency" and "sensibility" usually mean "sensitivity."

Page 42: Another way of looking at the evolution of Equation 3 and its offshoots is the following: let the fixed d.c. voltage source E_0 in series with the variable capacitor C and the resistor R be replaced by the same fixed d.c. voltage source E_0 in series with an additional voltage source, the variable a.c. voltage source e_{ac}, in series with only the fixed capacitor C_0 and the resistor R. This is made possible by a modified version of the compensation theorem. Then see the following note.

Page 42: Equation 3 and its offshoots can be seen intuitively if we work with charges Q_0, and Δq, operate at quasi-d.c. frequency (to remove the effects of R, which then acts as if it had the value zero), and throw away the distortion terms. Then, since

$$E_0 C_0 = Q_0 \tag{1}$$

$$E_0 (C_0 + \Delta C) = E_0 C_0 + E_0 \Delta C = (Q_0 + \Delta q), \tag{2}$$

then

$$E_0 \Delta C = \Delta q, \tag{3}$$

and

$$E_0(\Delta C/C_0) = \Delta q/C_0 = \Delta E, \text{ or } e_{ac}\,. \tag{4}$$

Now this is merely a variation of the compensation theorem that says in effect: If, in a circuit, where $E_0 = I_0 Z_0$, Z_0 becomes $Z_0 + \Delta Z$ (and I_0 becomes $I_0 - \Delta I$), then the effect of ΔZ is to create a differential voltage generator $\Delta E = I_0 \Delta Z$. ("Differential" in this usage means nothing but "relatively small.") One corollary is that $I_0 \Delta Z \cong Z_0 \Delta I$, another form of ΔE. This form implies that the differential generator ΔE has the internal impedance Z_0 (by way of $\Delta E/\Delta I$). Then from Equation (4) above,

$$\Delta E = E_0 (\Delta C/C_0) = \omega \Delta q \cdot 1/ \omega C_0, \tag{5}$$

which is merely $Z_0 \Delta I$. This implies that the differential voltage generator ΔE has the internal impedance $1/\omega C_0$, and this is precisely what shows up in Wente's Equation (4).

If we now let the voltge source ΔE or e_{ac} drive the series combination of $1/\omega C_0$ and R at *any* frequency, it turns out (by reworking Wente's exact equation, Equation 3), that e_{ac}, which is $E_0 \Delta C/C_0$ when $R = 0$ is practically independent of

R even at frequencies much higher than the quasi-d.c. frequency we started with, so long as $\Delta C/C_0$ is small. And so we can treat it as the exact a.c. generator.

Then

$$i = e_{ac}/\sqrt{(1/\omega C_0)^2 + R^2} = (E_0 \Delta C/C_0) / \sqrt{(1/\omega C_0)^2 + R^2}. \quad (6)$$

Then the load voltage is

$$e_R = i R = (E_0 RC_1 \sin \omega t/C_0) / \sqrt{(1/\omega C_0)^2 + R^2}. \quad (7)$$

And this is Wente's Equation (4) when we replace ΔC by its equivalent, $C_1 \sin \omega t$.

Page 47, paragraph 2: Wente gives capacitance as $C = (1/k)\pi R^2/4\pi d$ or $(1/k)$ Area$/4\pi d$, where the omitted $1/k$ is 1 stat-coulomb²/dyne-cm², and C is in stat-coulomb²/dyne-cm, which equals constant times farads. Nowadays we would use MKS units and express C as ϵ_0Area$/d$ in coulomb²/newton-meters, which equals another constant times farads. Note that $\epsilon_0 = 1/4\pi k$. We can check things dimensionally by noting that Energy $= 1/2\,(q^2/C)$. Hence $C \sim (q^2/\text{Energy})$.

Page 48, first equation: Note that this sensitivity is 6 dB above 1 millivolt or –54 dB versus 1 volt per dyne/cm², which is very acceptable even today.

Page 54: In Equation 13, the misprinted term should read $\rho\omega^2/T$. This then can be written $\omega^2/(T/\rho)$ or ω^2/c^2 or $(2\pi/\lambda)^2$. An equation similar to this is treated by Morse, *Vibration and Sound*, 2nd ed., McGraw-Hill, New York, p. 187 (1948), Equation 19.3. In the next few lines, note that I has the dimensions 1/length, like $2\pi/\lambda$.

Page 55, last paragraph: We are dealing with frequencies such as 6,350 Hz; 17,000 and 19,000 Hz; and 35,000 and 9,000 Hz; and we could get lost. Wente is making three separate points. First, the membrane or diaphragm in free space has a natural frequency or lowest resonance f_0 of 6,350 Hz. Second, when the diaphragm is spaced about 0.001 inches from the plate, the air or hydrogen in the gap becomes laterally trapped, and the gap acts like a closed cavity. The cavity stiffness then raises f_0 to 17,000 or 19,000 Hz. This shift in frequency is the same for air or hydrogen because the cavity stiffness is given by $\rho c^2/V$, whose other form is $\gamma P_0/V$; and since γ chances to be 1.4 for either gas and P_0 is merely the atmospheric pressure, the cavity stiffness must be the same for air or hydrogen. Third, there is a "cylindrical enclosure," part A in Figure 2 (also Fig. 8), which is a sort of mouthpiece. It will have a number of radial resonances, with perhaps the lowest being tightly coupled to the resonant transducer when the mouthpiece is closed by the thermophone (Fig. 8). This lowest resonance can be thought of as being due to a "breathing ring" of hydrogen (or air), this ring having some effective circumference $\pi(\alpha D)$ of maybe one-half to one-fourth the circumference πD of the cylinder. The ring circumference $\pi(\alpha D)$ must always equal one wavelength, or λ. The factor α for Wente's cylinder turns out to be 1/3.83 or 0.26, and the effective ring circumference is thus 3.58 cm, which must equal λ for every gas resonating in the lowest mode. (In the formula for f_0, middle of the page, the printer dropped the π from the denominator.) When hydrogen is used, having a velocity of 1,270 m/sec, f_0 should be Wente's value of 35,000 Hz. When air is used, having a velocity of 344 m/sec, f_0 should be 9,600 Hz (confirmed by experiment). Note, however, that when the thermophone is removed, leaving a shallow open mouthpiece, this radial resonance will be very weak and probably not even measurable.

Page 58, paragraph 2: Note that this discussion has occurred in the section entitled "Possibilities of Tuning." The discussion implies that the doubly tuned transducer will produce a tunable "wave analyzer," roughly analogous to Webster's phonometer. However, Wente never did find it practical to retune the mechanical resonance frequency throughout the audio band. The subject is mentioned again, more optimistically, in the conclusion of Wente's second (1922) paper.

Page 60, section 9: In this section Wente is discussing the reversing of his microphone to act as a loudspeaker, but he does not invoke the reciprocity principle. One reason is that the condenser microphone (a generator) will not work if an a.c. voltage is to be generated in the absence of a d.c. polarizing voltage. Yet it will work as a loudspeaker (a motor) without a d.c. polarizing voltage, albeit with frequency doubling. This is touched upon in the paper by R. K. Cook on the reciprocity method of absolute calibration, in Part IV of this volume (Paper 32). It should be noted that the condenser microphone is a modulation device that requires a carrier, either d.c. in Wente's case or a.c. in Riegger's case (H. Riegger, *Zeitschr.*

Tech. Phys. 5:577, 1924). Without the carrier no electric signal can be generated by the sound wave.

Wente could have carried through the remaining analysis using d.c. plus a.c. or $E_0 + e_{ac}$, applied to the loudspeaker. In this case the reciprocity principle would hold. However, he was apparently more concerned with designing and using the simplest standard source of sound, and a plain a.c. signal was all he needed for that.

Page 61, following Equation 16: It might be well at this point to review the various ingredients that Wente has been using in his equations. We will use CGS units for mechanical quantities and electrostatic units for electrical quantitites. Our aim is to check the dimensionality of some of Wente's relationships.

Wente's capacitance C is always $d(\text{cm})/4\pi k$, where $1/k$ is 1 esu or 1 stat-coulomb²/dyne-cm². In Wente's day it was customary to omit the factor $1/k$ or 1 esu so that the dimension of C was said to be length (in centimeters). Of course it should have been coulomb²/dyne-cm, which is the dimension of the farad. However, for his final equations, Wente always switches over from centimeters to farads, so his final equations always check out; only the intermediate ones do not. Thus,

$$P \cdot \text{Vol} = Pd^3 = \text{Energy} = 1/2\ CE^2.$$

Then

$$P = \frac{1/2\ [E^2C(1 - \cos 2\ \omega t)]}{d^3} = \frac{1/2\ [E^2d(1 - \cos 2\ \omega t)]}{d^3 4\pi k}\ .$$

Wente then omits k and says

$$P = \frac{E^2(1 - \cos 2\ \omega\ t)}{8\ \pi\ d^2}\ .$$

This is his Equation 15, and of course it does not check out dimensionally.

If we were to calculate the volume displacement of a freely sliding rigid piston, opposed only by cavity stiffness K_{acoustic}, then

$$\Delta V = P/(\gamma P_0/V_0) = P(Ad/\gamma P_0).$$

Then, using the capacitance version above of the equation for P:

$$\Delta V = (E^2/2d^3)(C/\gamma P_0)(1 - \cos 2\omega t)Ad.$$

Note that a quantity similar to $C/\gamma P_0$ is what Wente, in his Equation 16, calls C_f, the change in capacity per unit pressure. Then linear displacement u, or $\Delta V/A$, is

$$u = (E^2/2d^2)(C/\gamma P_0)(1 - \cos 2\omega t).$$

This is somewhat similar to Equation 18, and it checks out dimensionally if one measures C in farads, as Wente sometimes does (see p. 45, top).

The factor 8π has shown up (above, Equation 15, and other places) because in the electrical equations $1/2(C) = 1/2(d/4\pi k) = d/8\pi k$. But by coincidence, another factor 8π shows up in the mechanical equation for the flexing membrane. If we wish to calculate u for a flexing membrane when no cavity stiffness is present, $u = F/K_m$. Here, K_m, or $K_{mechanical}$, is the effective membrane stiffness in dyne/cm that prevails below the first resonance. But $K_m = 8\pi T$ dyne/cm (see Morse, p. 202, Equation 20.10). Hence $u = PA/8\pi T$. This is somewhat similar to Equation 17.

Combining the two types of stiffness, cavity and membrane,

$$u = \frac{PA}{8\pi T + K_{acoustic} \cdot A^2} = \frac{1/2[E^2C(1 - \cos 2\omega t)]A}{d^3[8\pi T + (\gamma P_0/Ad)A^2]}.$$

Then

$$u = \frac{1/2[E^2 (1 - \cos 2\omega t)]\pi a^2}{4\pi d^2(8\pi T + \gamma P_0 \cdot A/d)}.$$

This is somewhat similar to Equation 19 if we agree to omit $1/k$ from the factors composing C (as Wente did in Equation 15) and thus measure C as $d/4\pi$ centimeters. Note that Wente here uses both 8π factors simultaneously.

13

Reprinted from page 503 of *Phys. Rev.* **19**:498–503 (1922)

THE SENSITIVITY AND PRECISION OF THE ELECTROSTATIC TRANSMITTER FOR MEASURING SOUND INTENSITIES

E. C. Wente

[*Editor's Note:* In the original, material precedes this excerpt.]

Concluding Remarks.

Little has been said about amplifiers in this paper although the electrostatic transmitter has little practical value unless it is used with an amplifier. Experiments have shown that the transmitter changes but little with time and atmospheric conditions. The amplifier, when properly designed and if the vacuum tubes are carefully chosen, will maintain the same value of amplification for a long period, provided the plate voltage and the filament current are kept at a constant value. An amplifier used in most of the experiments described above did not vary by more than a few per cent. during the course of several months. A combination of electrostatic transmitter and amplifier can thus be used as an absolute phonometer, the readings of which are dependable from day to day.

In the preceding discussion it has already been pointed out that one of the chief merits of the electrostatic transmitter is the fact that it has a very uniform response and so is especially adapted for measuring sounds of complex wave form or for comparing the intensities of tones of different frequencies. However, in some classes of problems it is desirable to have an instrument which is sharply tuned so that it will respond to a tone of one frequency and be unaffected by any other tones that may be present at the same time. If an electrostatic transmitter is connected to a tuned amplifier the output of which goes to a vibration galvanometer, selectivity of a very high order may be obtained and an amplifier of sufficient amplification may be used to give the combination almost any desired sensitivity.

Research Laboratories of the
American Telephone and Telegraph Company,
and Western Electric Company, Inc.

14

Reprinted from *Phys. Rev.* **10**:22–38 (1917)

THE THERMOPHONE AS A PRECISION SOURCE OF SOUND.

By H. D. Arnold and I. B. Crandall.

THE acoustic effect accompanying the passage of an alternating current through a thin conductor has been known for some time, but as far as we are aware, no use has been made of the principle involved for the production of a precision source of sound energy, or standard phone. In 1898 F. Braun[1] discovered that acoustic effects could be produced by passing alternating currents through a bolometer in which the usual direct current was also maintained. An article by Weinberg[2] describes the old experiments of Braun, and also some experiments of Weinberg, in which acoustic phenomena were observed with other electrically heated conductors, rheostats, etc., through which large alternating currents were passed. A more recent application of the same principle is described by de Lange[3] in his article on the thermophone.

The writers have found that the thermophone together with a suitable supply of alternating current can be used very conveniently as a precision source of sound energy. On account of the fact that the published material on this electrical-acoustic effect is largely of a qualitative character it has been necessary to work out a quantitative theory; and it is the purpose of this paper to give the theory and show how the instrument can be adapted to acoustic measurements.

When alternating current is passed through a thin conductor, periodic heating takes place in the conductor following the variations in current strength. This periodic heating sets up temperature waves which are propagated into the surrounding medium; the amplitude of the temperature waves falling off very rapidly as the distance from the conductor increases. On account of the rapid attenuation of these temperature waves, their net effect is to produce a periodic rise in temperature in a limited portion of the medium near the conductor, and the thermal expansion and contraction of this layer of the medium determines the amplitude of the resulting sound waves. To secure appreciable amplitudes with currents of ordinary magnitude it is essential that the con-

[1] Ann. der Physik. 65, 1898, p. 358.

[2] Elektrot. Zeit. 28, 1907, p. 944. See also A. Koepsel, Elektrot. Zeit. 28, 1907, p. 1095.

[3] Proc. Royal Soc. 91A, 1915, p. 239.

ductor be very thin; its heat capacity must be small, and it must be able to conduct at once to its surface the heat produced in its interior, in order to follow the temperature changes produced by a rapidly varying current.

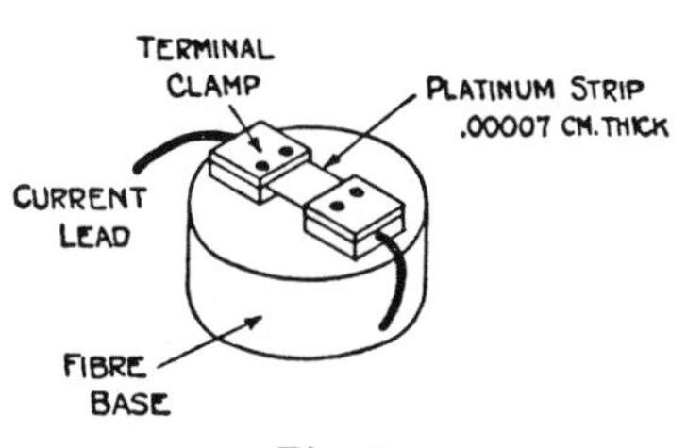

Fig. 1.
Simple Thermophone.

A simple form of instrument which we have used is shown in Fig. 1. There are two ways in which the strip may be supplied with electrical energy in order to produce sound waves, (*a*) with pure alternating current and (*b*) with alternating and direct current superimposed. If an alternating current $I \sin pt$ is supplied, the heating effect is proportional to

$$RI^2 \sin^2 pt = \frac{RI^2}{2}(1 - \cos 2pt), \quad (1)$$

so that the acoustic frequency is double the frequency of the applied alternating current. If it is desired to make the acoustic wave follow the alternating current wave, without introducing the double frequency effect, resort must be had to a superimposed direct current whose value is several times as large as the maximum value of the alternating current. If a direct current I_0 and an alternating current $I' \sin pt$ are used to heat the strip, the heating effect is proportional to

$$\begin{aligned} R(I_0 + I' \sin pt)^2 &= RI_0^2 + 2RI_0I' \sin pt + RI'^2 \sin^2 pt \\ &= R\left(I_0^2 + \frac{I'^2}{2}\right) + 2RI_0I' \sin pt - \frac{RI'^2}{2} \cos 2pt \quad (2) \end{aligned}$$

from which it is evident that the double frequency term can be made negligible by suitable choice of I_0 and I'.

When pure alternating current is used, the mean temperature of the strip is determined by the term $\frac{1}{2}RI^2$; when direct current is used with the alternating current, the mean temperature is determined by the term RI_0^2. The mean temperature of the conductor is one of the factors which sets a limit on the maximum amount of electrical energy used and hence on the maximum amount of acoustic energy that can be obtained. If only a small quantity of alternating current energy of suitable frequency is available, it is clear, from a comparison of equations (2) and (1) that more acoustic effect will be realized if direct current energy is added up to the limit that the strip will bear; for example, if I_0^2 is as large as I'^2, the product term in (2) is four times as large as the second term in (1).

Suppose now that an indefinite quantity of alternating current energy of any frequency is at hand; we desire to find the most effective way to actuate the element. Equating the terms in (1) and (2) which are proportional to the (limiting) mean temperature in each case

$$\left[\frac{RI^2}{2}\right]_{max} = \left[R\left(I_0^2 + \frac{I'^2}{2}\right)\right]_{max} \tag{3}$$

we can compute the maximum amplitude $2RI_0I'$ of the product term in (2) and compare this with the amplitude $RI^2/2$ of the periodic term in (1). The maximum value of the product $2RI_0I'$, consistent with condition (3) is $RI^2/\sqrt{2}$ and implies the relations

$$I = \sqrt{2}\, I' = 2I_0.$$

The amplitude $RI^2/\sqrt{2}$ in the second case is only slightly larger than the amplitude $RI^2/2$ which we should have according to (1); and in the second case there is the double frequency term of amplitude $RI^2/4$ which in most cases would be inconveniently large.

The conclusion from these calculations is that for sounding a pure tone of a given frequency it is better to actuate the strip wholly by alternating current of half that frequency. However, if it is desired to make the sound waves reproduce the electrical waves in both frequency and form, it is necessary to use in addition a direct current whose relative value is large. In this case the thermophone element is worked somewhat below maximum efficiency for the sake of minimizing the double-frequency effect.

Using the first method of excitation, it is necessary, if a pure tone is desired, that the alternating current used be a pure sine wave, absolutely free from harmonics. In order to show the acoustic effect of harmonics in the alternating current supply, consider an exciting current of the form

$$\sum_{k=1}^{n} a_k \sin kpt.$$

The heating effect produced is proportional to

$$\left(\sum_{k=1}^{n} a_k \sin kpt\right)^2 = \sum_{k=1}^{n} a_k^2 \sin^2 kpt + \sum_{\substack{j=1 \\ j \neq k}}^{n-1} \sum_{k=2}^{n} a_j a_k \sin jpt \sin kpt$$

$$= \sum_{k=1}^{n} \frac{a_k^2}{2} - \sum_{k=1}^{n} \frac{a_k^2}{2} \cos 2kpt + \sum_{\substack{j=1 \\ j \neq k}}^{n-1} \sum_{k=2}^{n} \frac{a_j a_k}{2} \cos (j-k)pt$$

$$- \sum_{\substack{j=1 \\ j \neq k}}^{n-1} \sum_{k=2}^{n} \frac{a_j a_k}{2} \cos (j+k)pt.$$

which shows that two series of combination-tones result in addition to the series of tones whose frequencies are double those of the applied fundamental and harmonics. One particular case is of practical importance: the case in which the alternating current wave consists of a fundamental and an appreciable second harmonic. In this case, besides the tones of double and quadruple frequency there are combination tones of *single* and triple frequency, a paradoxical result that is very easily verified by experiment. The importance of a pure alternating current supply is clear from the considerations given.

The Periodic Temperature Change in a Thin Flat Conductor Supplied with Alternating Current.

Consider first the case of a strip supplied with both direct and alternating current. Equating the rate of production of heat by the electric current to the rate of transfer of heat to the surrounding medium, plus the rate of storage of heat in the strip, the fundamental equation may be written:

$$0.24(I_0 + I' \sin pt)^2 R = 2a\beta T + a\gamma \frac{dT}{dt}, \qquad (4)$$

in which the unit is the calorie per second, and the constants are chosen as follows:

I_0 = direct current in amperes.
I' = maximum value of A.C. in amperes.
p = $2\pi f$; f = frequency.
R = instantaneous resistance of the strip.
T = temperature of strip above surroundings.
a = area of one side of strip.
β = the rate of loss of heat per unit area of the strip (due to conduction *and* radiation) per unit rise in temperature of the strip above that of its surroundings; it is equal to the product of the temperature gradient per degree rise, into the conductivity of the medium. It can be determined experimentally, and is a constant if only conduction is considered; if it is desired to take account of radiation a modified value of β for any value of T may be obtained which is sufficiently accurate for the purposes of calculation. The rate of radiation is not great at low temperatures, and only becomes equal to the rate of conduction at about 500° C.
$a\gamma$ = the heat capacity of the strip, γ being equal to the product of the thickness of the strip by the specific heat per unit volume.

The factor is analogous to the mass in vibratory mechanics, and the inductance in alternating current calculations.

The equation for T_0, the mean temperature above surroundings is:

$$0.24\left(I_0^2R + \frac{I'^2R}{2}\right) = 2a\beta T_0. \tag{4a}$$

Combining equations (4) and (4a) we have the following, which contains only factors which vary with the time:

$$0.24R\left(2I_0I' \sin pt - \frac{I'^2}{2}\cos 2pt\right) = 2a\beta(T - T_0) + a\gamma\frac{dT}{dt}. \tag{5}$$

In obtaining a solution for $T - T_0$ we shall neglect transient effects, also the double frequency term. The double frequency effect is the principal effect in the case of a pure alternating current supply as given below; but here we simply remark that we can make the double frequency term as small as we please by a suitable choice of the ratio $I_0 : I'$.

The solution of the equation

$$.48RI_0I' \sin pt = 2a\beta(T - T_0) + a\gamma\frac{dT}{dt} \tag{6}$$

is, neglecting transient effects,

$$T - T_0 = \frac{.48RI_0I'}{a\sqrt{4\beta^2 + \gamma^2p^2}} \sin\left(pt - \tan^{-1}\frac{\gamma p}{2\beta}\right), \tag{7}$$

which gives the periodic temperature variation of the strip. Note that if i is the effective (measured) value of the A.C., $\sqrt{2}\, i$ must be written in place of I' in (7).

If the strip is supplied with alternating current only, the fundamental equation becomes

$$0.24RI^2 \sin^2 pt = .12RI^2(1 - \cos 2pt) = 2a\beta T + a\gamma\frac{dT}{dt}. \tag{4'}$$

The mean temperature in this case is defined by

$$.12RI^2 = 2a\beta T_0 \tag{4'a}$$

and the differential equation which $T - T_0$ must satisfy is

$$.12RI^2 \cos 2pt = 2a\beta(T - T_0) + a\gamma\frac{dT}{dt}. \tag{5'}$$

The solution of this equation is, neglecting transient effects

$$T - T_0 = \frac{.12RI^2}{2a\sqrt{\beta^2 + \gamma^2p^2}} \cos\left(2pt - \tan^{-1}\frac{\gamma p}{2\beta}\right). \tag{7'}$$

Having found the magnitude of the temperature variation in the strip, we go on to calculate the magnitude of the effect in the surrounding medium.

Theory of the Effect in the Medium.

Consider an infinite plane metal plate with a column of gas extending normally from either face of a certain portion of the plate; this is equivalent, mechanically, to the strip conductor if terminal conditions are neglected. If the temperature of the plate is a sine function of the time, temperature waves will be propagated into the atmosphere on either side; and calculation will show that these waves are so heavily damped that they are practically extinguished after one wave-length has been traversed. Within this region there is a rise and fall of temperature of the medium with every cycle, and the resulting expansion and contraction of this narrow film of the medium near the source accounts for the sound vibration produced.

In the derivation of equations (7) and (7′) it has been tacitly assumed that no electrical energy was spent in expanding the strip, as this effect would be relatively very small. It is evident from conditions of symmetry that there is no force on the strip tending to make it vibrate; hence no energy can be used mechanically. In calculating the effect on the medium we shall consider two cases:

1. In which the periodic rise in temperature of the strip is allowed to produce a continuous stream of sound energy, propagated away from the strip as plane waves. It is an easy matter to modify this treatment to fit the case of diverging waves in the open atmosphere.

2. In which the strip is placed in a small cavity for the purpose of producing pressure changes; these pressure changes being used to actuate the ear, or some mechanical phonometer which constitutes one wall of the enclosure.

The reason for giving separate treatment to these two types of action, is that in the first case we can speak of a definite amplitude and particle velocity, and a corresponding propagation of energy; whereas in the second case, amplitude and velocity are indefinite terms, and pressure change is much more readily calculated. It is by virtue of pressure change that the acoustic energy generated makes its effect on the bounding wall, and if the dimensions of the cavity are small compared to the acoustic wave-length, the pressure change produced at the strip is quickly distributed over the whole enclosure.

First Case: Wave Propagation from the Strip.—Assume that the periodic temperature variation in the strip results in the expansion and contraction against constant atmospheric pressure of a certain layer of air next to the source. This implies that the very small pressure changes that do arise at the boundary of the layer (as the result of rapid change in volume) are propagated into the atmosphere with such high velocity that they do not react appreciably on the expansion of the layer. This condition is realized in practice because the velocity of sound in air is so much greater than the velocity of the vibrating boundary which produces the sound.

In treatises on the conduction of heat it is shown that the temperature at any point of the medium distant $\pm x$ from a plane source of temperature, varying periodically as in equation (7), may be expressed as the following function of space and time:

$$T_x' = T'\epsilon^{-\alpha x} \sin (pt \pm \alpha x), \tag{8}$$

in which $\alpha = \sqrt{p/2k}$, p being $2\pi \times$ frequency, and k the "diffusivity" of the medium, or the ratio of the thermal conductivity to the specific heat per unit volume. The value of this constant for air at 0° centigrade, using the specific heat at constant pressure is 0.17 C.G.S. units.

It is necessary to know the effect of the temperature of the medium on k and this can be found by considering separately the conductivity and the specific heat. The former is proportional to the square root of the absolute temperature; the specific heat per unit mass is practically independent of temperature thus making the specific heat per unit volume proportional to the reciprocal of the absolute temperature. Since k is the ratio, we may write

$$k = 0.17 \left(\frac{\theta}{273}\right)^{3/2} \tag{9}$$

where θ denotes the absolute temperature of the medium.

The velocity of propagation of the temperature wave is, from (8)

$$v = \frac{p}{\alpha} = \sqrt{2pk}$$

and the wave-length

$$\lambda = \frac{2\pi}{\alpha} = \sqrt{\frac{8\pi^2 k}{p}}. \tag{10}$$

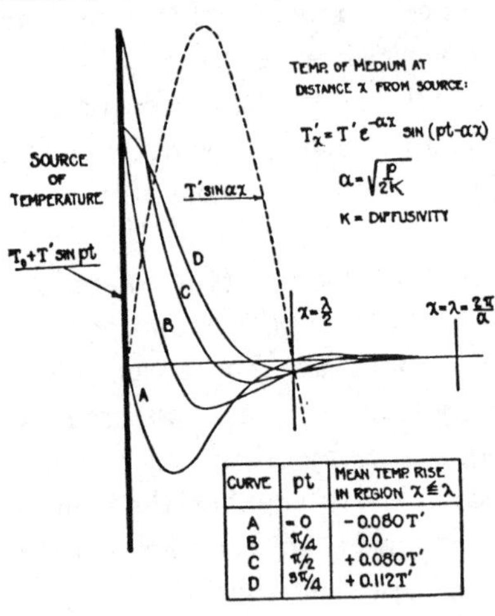

Fig. 2.

If the wave-length is taken as a unit, it is easy to plot the course of T_x' as a function of x for any given time, t, as is shown in Fig. 2. This shows clearly the enormous damping of these waves of acoustic frequency; it also shows that practically all of the expansion effect due to periodic rise in temperature takes place within the region bounded by the plane

$$x = \lambda = \frac{2\pi}{\alpha}.$$

In order to compute the amount of the periodic expansion, we desire to know the mean value of the temperature rise in this region as a function of the time: that is,

$$\frac{T'}{\lambda}\int_0^{\lambda} \epsilon^{-ax} \sin(pt - \alpha x)dx$$

$$= \frac{T'(1 + \epsilon^{-2\pi})}{2\lambda\alpha}(\sin pt - \cos pt) = \frac{T'}{2\sqrt{2}\pi}\sin\left(pt - \frac{\pi}{4}\right) \quad (11)$$

i. e., its maximum value is $.112T'$, and it lags the varying temperature of the strip by the angle $\pi/4$.

If the mean absolute temperature of the air film is θ, the maximum expansion will be,

$$\frac{d\theta}{\theta} = .112\frac{T'}{\theta} \text{ per unit volume,} \quad (12)$$

or per unit length, if expansion is considered to take place in only one direction. The length in question is a wave-length, and this by equations (9) and (10) is, at $\theta°$

$$\lambda = 2\pi\sqrt{\frac{2k}{p}} = 0.82\sqrt{\frac{\pi}{f}}\left(\frac{\theta}{273}\right)^{3/4}. \quad (13)$$

Multiplying (12) and (13) we obtain for the absolute increase in length due to expansion

$$\xi = \frac{.16T'}{\sqrt{f}\,\theta^{\frac{1}{4}}(273)^{\frac{3}{4}}}. \quad (14)$$

This may be considered as the maximum "amplitude" of a sound wave leaving the plane $x = \lambda$, if the effect of the expanding and contracting air film on the surrounding air is the same as that of a solid moving piston—assuming also that the amplitude of the sound produced by a moving piston is equal to the amplitude of the motion of the piston itself.

If the thermal conductivity were proportional to the first power of the absolute temperature, instead of to the square root, we should have had, instead of (12)

$$\xi_1 = \frac{0.16T'}{\sqrt{f}\,.273}. \quad (14a)$$

The departure of (14*a*) from (14) is not serious, if the temperature of the film is 300° C. or less, as in this case $\theta^{1/4}\cdot(273)^{3/4}$ is less than 330. The air film is always considerably cooler than the strip, so that the strip might have a temperature of (say) 500° without causing more than a 20 per cent. discrepancy between ξ and ξ_1.

In order to have the amplitude of the sound wave in terms of the alternating current supplied to the strip, we make use of equations (7) and (7′) which give the variation in temperature of the strip.

Using (7) and (14*a*), we have for a strip supplied with direct current I_0 and alternating current of effective value i, the acoustic amplitude

$$\xi = \frac{4.0 \times 10^{-4} R I_0 i}{a\sqrt{f}\sqrt{4\beta^2 + \gamma^2 p^2}} \sin\left(pt - \tan^{-1}\frac{\gamma p}{2\beta} - \frac{\pi}{4}\right). \tag{15}$$

Using (7′) and (14*a*) we have for a strip supplied with alternating current only

$$\xi' = \frac{7.0 \times 10^{-5} R i^2}{a\sqrt{f}\sqrt{\beta^2 + \gamma^2 p^2}} \cos\left(2pt - \tan^{-1}\frac{\gamma P}{\beta} - \frac{\pi}{4}\right). \tag{15'}$$

These two equations contain no transient terms; they are solutions for the state of steadily maintained vibrations. The acoustic amplitude ξ' (Equation 15′) is of double the frequency of the applied alternating current.

Using either method of actuating the strip, there is a low critical frequency above which the factor γp (which represents thermal inertia) is so much greater than β (which represents conduction loss, or dissipation) that the latter can be neglected. (This frequency is in the neighborhood of 100 for platinum strip 1 micron thick.) Neglecting β, (15) can be written

$$\xi = \frac{6.4 \times 10^{-5} R I_0 i}{a\gamma f^{3/2}} \sin\left(pt - \frac{3\pi}{4}\right), \tag{15a}$$

and instead of (15′) we have

$$\xi' = \frac{1.1 \times 10^{-5} R i^2}{a\gamma f^{3/2}} \cos\left(2pt - \frac{3\pi}{4}\right). \tag{15'a}$$

In considering how the efficiency of the process depends on the constants of the strip, we note that it is advantageous to make the resistance R as large as possible, and the heat capacity $a\gamma$ as small as possible. The advantage of thinness is plain.

In calculating the intensity of a sound wave, or the rate of flow of energy in the medium it is necessary to know the square of the particle velocity; and this is, from (15*a*) (using superimposed direct current)

$$\dot{\xi}^2_{\max} = p\xi_{\max} = \frac{4.0 \times 10^{-9} R^2 I_0^2 i^4}{a^2\gamma^2 f^3}. \tag{16}$$

Similarly from (15′*a*), for alternating current,

$$\dot{\xi}'^2_{\max} = \frac{1.2 \times 10^{-10} R^2 i^4}{a^2\gamma^2 f^3}. \tag{16'}$$

These equations enable us to find the strength of the source; and knowing this, we can calculate the intensity of the sound at any distance from the source, in the ideal case in which energy is propagated in the form of spherical waves in a homogeneous medium.[1]

Since the dimensions of the source are small compared with the wavelength of sound, we may consider the strip as equivalent to a small sphere of the same area ($2a$) and which produces the same fluid velocity $\dot{\xi}$ at the surface. The velocity potential for the resulting spherical distribution of sound waves is

$$\varphi = -\frac{2a\dot{\xi}_{max}}{4\pi r}\cos\left(pt - \frac{2\pi r}{\lambda}\right), \tag{17}$$

in which $2a\dot{\xi}_{max}$ is the strength of the source, or maximum rate of emission of fluid at the source. In order to calculate the intensity of the sound produced, we make use of the two following equations

$$\text{Intensity} = \frac{W}{t} = \frac{\Pi^2_{max}}{2\rho_0 c}, \tag{18}$$

$$\Pi = -\rho_0 \frac{d\varphi}{dt}, \tag{19}$$

in which Π is the pressure change at any point in the field, c the velocity of sound, and ρ_0 the mean density of the medium. Substituting (17) in (19) we obtain for Π in terms of $\dot{\xi}_{max}$

$$\Pi = \frac{-\rho_0 a\dot{\xi}_{max}\cdot 2\pi f}{2\pi r}\sin\left(pt - \frac{2\pi r}{\lambda}\right),$$

and for the intensity, according to equation (18)

$$\frac{W}{t} = \frac{\rho_0 a^2 \dot{\xi}^2_{max}\cdot f^2}{2cr^2}, \tag{20}$$

or finally in terms of the electrical energy used in the strip (direct current case)

$$\frac{W}{t} = \frac{2 \times 10^{-9} R^2 I_0^2 i^2 \rho_0 f}{cr^2\gamma^2}, \tag{21}$$

or, for alternating current,

$$\frac{W}{t} = \frac{6.0 \times 10^{-11} R^2 i^4 \rho_0 f}{cr^2\gamma^2}. \tag{21'}$$

[1] The solution here given for intensity in the case of ideal spherical distribution may easily be applied to the more practical case in which the small thermophone element is placed close to an infinite rigid plane wall. In this case, the velocity potential on the thermophone side of the wall will be twice as great as given by (17) and the intensity four times as great as given in (20).

Thus the actual intensity at any point some distance away from a thermophone whose power input is constant should increase with the first power of the frequency, and decrease with the square of the distance r. It is independent of a, the area of the strip.

Second Case: Production of Pressure Changes in Small Enclosure.—Let us assume that the strip is placed in an enclosure the dimensions of which are small compared with the acoustic wave-length, and further that the shortest distance from the strip to the boundary is large compared to the wave-length of the heat wave originating at the surface of the strip. These conditions are readily satisfied for all ordinary acoustic frequencies. If the temperature variation of the strip is given by

$$T' \sin \omega t$$

the temperature variation at any near-by point in the enclosure is

$$T_x' = T' \epsilon^{-\alpha x} \sin (\omega t \pm \alpha x). \tag{8}$$

We can consider that both sides of the strip, each of area a, give rise jointly to the temperature wave; also that this temperature wave travels a mean distance x before striking boundary defined by the equation

$$\bar{x} = \frac{V_0}{2a},$$

where V_0 is the volume of the enclosure. The alternating temperature averaged over the whole enclosure is then

$$\delta T = \frac{2a}{V_0} \int_0^{\bar{x}} T_x' dx = \frac{2aT'}{V_0} \int^{\bar{x}} E^{-\alpha x} \sin (\omega t - \alpha x) dx. \tag{22}$$

The thermal conductivity of the gaseous medium varies as the square root of the absolute temperature, while the specific heat per unit volume is practically constant at constant volume, so that the diffusivity is

$$K = K_0 \sqrt{\frac{\theta_1}{273}}.$$

In terms of K_0, the diffusivity at 0° Centigrade, θ_1 is the absolute temperature of the gas near the element, this being approximately the same as the temperature of the element itself.

We then have

$$\alpha = \sqrt{\frac{\omega}{2K_0} \sqrt{\frac{273}{\theta_1}}}. \tag{23}$$

As α varies only as the fourth root of $\theta_1/273$, and conditions are easily arranged so that the temperature of the gas is not excessive, α may be considered constant in the evaluation of the integral in (17). Integrating,

$$\delta T = \frac{aT'}{V_0\alpha}\left[\epsilon^{-(V_0\alpha/2a)}\left\{\sin\left(\omega t + \frac{V_0\alpha}{2a}\right) - \cos\left(\omega t + \frac{V_0\alpha}{2a}\right)\right\} - \sin\omega t - \cos\omega t\right].$$

Now K_0 is of the order of unity ($K_0 = 1.5$ for hydrogen and .23 for air, using specific heat at constant volume) so that α (equation 23) is large for all acoustic frequencies. We may, therefore, neglect $\epsilon^{-(V_0\alpha/2a)}$ and write

$$\delta T = -\frac{\sqrt{2}\,aT'}{V_0\alpha}\sin\left(\omega t - \frac{\pi}{4}\right)$$

and, substituting the value of α from (23)

$$\delta T = -\frac{2aT'}{V_0\sqrt{\omega}}\sqrt{K\sqrt{\frac{\theta_1}{273}}}\sin\left(\omega t - \frac{\pi}{4}\right). \qquad (24)$$

If the walls of the boundary are rigid, we have for a perfect contained gas, $\delta V = 0$ and the pressure change in terms of temperature change is

$$\Pi = \frac{P}{\theta_2}dT$$

if P = total pressure and θ_2 is the mean temperature of the gas. Substituting δT from (24) we have for pressure change in the enclosure

$$\Pi = \frac{2aT'P\sqrt{K_0\sqrt{\theta_1/273}}}{\theta_2 V_0\sqrt{\omega}}\sin\left(\omega t - \frac{\pi}{4}\right), \qquad (25)$$

in terms of temperature variation in the strip. When direct current is used with the A.C. this is given by (7); substituting this expression for T' and dropping the dissipation factor β, we have, ($\omega = p$)

$$\Pi = \frac{.086RI_0iP\sqrt{K_0\sqrt{\theta_1/273}}}{\gamma V_0\theta_2 f^{3/2}}\sin\left(pt - \frac{3\pi}{4}\right) \qquad (26)$$

and when the strip is actuated only by alternating current, we have from (25) and (7′), dropping β as before, and noting that $\omega = 2p$,

$$\Pi = \frac{.0106Ri^2P\sqrt{K_0\sqrt{\theta_1/273}}}{\gamma V_0\theta_2 f^{3/2}}\cos\left(2pt - \frac{3\pi}{4}\right). \qquad (26')$$

In (26′) f is the frequency of the alternating current and half the acoustic frequency.

Equations (26) and (26′) are in the most convenient form for calculating the stress exerted on any part of the boundary, which may be the exposed face of a sound detecting mechanism, as for example the ear. The intensity of the sound produced in the enclosure can easily be computed from the usual equations

$$\frac{W}{t} = \tfrac{1}{2}\rho_0 c^3 s^2 = \tfrac{1}{2}\frac{\Pi^2}{\rho_0 c}, \tag{27}$$

in which s = maximum condensation (Π/P), Π = maximum pressure change, ρ_0 = mean density, and c = velocity of sound in medium. Substituting the value of Π from (26) in (27), the intensity is, in the case of direct current operation

$$\frac{W}{t} = \frac{3.7 \times 10^{-3} R^2 I_0^2 i^2 P^2 K_0 \sqrt{\theta_1/273}}{\rho_0 c \gamma^2 V_0^2 \theta_2 f^3}, \tag{28}$$

and in the case of alternating current only, from (26′)

$$\frac{W}{t} = \frac{1.1 \times 10^{-4} R^2 i^4 P^2 K_0 \sqrt{\theta_1/273}}{\rho_0 c \gamma^2 V_0^2 \theta_2^2 f^3}. \tag{28′}$$

It is seen from these equations that the intensity in this case is inversely proportional to the cube of the frequency. The temperature θ has been retained in equations (28) and (28′), and the calculation has been carried through to a determination of the intensity; but there is not much difference between equations (21), (21′) which deal with the intensity in the first case, and (28), and (28′) which deal with the intensity in the second case, except the frequency-variation law.

In all cases the temperatures of gas and of strip must be taken into account; and in most cases it is possible to arrange experimental work and calculation so that this can be done in a very simple way.

Experimental Tests.

The first test that was made was a rough verification of equations (15*a*) and (26) to see if the computed effect was of the right order of magnitude. The method used consisted in setting the thermophone and an electro-mechanical source (ordinary telephone receiver) for equal intensity at the same pitch, and measuring the electrical input into each instrument. The setting for equal intensity was made with the unaided ear, for simple experiments have shown that the ear judges equality between two tones of the same pitch to within 4 or 5 per cent.[1] The telephone receiver had previously been calibrated as a sound generator by measuring the motion of the diaphragm with a microscope when a known value of alternating current was sent through it. In the case of the vibrating telephone diaphragm, the motion of the diaphragm is greatest near the center, falling off to zero at the edge. The law of

[1] Or to one per cent. under favorable conditions. The ear seems to be about as good in these measurements as the eye is in the analogous photometrical case.

distribution of amplitude over the diaphragm is, for small vibrations (at the particular frequency used), such that the bowed diaphragm may be considered from the standpoint of air displacement as replaced by a piston whose area is 0.306 that of the diaphragm, and which moves back and forth with an amplitude equal to the amplitude of the diaphragm at the center.

The data of this experiment were:

Frequency, 800.

Constants of telephone receiver:

Area of diaphragm, 18.3 sq. cm.

Effective area, 5.5 sq. cm.

800-cycle current, 1.7×10^{-5} amp.

Amplitude at center of diaphragm 1.85×10^{-6} cm.

Constants of thermophone element:

Material, platinum, of thickness 7×10^{-5} cm.

Area $a = 0.8$ sq. cm.

Effective area $2a = 1.6$ sq. cm.

$\gamma =$ (thickness times specific heat per unit volume) $= 5 \times 10^{-5}$

Resistance 1.0 ohm

Direct current $I_0 = 1.2$ amperes.

800-cycle current $= 5.6 \times 10^{-2}$ amp.

The amplitude (ξ_{max}) is computed from (15a) corrected for temperature as per (14):

$$\xi_{max} = \frac{6.4 \times 10^{-5} R I_0 i \cdot \sqrt[4]{273}}{a \gamma f^{3/2} \sqrt{\theta}}. \qquad (15b)$$

Allowing for a temperature of about 150° centigrade ($\theta = 423$), we compute

$$\xi_{max} = 4.2 \times 10^{-6} \text{ cm.}$$

In comparing the acoustic outputs from these two sources, we shall assume that they are two pistons which communicate their amplitudes of motion to the adjacent medium. The strength of each source should be proportional (at fixed frequency) to the area of the piston times the amplitude of its motion. In the case of the telephone receiver, this quantity is $5.5 \times 1.85 \times 10^{-6} = 1.02 \times 10^{-5}$ cm.^{3}; and in the case of the thermophone element, $1.6 \times 4.2 \times 10^{-6} = 0.67 \times 10^{-5}$ cm.^{3} In these experiments the thermophone element was fitted into a receiver case, similar to that of the telephone receiver, and both instruments were held *loosely* to the ear. Assuming them to be *tightly* held it would be more correct to compute, instead of displacement, the relative pressure changes in the enclosed volume of air, (V_0) in order to compare the two sources. In the case of the telephone receiver the pressure change would be

$$\frac{\Pi}{P} = \frac{1.02 \times 10^{-5}}{V_0}$$

and for the thermophone, using equation (26)

$$\frac{\Pi}{P} = \frac{0.89 \times 10^{-5}}{V_0}.$$

The agreement between the two values, computed in either way is fairly good, considering the number of factors that have to be taken into account in making the comparison.

A second experimental test was made for the purpose of verifying the intensity-frequency relation given in equation (28). Ear comparison of intensities was again resorted to, the energy from the strip conductor being compared with that from a special telephone receiver at various frequencies. (The dynamical characteristics of the telephone receiver had been roughly determined so that it was possible to regulate it for equal acoustic output at various frequencies by adjusting the alternating current input.) The A.C. power input i^2R in the strip was measured for equal intensity at several frequencies, and the results are shown in Fig. 3.

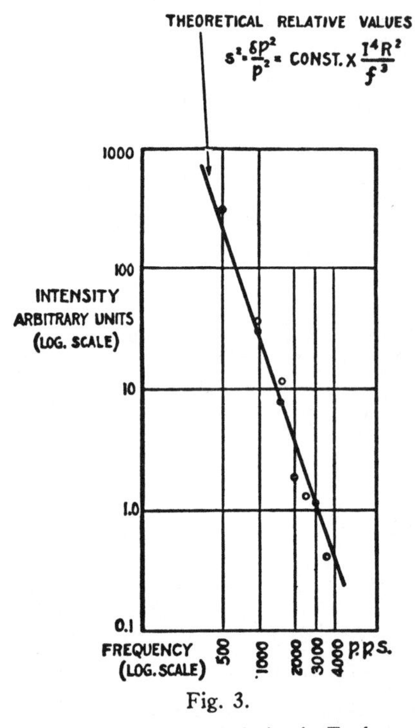

Fig. 3.

Intensity-Frequency Relation in Enclosure.

The points represent the relative intensity at different frequencies for equal A.C. power input, and are proportional to the reciprocal of the power input for equal intensity at each frequency. The curve represents the theoretical decrease in intensity according to the cube of the frequency, and the general result is a confirmation of this relation.

The writers are indebted to Mr. E. C. Wente of this laboratory for an experimental method and data which afford a much more accurate and satisfactory test of the theory than the two experiments given above.[1] The thermophone element was placed in an enclosure whose

[1] This experiment was carried out by Mr. Wente in connection with work on the theory and calibration of a new phonometer which is reported on in the paper immediately following.

volume V_0 was about 45 cubic centimeters; one of the walls of which consisted in a phonometer or pressure-measuring instrument as shown in Fig. 4*a*. This wall yielded so little that the experiment can be considered as carried out rigorously under constant volume. The pressure change in this case, if only alternating current is used to actuate the strip, is given by equation (26′). The experiments were made at a frequency of 20 cycles, the (platinum) strip being made sufficiently heavy to give a large value of thermal inertia γp so that the dissipation term β could be neglected. In order to eliminate an absolute calibration of the phonometer, a second experiment was made, using the piston apparatus shown in Fig. 4*b*, at the same frequency. The maximum pressure change Π as produced by the piston is easily calculated from mechanical considerations, and the comparison is easily made.

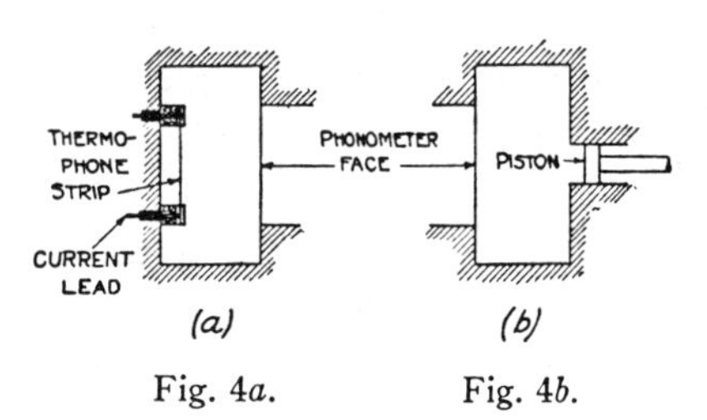

Fig. 4*a*. Fig. 4*b*.

When the piston apparatus was used, the ratio of phonometer reading to calculated pressure increase was 2.02 arbitrary units; and when the strip conductor was used, the ratio of the phonometer reading to pressure change as calculated from (26′) was 1.92 on the same scale. This confirmation to within 5 per cent., was the best we have had of the theory given in this paper.

The results obtained with platinum show that good quantitative work can be done with the thermophone when this material is used for the element. However, it is possible to obtain other materials, such as gold leaf, which are much thinner than bolometer platinum—and which are therefore very useful in cases where higher efficiency is needed. Caution should be used in applying the theoretical formulæ to elements of gold leaf since the heat capacity of gold leaf seems to be very different in different samples. Any such variations, due perhaps to absorbed gases, may be cared for (as shown by E. C. Wente in the following paper) if a check can be made against a platinum element in the same atmosphere. The correction factor thus obtained should hold for all frequencies so long as the gold foil is not unduly heated.

Comparative Value of the Thermophone as a Laboratory Source of Sound.

With regard to efficiency the thermophone compares favorably with electromagnetic and electrostatic devices except in the vicinity of their natural frequencies. In certain work it is essential that the response should be as nearly uniform as possible over a wide range of frequencies and that the relative response should be easily determinable. For such work the advantages of the thermophone are evident, for while its response diminishes with increasing frequency the law of variation is simple. When sound of indeterminate loudness and of one frequency only is desired the volume obtainable from the thermophone does not compare favorably with that from resonant mechanical devices.

The thermophone is particularly adapted to laboratory purposes because it requires no adjustment. It is extremely simple in structure and the units are readily reproducible. The determination of the acoustic effect of the thermophone depends principally upon the thermal properties of materials and is remarkably simple as compared with corresponding determinations for resonant apparatus, which usually involve motions of complicated mechanical systems. In addition, the response of the thermophone is uniform through indefinite periods of time and is not subject to the trouble of accidental detuning, which so often occurs in resonant apparatus.

Possibly even more important than the ease of determination of the sound effects in the air close to the element is the fact that these sound effects cannot react appreciably upon the source of energy whence they arise. Whenever a vibratory system is used it is always subject to reactions which may present serious complications. The thermophone seems the nearest equivalent to an ideal piston source at present obtainable.

Various modifications of size, shape and electrical resistance of the thin conductor employed may be necessary in experimental work. These need change the theory given in no essential way. On account of its simplicity from theoretical and practical points of view we believe that the thermophone in conjunction with a suitable supply of alternating current will be of material value as a precision source of sound.

Summary.

1. A description of a simple thermophone structure is given together with the theory of its operation.

2. An account is given of experimental tests the results of which are substantially in accord with the theory.

3. The thermophone is adapted to two classes of service (*a*) as a precision source of sound at any frequency (*b*) as a source of sound of known relative loudness at different frequencies throughout the acoustic range.

Research Laboratory of the American
Telephone and Telegraph Co. and Western Electric Company, Inc.

EDITOR'S NOTES

Page 23, Equation 2: For example, in the second line let $I_0 = 1$ and $I' = 0.1$. Then the third term divided by the second term is 1:40 or 2.5%, or −32 dB on a 20 log scale.

Page 25: The heat capacity $a\gamma$ (third term of Equation 4), often called c_H, has the dimensions calories/degree. The heat treatises tell us that $c_H = Q$ calories/ΔT degrees; so $c_H d\Delta T/dt = dQ/dt$ calories/sec. This tells us the rate of *storage* of heat in the strip. (Note that we will usually use ΔT where the authors use simply T for temperature difference.)

Arnold and Crandall suggest (p. 26) treating c_H as analogous to electrical inductance L. This then suggests use of the relation $L\,di/dt = E$. Here current i corresponds to temperature difference T, and voltage E corresponds to calories per second, or dQ/dt. T is not dimensionally equivalent, however, but merely roughly analogous to i or i^2 or to E or E^2, for that matter.

Furthermore, we could substitute for $L\,di/dt = E$ the equally valid electrical analog $C_0 dE/dt = dq/dt$. Here, electrical capacitance C_0 corresponds to heat capacity c_H; voltage (or potential difference) E applied to the capacitor corresponds to temperature difference T; and charge q in coulombs corresponds to heat Q in calories. Note again that these "correspondences" are not dimensionally valid. This can be seen by inspecting the left side of Equation 4, which is basically I^2R watts or joules/sec. This is not dimensionally the same as $C_0\,dE/dt$ or dq/dt coulombs/sec, which is I amps. Since 4.186 joules equals 1 calorie, 1 joule/sec equals 0.24 cal/sec. Moreover the dimension of each term in Equation 4 is calories/sec.

Page 25: There remains the term $2a\beta T$ (Equation 4). The factor $a\beta$ must have the dimensions calorie/sec-degree. When conduction alone is considered, we have from the heat treatises, $dQ/dt = (KA/L)\Delta T$. This tells us the rate of *loss* of heat from the strip. Here Q is heat in calories, KA/L is thermal conductance in calorie/sec-degree, and ΔT is temperature difference in degrees across the thickness L. Note that K is specific conductivity, A is area, and L is thickness of the metal strip. An extension of our analogy may be helpful here. Consider the flow of electric current through a resistor: dq/dt, or $i = (KA/L)E$. Here q is charge in coulombs, i is current in amps, KA/L is electrical conductance in mhos or coulomb/sec-volt, and E is the voltage drop or potential difference across the thickness L from one face A of a strip to the other. Again q corresponds to Q, and E corresponds to ΔT, as above. So just as Equation 4 describes a temperature difference T (which is our ΔT) driving calories into a storage unit and a loss unit in series, the electric analog describes a potential difference E driving coulombs into a capacitor and a resistor in series.

Note that Equation 4 makes use of the first statement of Equation 2 (left side), whereas Equation 4a makes use of the second statement of Equation 2 (right side, first term). Equation 4a contains only d.c. terms; Equation 5 contains only a.c. terms (obtained by subtracting Equation 4a from Equation 4).

Page 27, paragraph 1: The reader might wish to imagine a child's cylindrical balloon closely surrounding the strip conductor. A temperature wave having thickness equal to about λ is confined within the boundaries of the balloon—that is, there is no temperature transmission through the balloon walls. These walls then expand and contract to produce acoustic waves, just as does that ideal acoustic source, the breathing sphere.

Page 28: Equation 8 is the solution to the diffusion equation in heat, to the skin-effect equation in electromagnetic wave theory, and to the submarine cable telegraphy equation in communication engineering. For further reading, see Pipes (1946), Ramo and Whinnery (1944), and Heaviside (1950), as listed at the end of Part II in the bibliography. Note that α shows up twice: as an attenuation factor and as the wave number of a traveling wave (Equations 8 and 10).

Figure 2, showing diffusion waves of temperature, can be understood from two different electrical analogies. The first analogy is the skin-effect phenomenon. Imagine the space to the left of the y axis to be a vacuum and to have incoming electromagnetic waves with maximum amplitude equal to that of curve C. The space to the right (which in Figure 2 is actually heated air) is to be a semi-infinite block of copper. The four incoming traveling waves (which are actually four time-snapshots of a single incident wave) can be found by extending curve A to the left as an unattenuated sine wave; likewise curves B, C, and D. We already have (in Fig. 2) four snapshots of the diffusion wave at four different times. To the right of the y axis we see that the wave continues to propagate within the copper block but only to a small distance X_{Δ}, the skin depth. The attenuation versus distance is very great, as shown by the factor $\epsilon^{-\alpha x}$ and as seen in Figure 2, and gets greater with frequency. Treating A, B, C, and D as four time positions on *one* incoming wave, and labeling from right to left, we see that position A (to be called the A-curve) is the first to cross the x-axis or T_0. The B-curve crosses this axis 45° later; the C-curve 90° later; and the D-curve 135° later. The D-curve has tried to maintain its original C-max value (to the right of the y-axis) but the attenuation has knocked down this peak.

The second analogy is the submarine cable telegraphy situation. We replace the heated air to the right of the *y* axis by an inductanceless transmission line comprised of *n* sections of a series resistance *R* and a shunt capacitance *C*. We will try to send a square pulse down this line. With a battery we suddenly charge up the first *RC* section to a unit positive value $+q_1$ and then abruptly remove the battery. The charge will propagate, as the positive half cycle of one square wave, down the entire length of the line but with tremendous attenuation and smearing. Now at a short time Δt later, we suddenly charge up the first *RC* section to a unit negative value $-q_1$ and then abruptly remove the battery. This charge also will propagate, as the negative half cycle of one square wave, down the entire line. However, the sum of the two square impulses at the receiving end will be greatly smeared, showing a slow build-up and a long tail or wake. Some of this is suggested in Figure 2 and also in Heaviside (see Annotated Bibliography).

[Heaviside then taught how to sharpen up the received electric impulse (that is, remove the smearing) and how to fight the attenuation, by adding a series inductance *L* to each of the *n* meshes. However, there is no analog to inductance *L* in thermal conduction, which is one way of showing that the diffusion equation of heat can never be "upgraded" to become the wave equation. But of course the enormous attenuation is very desirable in the thermophone problem.]

Perhaps the most useful information given by Equation 10 is that the "balloon boundary" where $X_\Delta = \lambda$ is proportional to $1/\sqrt{f}$. (This shows up in Equation 15a.)

Pages 29–31: In pages 29–30 the authors are leading up to Equation 15a, which shows that the particle displacement ξ follows an inverse 3/2 power law with respect to frequency, which is a −9 dB/oct law. [Note the printer's error in Equation 14: the radical symbol should be erased.] Thus from Equations 6 and 7 we get a $1/f$ term due primarily to the varying temperature's time constant, as governed by the γp term, where p is angular frequency. From Equation 10, or Equations 13 or 14, we get a $1/\sqrt{f}$ term due to the frequency dependence of the distance term X_Δ, defining the skin depth or the balloon boundary (heat diffusion). Note that in Equation 16, $p\xi$ should read $(p\xi)^2$ and that a factor p^2 or ω^2 should be multiplied with the numerator, making velocity squared $\dot{\xi}^2$ be $\sim 1/f$.

Page 31, Equation 17: The velocity potential ϕ is proportional to $\dot{\xi}$, which is $\sim 1/\sqrt{f}$. Hence in Equations 19 et seq. the free-field acoustic pressure $\Pi \sim \omega\,\dot{\xi}$, or $\sim\sqrt{f}$, which is a +3 dB/oct law. The square root of Equation 21 (which itself derives from a corrected Equation 16) says the same thing—namely, that $\Pi \sim iI_0 R\sqrt{f}$.

Page 32, second case: We now turn to the closed-cavity case where the authors are leading up to Equation 26, which shows that here acoustic pressure Π falls off as $f^{-3/2}$, which is a –9 dB/oct law. An important equation is Equation 24, which is analogous to Equations 11 plus 13. Equation 24 shows explicitly the $1/\sqrt{f}$ contribution of $1/\alpha$. In addition, however, the T' factor is reputed to be identical with Equation 7, which supplies the $1/f$ contribution. If then we rewrite the next relation as $\Delta p/P = \delta T/\theta_2$ (where Δp is called Π in the text), we see that Equation 25 is reasonable. Then Equation 26 follows, giving the pressure law that Π or Δp drops as $f^{-3/2}$ or –9 dB/oct.

Now if we were to calculate displacement ξ, we would find that, just as in the free-field case, it drops by –9 dB/oct. So the pressure "tracks" the displacement in the closed-cavity case but surpasses it by +12 dB/oct in the free-field case (giving finally +3 dB/oct). In other words, the difference between the Δp law for a free field (proportional to acceleration) and the Δp law for a closed cavity (proportional to displacement) is always +12 dB/oct. This is a universal law, analogous to Schottky's 6 dB/oct reciprocity law (see Part IV, this volume).

Page 35, Equation 15b to end: In the "loosely held example," the required thermophone source strength or volume velocity is lower than its rival by about 3.5 dB. Why? Let us guess that at 800 Hz the shunt leakage inertance of the thermophone against the ear resonates with the "effective ear volume" V_E of the listener. Then a constant-displacement generator ξ might produce a 3.5 dB boost across this "tank circuit."

When the standard telephone receiver, however, is held against the ear with the identical leakage, this resonance at 800 Hz could well be absent because the "effective ear volume" now is V_0, somewhat smaller than V_E—that is, the large V_E equalled V_0 plus V_{cav}, the cavity in which the thermophone is inserted. In a standard receiver this cavity V_{cav} is sealed off by a diaphragm. (In fact, its impedance is in series with that of V_0, but this goes beyond the present paper.) Conversely, in the thermophone example with the diaphragm removed, the two impedances are in parallel, proportional to $1/(V_0 + V_{cav})$, and hence the net impedance (also the net stiffness) is smaller than that of V_0 alone. Hence, if $(V_0 + V_{cav})$ causes a leakage resonance at 800 Hz, V_0 alone might cause its own leakage resonance to occur in the neighborhood of 1,000 Hz and so not be found too prominently at the lower frequency.

In the "tightly held example," the authors calculate pressure directly (using Equation 26 for the thermophone case). Probably they should have used $\Pi/1.4P$, where 1.4 is the ratio of specific heats. This would lower the numbers to $0.73 \cdot 10^{-5}$ and to $0.64 \cdot 10^{-5}$. These numbers are actually the volume displacements ΔV. Alternatively, it is possible to derive the pressure without going back to Equation 26, by using Equation 15b for the source strength or volume velocity, $\omega\xi \cdot$ Area. Then $\Pi/1.4P = \omega \Delta V/\omega V_0 = 0.67 \cdot 10^{-5}$ cm^3 divided by V_0. This compares quite well with $0.64 \cdot 10^{-5}$ (above). However, there are still some loose ends in both examples.

Page 36: Note that in Figure 3 the intensity falls off at about −9 dB/oct. This is merely an f^{-3} law on a 10 log scale, as the theory predicts for intensity (Equation 28). When we discussed pressure and the $f^{-3/2}$ law we used a 20 log scale and hence also got −9 dB/oct.

15

Reprinted from pages 305–307 of *Speech and Hearing,* Van Nostrand, New York, 1929, 331 p.

APPENDIX A

H. Fletcher

A DRAWING showing the arrangement for calibrating a condenser transmitter by means of a thermophone is given in Fig. 1. Hydrogen gas is sent into one of the capillary tubes shown and out of the other until the enclosed chamber is filled with this gas. Capillary tubes are used for this purpose so that from an acoustic standpoint the chamber may be considered completely closed. The measurements are made with hydrogen gas so that the wavelength will be as large as possible compared to the size of the chamber. This is necessary, for in

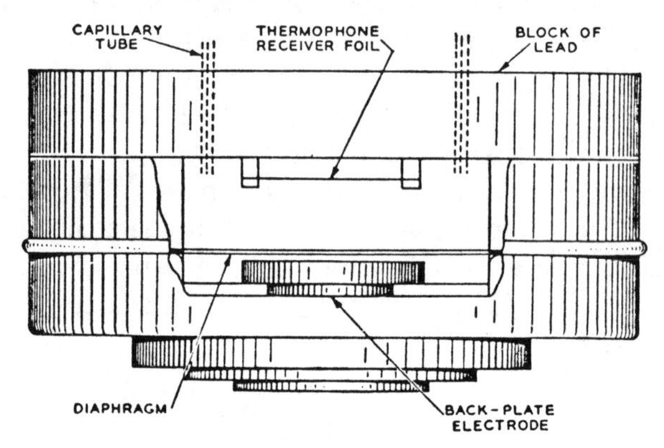

APPENDIX A. FIG. 1.—ARRANGEMENT FOR CALIBRATING ELECTROSTATIC TRANSMITTER.

the development of the formula given below it was assumed that variations of pressure at different positions within the chamber are all in phase. When the wavelength is comparable to the size of the chamber, standing wave patterns are set up. Under such conditions this formula does not hold.

A direct current I_0 is sent through the gold foil strip, heating it to a temperature θ_0. The final temperature which it assumes depends upon I_0 and also upon the heat capacity of the foil, the properties of the gas, and the size of the chamber. It can be calculated from the change in resistance of the foil. An alternating current $I_1 \cos \omega t$ is then superimposed upon the direct current. It causes fluctuations in the temperature of the gold strip and also in the gas immediately surrounding it. This in turn causes fluctuations in pressure which depend upon the rate at which the heat is conducted and radiated away from the gas layer next to the strip. The following formula was deduced by E. C. Wente of Bell Telephone Laboratories, and its validity has been demonstrated experimentally. The pressure variation p is given by

$$p = \frac{.478 R I_0 I_1 \cos(\omega t - \Phi) M}{\left[\left(GQ - UF - \frac{Aa\omega}{\alpha}\right)^2 + \left(FQ + UG - \frac{Aa\omega}{\alpha}\right)^2\right]^{1/2}}$$

R = resistance of thermophone strip;
I_0 = direct current;
I_1 = amplitude of alternating current;
ω = 2π times the frequency of variation;
Φ = phase between pressure and current;
M = correction factor which is nearly unity and is given by

$$M = \left(1 - \frac{S}{V_0\alpha} + \frac{S^2}{2V_0^2\alpha^2}\right)^{1/2}\left(1 - \frac{2}{\alpha' l} + \frac{2}{(\alpha' l)^2}\right)^{1/2}$$

S = area of the walls of the chamber:
V_0 = volume of the chamber;
α is related to the heat diffusibility and is given by

$$\alpha = \sqrt{\frac{\rho_0 \omega C_p}{2K}}$$

ρ_0 = density of the gas;
C_p = specific heat of the gas at constant pressure;
K = the heat conductivity of the gas;
α' is the same as α except that the constants involved refer to the thermophone strip.

$$G = \frac{V_0 \alpha T_a}{2ap_0}\left[1 - \frac{k-1}{k}\frac{\theta_0}{T_a}\left(1 - \frac{2a}{\alpha V_0}\right)\right]$$

$$Q = 2a\alpha K + 8aC\theta_0^3$$
$$U = a\gamma\omega + 2a\alpha K$$
$$F = \frac{V_0\alpha T_a}{2ap_0}\left(1 - \frac{k-1}{k}\frac{\theta_0}{T_a}\right)$$

k = ratio of the specific heats of the gas at constant volume and at constant pressure $= \frac{c_p}{c_v}$.
A = the reciprocal of the mechanical equivalent of heat;
a = area of one side of the thermophone strip;
θ_0 = temperature of the thermophone strip due to I_0;
γ = heat capacity per unit area of the strip;
T_a = average temperature of gas within the enclosure;
p_0 = average pressure within the enclosure;
C = Planc's radiation constant.

The phase θ is determined by $\tan\theta = \dfrac{FQ + GU - \frac{Aa\omega}{a}}{GQ - FU - \frac{Aa\omega}{a}}$.

For frequencies above 50 cycles the effect of radiation can be neglected and the above formula is simplified thus:

$$p = \frac{.478 I_0 M}{\sqrt[N]{1 + (B + D)^2}} IR$$

where p and I are either maximum or effective values and

$$N = \frac{\gamma\omega V_0\alpha}{2p_0}\left(T_a - \frac{k-1}{k}\theta_0\right)$$

$$B = 1 + \frac{4K\alpha}{\gamma\omega}$$

$$D = \frac{2a}{V_0\alpha}\,\frac{\frac{k-1}{k}\theta_0}{T_a - \frac{k-1}{k}\theta_0}.$$

Neglecting the radiation effect causes a 2 per cent error at 32 cycles. The correction factor M varies from .7 at 32 cycles to .98 at 1000 cycles.

EDITOR'S NOTES

Fletcher's first equation for the sound pressure in a closed cavity is Wente's Equation 7, which thus allows the reader to jump into the middle of Wente's paper "The Thermophone" (*Phys. Rev.* **19**:333–335, 1922). See also Ballantine's discussion of the thermophone, in which his Equation 4 is similar to Fletcher's first equation ("Technique of Microphone Calibration," *Acoust. Soc. Am. J.* **3**:319–360, 1932). In Fletcher's second equation, a simplified formula for p, on page 307, the reader can correct the printer's error by moving N one centimeter to the left of the radical sign. N is then a simple multiplier. N is approximately UF, which is proportional to $\omega\alpha$ or $\omega^{3/2}$. Hence we see that p is approximately proportional to $\omega^{-3/2}$, which is a –9 dB/oct law.

Part III

DIRECT VISUAL OBSERVATION OF SOUND WAVES

Editor's Comments on Papers 16 Through 24

It has been pointed out in the optics literature that we cannot see the traveling wavefront of an expanding spherical wave of light. The reason is that any stroboscopic observation would require that the probing wave travel much faster than our light wave in order to freeze the light wave momentarily and allow its head to be photographed. This, of course, is not possible. Likewise, a probing wave of sound—for example, a narrow beam of ultrasound—is unable to accomplish a stroboscopic observation of an expanding spherical wave of sound. However, a probing wave of pulsed light should be able to freeze an expanding spherical wave of sound since the velocity ratios are about one million to one. August Toepler recognized this and invented a process, the schlieren method, for seeing a traveling sound wave directly. Toepler was a highly inventive applied physicist, too innovative for the *Annalen der Physik,* which in 1864 rejected his startling paper on the schlieren method. The paper was issued, nevertheless, as a privately printed brochure. (The *Annalen* editor, Poggendorff, did not reject Toepler's follow-up papers!)

The schlieren method is a means for "amplifying" small differences in the optical index of refraction of the medium through which the sound wave travels. This amplification thus increases the contrast between transparent objects having, for example, a difference in indexes of one part in a million.

Papers 16 and 17 consist of nearly all the acoustics-related passages from Toepler's 1864 brochure and from the 1867 papers in the *Annalen der Physik*.

Toepler used a geometrical-optics point of view in all his analysis. The first analysis of Toepler's schlieren method from a wave-optics point of view seems to have been done by Frits Zernike in the period 1942–1946. Paper 18 consists of five pages from Zernike's 1946 paper on phase-contrast microscopy (Zernike was awarded the Nobel prize in 1953 for his phase-contrast work). The paper was apparently written in 1942 but was kept hidden from the Germans by the physicists of the Netherlands until the day World War II ended. As noted, Zernike analyzed Toepler's schlieren method from a wave-optics point of view and then compared the schlieren method's operation in enhancing small differences in index of refraction with his own phase-contrast method. Zernike's high fidelity method is limited to microscopy; Toepler's lower fidelity method is not thus limited.

Paper 19 presents V. Dvorak's simplification of Toepler's schlieren method. For some reason it seems to have been overlooked by later investigators and had to be reinvented (by A. L. Foley). The method was extensively used by W. C. Sabine (see Paper 22), who attributed it vaguely to Toepler and Foley.

Paper 20, by A. L. Foley and W. H. Souder of Indiana University, reinvented in 1912 the shadow method of Dvorak (Foley was clearly unaware of Dvorak's paper). Foley carried Dvorak's work further by contributing a series of remarkable shadow photographs of sound waves; just as R. W. Wood carried Toepler's work further with a series of remarkable schlieren photographs of sound waves (see Annotated Bibliography). We present a few nonconsecutive pages of the Foley and Souder paper.

Paper 21 presents a few pages from the 1945 article by Barnes and Bellinger. In this article they give a good discussion of some subtle differences between schlieren pictures and shadowgraph pictures in addition to some valuable experimental tips.

We present, as Paper 22, Figure 14 from W. C. Sabine's 1913 paper in the *American Architect.* (This paper is given in its entirety in Northwood, 1977. Sabine uses the shadowgraph method, with credit to his active contemporary, Foley. Both men were apparently unaware of Dvorak's work.

As Paper 23 we present Figure 34 from W. E. Kock's 1965 book, *Sound Waves and Light Waves,* wherein Kock demonstrated his method of "seeing" audio sound waves. (Kock's earliest photographs were published in the *Bell Laboratories Record* in the 1940s). Kock's method is not the direct ("unmittelbar") visual observation method that Toepler insisted on. Rather, Kock employs indirect visual observation using a scanning microphone coupled with a neon light; but the results are startling.

Finally, in recognition of a famous lecture demonstrator, we present two nonconsecutive pages from the third edition of John Tyndall's *Sound* (Paper 24). Both Toepler and Dvorak made reference to Tyndall's sensitive smoke jets. Even Rayleigh (Strutt, 1876) found it convenient to use Tyndall's sensitive flame as an indicator of sound pressure in a reciprocity experiment.

REFERENCES

Northwood, T. D., 1977, *Architectural Acoustics,* Dowden, Hutchinson & Ross, Stroudsburg, Pa., pp. 120–132.

Strutt, J. W. (Lord Rayleigh), 1876, On the Application of the Principle of Reciprocity to Acoustics, *R. Soc. London Proc.* **25**:118–122. (Paper 36 in *Acoustics: Historical and Philosophical Development,* R. B. Lindsay, ed., Dowden, Hutchinson & Ross, Stroudsburg, Pa., 1974, pp. 402–406. See especially pages 404–405).

ANNOTATED BIBLIOGRAPHY

Barnes, R. B., and C. J. Burton, 1949, *J. Appl. Phys.* **20**:286–294.
Good schlieren photographs.

Bergmann, L., 1942, *Der Ultraschall,* 3rd ed., VDI Press, Berlin, pp. 36, 37, 94–103.
Famous German reference with good photographs of sound waves.

Born, M., and E. Wolf, 1959, *Principles of Optics,* Pergamon Press, New York.
See especially Section 8.6.3, pp. 420–429.
Goodman, J. W., 1968, *Introduction to Fourier Optics,* McGraw-Hill, New York.
See especially Chapter 7 and the discussion of Figure 5-5(c).
Holder, D. W., and R. J. North, 1963, *Schlieren Methods,* National Physical Laboratory, Notes on Applied Science No. 31, Her Majesty's Stationery Office.
This monumental treatise of more than 100 pages contains some startling schlieren photographs pertaining to airfoils.
Kingslake, R., ed., 1965, *Applied Optics and Optical Engineering,* vol. 4, Academic Press, New York.
See Chapter 10 by G. E. Fisher, "Refractometry," especially Sections D and E (pp. 378–380).
Martin, L. C., 1950, *Technical Optics,* Pitman, London.
See especially Chapter 3, pp. 102–128 on Abbe's work.
Neubauer, W. G., 1969 *Acoust. Soc. Am. J.* **45**:1134–1144.
Fine photographs of sound waves in a liquid.
Strong, J., 1958, *Concepts of Classical Optics,* W. H. Freeman & Co., San Francisco.
See especially Appendix K, "The Wave Theory of Microscopic Image Formation," by F. Zernike.
Willard, G. W., 1947, *Bell Lab. Rec.* **5**:194.
Good schlieren photographs.
Wolf, E., ed., 1961, *Progress in Optics,* vol. I. North-Holland Publishing Co., Amsterdam.
See Chapter IV by D. Gabor, "Light and Information," especially Section 4 (pp. 122–124).
Wood, R. W., 1899, Photography of Sound-Waves by the "Schlieren-Methode," *Philos. Mag.* **48**:218.
Wood, R. W., 1900, Photography of Sound-Waves and the Evolutions of Reflected Wave-Fronts, *Philos. Mag.* **50**:148.
Wood, R. W., 1901, Photography of Sound-Waves, *Philos. Mag.* **52**:589.
In these three papers the future author of *Physical Optics* (R.W. Wood) illustrates some phenomena of optical refraction and reflection by means of photographs of sound waves in gases.

16

OBSERVATIONS WITH A NEW OPTICAL METHOD

A Contribution to Experimental Physics

A. Toepler

These excerpts, translated by R. Bruce Lindsay, were edited and paraphrased expressly for this Benchmark volume by the volume editor from "Beobachtungen nach einer neuen optischen Methode: Ein Beitrag zur Experimentalphysik," in Ostwalds Kl. Exakten Wiss. ***157****:5–61 (1906)*

[*Editor's Note:* The superscripts in the text refer to the Editor's Notes at the end of the translation.]

There are many phenomena taking place in the interior of optically transparent media that are not open to direct observation by the eye or by the optical instruments so far in use because these phenomena frequently exert an almost vanishing effect on the passage of light rays traversing the medium. In this category belong the diffusion motions of bodies of different states of aggregation, whether produced by heat or by pressure, by sound vibrations, etc. Obviously what is needed to bring these phenomena into the realm of immediate observation is a more reliable method of making directly visible very small fluctuations of the optical refracting properties of the media.

As early as the year 1859 I was concerned with a closely related problem, which was indeed of more technical than scientific interest. This was: a simple method of recognizing in glasses intended for optical use and with plane or spherical surfaces, so-called schlieren. As is well known we give the name schlieren to the streaks and other portions of optical glass in which the density varies from that of the mass of glass as a whole. [*Editor's Note:* In this translation we shall use the word *schlieren* whenever it occurs since it is now in fairly common English usage.[1]] Such nonhomogeneous glasses, when used in the manufacture of telescope objectives of large focal length, give rise to distortion due to the irregular refraction of light, especially when large magnification is desired. [*Editor's Note:* He then goes into a digression on the problems of grinding lenses and the effects of schlieren therein.]

Now I succeeded, by means of a simple optical apparatus, in locating such nonuniformities in the refracting power. I recognized further that the same method, modified in suitable fashion, could also be applied to the domain of scientific investigations; and indeed wherever the matter concerns small variations in the density or the refracting power of the transparent media. The outcome corresponded so closely with my expectations that I was led to test the method in a rather large number of observations. Now that the apparatus has experienced many alterations in the course of time, I feel impelled to publish the details of my method in the hope that it might prove to be a useful tool in experimental physics.

PRINCIPLE OF THE METHOD OF OBSERVATION

The purpose of the present article is in the first place to provide a thorough description of the method and the necessary apparatus. If I here go into greater detail than seems strictly necessary for a mere understanding of the procedure, the reason is that the success of the scheme depends very closely on the correctness with which the operations are carried out and necessary adjustments made. I wish to avoid at once a drawback that often occurs in the publication of new observations, namely, that such observations may not be confirmed by later experimenters. In the majority of cases such differences as are found can be attributed to a deviation in the method of observation. The detailed description of my method will perhaps find its excuse in this circumstance. The conclusion of the article consists of the communication of some investigations that I have carried out by my method.

The foundation on which the method in question rests follows very simply from theory. In Figure 1 of Plate I the letter *a* represents a luminous point having no volume or surface extension. We further let *L* represent a convex lens of long focal length and large aperture, which focuses the rays emerging from *L* in *b*. On the opposite side of the lens *L* from *a*, an eye *O* can be brought so close to the focus *b* that the rays diverging again from *b* can all reach the retina through the pupil. (For the sake of clarity no attempt is made to preserve precise dimensions in the figure.) If the observer accommodates his eye corresponding to the distance from *O* to *L*, a distinct image of the lens *L* will be formed on the retina, which will appear to be completely and uniformly illuminated by the rays emanating from *a*. In the figure, *hc* denotes a solid opaque screen that for the moment we shall not consider to be present.

Now if at *gi* in the interior of the lens there is a small segment of arbitrary shape whose refractive index differs only slightly from that of the lens as a whole—in short, if there is a schliere (or streak) —corresponding light rays will not meet precisely at *b* but will pass around *b* in order to reach the retina. The irregularly refracted rays, for example, *gf* and *id*, will then reach the position *rs* in the retina image, corresponding to the schliere *ig*, because the eye is accommodated for the distance to the lens. [*Editor's Note:* This accommodation lies at the heart of the method.] In spite of the irregular refraction of the rays, the observer will see the whole field of view uniformly illuminated in case the diaphragm *hc* is absent. If, however, one inserts this in the direction indicated by the arrow in the figure, until the edge is close to point *b*, there exists a position for which certain irregular rays *id* are screened out. Thus in the uniformly illuminated field of view *mn* there must be produced a dark hole *s*. In fact, in this way in general one notes a dark indication of a schliere against a bright background.

If the diaphragm *hc* is lowered still more so that even the focus *b* is screened, the whole field of vision must suddenly become dark. However, in this situation other irregular rays *gf* underneath *b* get into the eye and produce for this position of the diaphragm a bright image of the schliere on the dark background.

It is now clear that the latter arrangement, in which *b* is fully screened out by the sharp edge *h* of the diaphragm, is the most sensitive position for making the schliere visible. The former dark image on the bright background easily eludes observation, since by intensive illumination of a large field of vision the sensitivity of the retina is much reduced; whereas with a bright spot on a dark background the most delicate schlieren can be readily detected.

The moment at which the gradually moving diaphragm actually reaches point *b* is very easily recognized by the fact that the bright field of vision suddenly becomes dark. Indeed the darkening takes place everywhere at the same instant, that is, assuming there are no schlieren. If the edge of the diaphragm does not quite cross *b* but passes a short distance in front of it, as at *a* in Figure 2, the eye will recognize a partial darkening of the image *mn*. If the edge of the diaphragm has then reached the optic axis, the field of view is half dark and half bright and indeed dark on the side opposite to the screen. It is readily seen that the converse takes place if the diaphragm is located between *b* and the eye. Neither position is usable for the detection of schlieren. The sensitive position is only that in which the edge of the diaphragm cuts across *b*. For brevity, in the future we shall refer to this case as the "sensitive arrangement."

Though in the practical course of investigation much will be found to take place other than what the above theoretical discussion indicates, we can at any rate recognize in the apparatus setup in Figure 1:

(1) That the schlieren in the interior of the lens *L* are not the only things made visible to the eye by the sensitive arrangement indicated. The path of the rays in the vicinity of *b* is also very much influenced by the condition of the transparent medium, e.g., the air, which is located before and after the lens *L* between points *a* and *b*. Since the eye *O* is accommodated to the distance *OL* (ordinarily of the order of 15 to 25 feet) it will also be accommodated to the layers of air on both sides of the lens so as to permit the detection of small changes in the index of refraction of the air in the same way in which the schlieren in the glass are observed. The same situation will hold (assuming that the lens is schlieren free) if a liquid contained in a vessel with plane-parallel transparent walls is placed either in front of or behind the lens. Every nonuniformity in the index of refraction of such a liquid can be directly observed. For convenience, in what follows the name schlieren will be used to denote *all* nonuniformities in the optical behavior of any otherwise homogeneous medium, whether solid, liquid, or gaseous.

(2) The eye will perceive directly not only such schlieren. Every error in the grinding of the lens surfaces will show up in similar fashion (as bright on dark, or the opposite). Moreover, irregularities in the surfaces of plane-parallel plates just before or behind *L* will also show up.

(3) We must also observe that for a certain position of the diaphragm *hc* schlieren are conceivable that the eye will not be able to detect. It is readily seen that this can happen in only one case. The above discussion has referred to irregular rays deviated in a plane normal to the edge *h* of the diaphragm. Let us assume that there is in the lens *L* a schliere extending in the form of a thread of cylindrical cross section in the direction from *p* to *q* (Fig. 1). This thread will clearly suffer deviations only at right angles to its axis; accordingly in planes that are parallel to edge *h*. In the arrangement in Figure 1 for this case, even when the moment of sensitive arrangement prevails, the schliere will not be visible.

In order to make all schlieren visible, whether in the lens or in the transparent medium surrounding it, the only way then is to rotate the diaphragm about point *b* in a plane normal to the optic axis *ab* and by experiment find the place corresponding to the sensitive arrangement. It turns out that, in general, to secure all schlieren it is necessary to locate the edge *h* in only two mutually perpendicular positions. In extremely delicate investigations, however, it is desirable to be able to locate the diaphragm in any arbitrarily chosen position.

It may now be appropriate to call attention to the great sensitivity provided by the method under favorable circumstances. A further examination of Figure 1 makes this clear. If irregular rays like *gf* pass below *b* by an immeasurably small amount, a position of *hc* is still possible for which *b* is screened out and the irregular rays are isolated for the eye. All that is needed is a mechanism to move the edge *h* back and forth through a very small range and thus provide a fine adjustment. Moreover, it is readily seen that, other things being equal, the sensitivity of the apparatus stands in a definite relation with the distance *Lb*, since the deviation of the irregular rays from *b* increases with increasing distance.

If now we wanted to carry out an investigation based on the above simple considerations, the results would be generally defective. Indeed, the procedure cannot be carried through in many cases. In reality the assumptions just made do not hold. For a more convenient and surer carrying out of scientific investigations the use of more complicated apparatus is unavoidable. Nevertheless, I have convinced myself that many of the investigations to be described later can be carried out successfully with a large telescope objective of good quality, a diaphragm held in the hand, and with the use of the eye alone. If for the time being we ignore the fact that in reality the source of light *a* as well as the image *b* possess spatial extension and that, further, in order to correct for chromatic and spherical aberration the large single lens must be replaced by a system of lenses, one of the principal difficulties is involved in the position of the eye. The light rays diverge from point *b* (Fig. 1) toward the eye. During the search for the sensitive arrangement, which has to be repeated for each observation, the field of vision oscillates between bright and dark. Hence the pupil of the eye contracts so strongly that with the smallest motion of the head or eye the entering rays are partially blocked out, so that it becomes impossible really to tell whether the observed change in the illumination of the field of vision results from the change in the position of the diaphragm or the change in the pupil.

In order to take care of this problem the eye must be brought close enough to *b* so that the cornea lies just before *b* and the bundle of rays entering the eye is as small as possible. In such a situation, however, holding the diaphragm in the hand becomes impossible; and the same is true of the inability of the eye, covered with tears, to recognize distant objects.

The idea then came to me that with the eye very close to *b* it might be possible to use the periphery of the pupil itself as the diaphragm. There is nothing theoretical against such a choice. It is, of course, difficult by mere motion of the head to make *b* fall on the periphery of the pupil and even more difficult in this case to realize the sensitive arrangement since, as has been mentioned, the pupil is in continual motion beyond the control of our will. Moreover, the periphery of the pupil is not precisely circular in form and indeed the iris at this boundary does not appear to be sufficiently opaque. Finally, direct observation with the eye, as is assumed in Figure 1, has the essential disadvantage that with the great distance from *O* to *L* (at least 10 to 15 feet) the lens or the object in front of it to be investigated appears too small to permit the recognition of fine details.

All these difficulties can be overcome by the use of a suitably arranged telescope. The arrangement, which I have used with surprising success in all the investigations to be described hereafter, is shown in Figure 3.

Here *a* represents the source of illumination, in most cases a very small circular opening in a thin black metal sheet. Directly back of this opening is a very bright flame, probably best provided by an Argand oil lamp of the best construction. What happens is thus obviously the same as if the rays diverging toward the lens *L* originated in *a*. *L* is a lens system of at least 2.5 to 4 feet focal length and of the largest possible aperture, providing at a distance of 10 to 25 feet from *a* a very clear image. At this distance of separation a small astronomical telescope *F* is placed on the axis of the lens system. (Note that in Fig. 3 no account is taken of the relative dimensions and distances.) The telescope is so placed that: (1) the distinct image of *a* falls on the near surface of the objective lens *b* of the telescope or directly before it; and (2) the telescope is adjusted for the distance of the lens system *LL'* in such a way that the eye *O* sees a distinct, enlarged, and inverted image *mn* of the nearer lens *L'*.[2] The distance of the eye from the ocular lens is to be adjusted as precisely as possible so that for a comfortable adjustment of the eye to the metal holder of the ocular, the optical image that the lens *s* forms of the objective *b* coincides with the opening of the pupil. It is clear that in this arrangement the converging rays from *L* to *b*, before they uniformly illuminate the field of vision *mn* in the eye, have a nodal point *p* in the eye itself and indeed within the pupil. If now an opaque screen *hc* is placed directly before the objective *b* it will suddenly screen out the whole bundle of rays, and a so-called sensitive arrangement will result. That which happens at *b* repeats itself precisely at *p*. Hence with this arrangement the result will be just as if we had set up the movable diaphragm in the eye itself at *p*.

At the same time we gain the substantial advantage of seeing the sought-for schlieren (in the lens system *LL'* or in its neighborhood) arbitrarily magnified, so that even very fine objects under investigation can be observed at distances appropriate to the method, in case they are perceptible at all. Moreover, for scientific work the application of the telescope has the incalculable advantage that one can combine qualitative observations with exact measurements, if one equips the ocular with cross hairs. [*Editor's Note:* The author then discusses some fine points of the method, and some nonacoustic applications, such as the mixing of slowly miscible transparent liquids, before returning to acoustics.]

VISUALIZATION OF SOUND WAVES IN ATMOSPHERIC AIR

After repeated determinations of the sensitivity of the method (as communicated above) I could have no doubt that the condensations and rarefactions of the air in the generation and propagation of intense sound waves could be made directly visible by the schlieren apparatus. If at a given moment a short sound wave is at an appropriate position with respect to the head of the apparatus and if at the same time we illuminate the field of vision, the sound wave will appear in its true form, a schliere, if the intensity of the sound is great enough to produce a sufficiently large change in the index of refraction to be detected by the apparatus.

I turned my attention in the first instance to regular oscillations, corresponding to musical tones. As an illuminator I used the spark of a Leiden jar, which at a given instant could be discharged between two metal plates separated by a very short distance. Naturally, I could operate with only very high-pitch tones in order to have at least one whole wavelength in the field of view. Thus, due to the small size of the head of my apparatus, I had to be satisfied with short wooden pipes about 2 to 4 cm in length. When strongly blown these emit a high-pitch tone. The mouthpiece of each pipe was connected to a glass tube, which in turn was connected to a bellows.

It will not cause surprise to be told that all investigations made with these small pipes led to negative results. The density variations in the oscillating air even right next to the sound generator were too small to be made visible with my apparatus. I therefore inserted the pipe between two plane glass schlieren-free discs and placed these in a parallel position before the head of the apparatus. In this arrangement the sound waves were kept from propagating as spherical waves, and as long as the waves moved between the plates, the decrease in intensity with distance was inverse linear instead of inverse square. I hoped that in this way I could detect visible wave schlieren at an appropriate distance from the sound source. But even this arrangement proved unsuccessful. Under certain circumstances I could detect vibration figures on the glass plates, but there was no sign of anything corresponding to the sound waves in the enclosed air.

The situation was otherwise in the following *indirect* procedure. In an earlier section I considered the vertical thin column formed above the opening of a tube out of which illuminating gas slowly streams. This thin column, or "thread" of gas, as it rises vertically in quiet air before the field of view, participates in the smallest motion of the surrounding air. If a sound is produced in the neighborhood and the field of vision is momentarily illuminated from time to time, the thread is systematically distorted and hence provides information about the oscillating motion of the air.[3] Since, however, I have not brought this sort of investigation to a conclusion, I must postpone further details about it.

In any case this scheme hardly satisfies my fundamental intentions since it provides only an *indirect* indication of the oscillating motion of the air. It thus seemed as if my attempt to provide the direct visualization of sound waves had run onto the rocks when at last I got the idea of using very intense electrical discharges as sources of sound. And indeed I was successful in this way in making the laws of sound propagation, previously determined only through calculation, directly available to observation by the eye. This turned out to be such a reliable procedure that I was able to employ it with my standard schlieren apparatus.

The little that I plan to communicate at the moment follows at the end of the paper. I wish at this point to present first some interesting relevant observations about electric sparks.

OBSERVATIONS ON ELECTRIC SPARKS IN ATMOSPHERIC AIR

If we use the customary lamp illuminator in front of the head of the schlieren apparatus and produce a series of electric sparks in the vicinity, we observe (when the sensitive arrangement is secured) for each discharge a schlieren cloud rise from the space traversed by the spark. This little cloud clearly originates from the air that has been heated by the passage of the spark. Such clouds appear (even if weak) when the spark is produced by an electrostatic machine. They are produced more strongly when the spark comes from the discharge of a Leiden jar; and most strongly of all by the use of the galvanic induction spark. [*Editor's Note:* The author then describes some detailed observations on these heat-created schlieren clouds before returning to his first goal: the direct visualization of sound waves.]

It was my intention to make visible the condensation wave[4] that as carrier of the sound produced by an electric spark, propagates in every direction in atmospheric air. With this aim in mind I allowed the sound-generating spark to discharge directly before the head of the apparatus [actually before L']; and a short time thereafter let a second spark illuminate the field of vision *in place of* the standard illuminator. This took place in such a short time that the sound wave was still in the field of view. Before I go into these observations I must put in a few words about the construction of the electrical illuminator. [*Editor's Note:* The author then discusses details of his spark-generating devices: Leiden jar, induction coil, spark gap, etc. He closes by promising further publication of observations on both electric sparks and acoustic phenomena.]

NOTES AND REFERENCES

1. Although *schliere* (plural: *schlieren*) translates literally into *streak,* a more useful translation would be *anomalous refractor,* or simply *refractor,* for all of Toepler's writings.
2. Note again that with all the things available for the telescope to focus on, only L' is chosen.
3. This is what Tyndall was independently doing in 1867. See *Acoustics: Historical and Philosophical Development,* R. Bruce Lindsay, ed., Dowden, Hutchinson & Ross, Stroudsburg, Pa., 1972, p. 352.
4. Note that the sound pulse continues to propagate long after the heat pulse has died out. This point is treated by Arnold and Crandall, in Paper 14 in this volume.

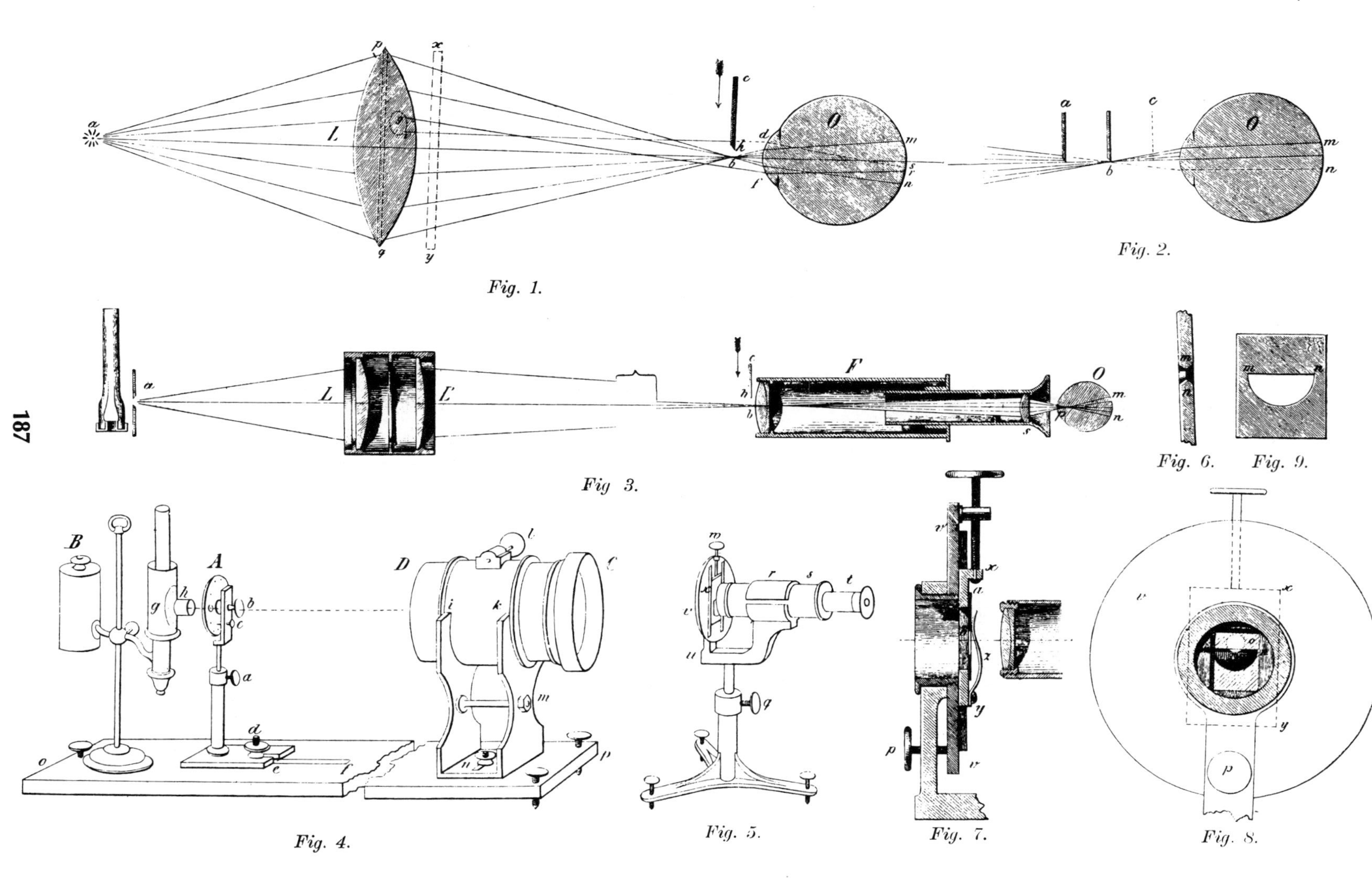
Fig. 1.
Fig. 2.
Fig 3.
Fig. 6.
Fig. 9.
Fig. 4.
Fig. 5.
Fig. 7.
Fig. 8.

17

OPTICAL STUDIES WITH THE SCHLIEREN METHOD

A. Toepler

These excerpts, translated by R. Bruce Lindsay, were edited and paraphrased expressly for this Benchmark volume by the volume editor from "Optische Studien nach der Methode der Schlierenbeobachtung," in Ostwalds Kl. Exakten Wiss. ***158**:24–102 (1906)*

[*Editor's Note:* The superscripts in the text refer to the Editor's Notes at the end of the translation.]

IMPROVED EQUIPMENT

The schlieren method* fulfills the purpose of making immediately visible small nonuniformities or changes in refractive index in apparently homogeneous media, so that at the same time many of the causes that produce such nonuniformities (such as small changes in density, temperature, elasticity, motion effects, etc.) can be made optically observable. Through the introduction of several improvements, particularly for instantaneous observation, the schlieren apparatus has achieved such a sensitivity that we are now in a position to see the sound wave in air resulting from the weakest spark from an electrostatic machine. It has developed further that we seek in vain for a means of finding an absolute value of this sensitivity, as the investigations to be mentioned below will attest. In what follows I shall describe a form of apparatus that even though of modest dimensions, can suffice for delicate investigations.

In order to avoid undue verbosity and lack of clarity in later communication of detail, may I now be permitted to present briefly once again the principle of the method and in the course of this call attention to certain points.

Figure 1, Plate I, shows in oblique projection the arrangement of the essential parts of the apparatus, as well as the path of the rays. Let *abc* be a uniformly illuminated surface, e.g., of triangular form, such as we can produce by means of an opening in an otherwise opaque screen with a brightly burning large flame behind it. The precise shape

* I derived the name, for the sake of simplicity, from the circumstance that the method was used as an extremely sensitive means of identifying with certainty nonuniformities in glasses (the so-called schlieren or streaks). In my "Vibroscopic Observations" (vol. 128 of these *Annalen,* p. 136), I mentioned the application of the method to the analysis of singing flames and also pointed out that Foucault had used a scheme for the testing of spherical mirrors, which in principle is close to the schlieren method but which was not used by him for further scientific purposes. [*Editor's Note:* Toepler's "Optical Studies with the Schlieren Method" was originally published in *Poggendorffs Annalen,* vol. 131, 1867.]

of the opening is immaterial. It is essential, however, that at least one side of the opening be as straight as possible, and this straight line in our case is the horizontal side *ab*. Under the given assumptions, all points of *abc* can be considered as self-luminous. Let a large lens, assumed initially to be entirely free of aberrations, whose vertical section is *mn*, produce at a rather great distance the real image *a'b'c'* as indicated by the path of the rays through *m* and *n*.

Directly behind the image *a'b'c'* let there be placed the objective lens of a telescope (in the figure only the forward part of the telescope is shown and for the sake of clarity is placed somewhat far from *a'b'c'*). Let this telescope be so arranged that an eye looking through it sees clearly all parts of the lens surface *mn* [rather than the triangle *a'b'c'*]. Let *m'n'* be the inverted image of the lens *mn* in the telescope, so that *m* and *m'* as well as *n* and *n'* represent corresponding points in the object and the telescope image. No other light is permitted to fall on the lens *mn* than that emitted by the surface *abc*. Finally, let an opaque screen (diaphragm) *BA* be so displaceable in the direction of the arrow that its straight lower edge moves precisely parallel to the boundary line *b'a'*. Thus the image *a'b'c'* can be gradually covered up, until the edge *BA* falls exactly on the line *b'a'*. It is thus readily seen that up to this limiting position of the screen the observing eye must see the lens *mn* as growing decreasingly bright. However, for any arbitrary position of the diaphragm the whole lens will still appear to be as a *uniformly illuminated disc,* as long as a vestige of the image *a'b'c'* is not covered up.

In order to recognize the correctness of the last claim, we need only note that every point of the image *a'b'c'* receives rays from the whole lens *mn* in uniform fashion. If, for example, through displacement of the screen *BA* the point *c'* is first covered up, the ray *mc'*, which in the telescope was directed toward *m'*, is entirely suppressed. Hence, at *m'* the telescope image must experience a weakening. The same holds also for *n'*, since for *c'* the ray *nc'* also suffers a blackout. In short, the telescope image *m'n'* will become uniformly dimmer at all points by the covering up of *c'*. The same conclusion also follows for all other gradually covered points of *a'b'c'*, so that the round telescope image *m'n'* will become continually darker, though at any instant it will represent a uniformly illuminated circle.

When the screen *BA* finally reaches the line *b'a'*, the eye will note a sudden transition from bright to dark. In conformity with previous notation, this position of the diaphragm *BA* will be called the "sensitive arrangement." [*Editor's Note:* The author then goes into a detailed discussion of blemishes of lenses, either on the surface *E* or within the lens at *pq*, and he discusses fine points about the shaded appearance of certain images, including the invisible zone of flames.]

Finally, it would not be difficult to see from Figure 1, Plate I on what factors the sensitivity of the method essentially depends. The sensitivity is the greater, the sharper the image *a'b'c'*, and accordingly the more perfect the lens *mn* is. Furthermore, observation becomes better the more precisely the two lines *AB* and *a'b'* in the sensitive arrangement can be made to coincide or, accordingly, the more nearly straight they are. Finally, other things being equal, the sensitivity increases directly with the distance of the image *a'b'c'* from the transparent object of observation, namely, the lens at *mn*. At large distances we need fear no trouble from the circumstance that the whole image *a'b'c'* no longer lies on the front surface of the telescope objective. The above conclusions all remain valid even when only a part of the line *a'b'* comes into operation on the middle of the objective lens of the telescope.

The above described procedure serves at once as a testing device for the lens *mn* as regards its surface quality as well as the homogeneity of the glass composing it. Moreover, every other defect of the lens is easily recognized at first glance. In my paper "Observations" [*Ostwald's Klassiker 157*], it was explained how spherical and chromatic aberration can be distinguished in their effects on the telescope image. Similarly, I also explained there how we can tell immediately from the telescope image whether the diaphragm actually goes through the plane of the image *a'b'c'* or lies before or behind it.

In my previous observations I used as luminous surfaces at *abc* small *circular* openings whose optical image likewise was completely covered by the screen before the telescope out to the outermost edges. Though with openings of this kind the mechanism of the apparatus as well as its operation is somewhat simpler than in the case of the straight line boundary of the luminous surface, it must not be overlooked that the latter must produce greater sensitivity for very difficult objects. For with the circular shape of the luminous opening, the edge of the screen in the sensitive arrangement touches the optical image in only a single point. For difficult objects, which in general can be seen only through this arrangement, only a few irregular rays can succeed in getting into the telescope; so the observations will therefore suffer from small light intensity. I shall indeed, in the following more precisely described arrangements of apparatus, call attention to many details that were overlooked in my earlier observations. [*Editor's Note:* Toepler then discusses in detail a manufactured version of his apparatus, shown partially in Figures 2 and 3, as exhibited at the Paris Exposition of 1867. He then reiterates some comments on the sensitivity of his apparatus, as demonstrated by the heating of a jar of water by immersing one's hand, and by the mixing of drops of alcohol with water. He concludes this installment by saying that in his next installment he will discuss phenomena exhibited by electric sparks.]

THE [SOUND] WAVES PRODUCED IN AIR BY ELECTRIC SPARKS.

Since the schlieren apparatus with constant illumination makes readily observable the slow streaming of air due to the unequal heating of the medium surrounding the lens system, we cannot be surprised at its ability to make [sound] waves in air visible by means of momentary illumination. We can, indeed, see a sound wave in the layer of air close to the lens with the use of the electric illuminator [i.e., the electric spark], if among several conditions one particularly is fulfilled, namely, that in the portion of the wave bounded by the field of view of the apparatus there exist sufficiently large variations in the density of the air to produce perceptible optical variations. In order to understand the necessity for this condition let us suppose that in the neighborhood of the apparatus a tone is produced that is intense but not of high pitch. For the sake of simplicity, let us further suppose that the fluctuating condensations and rarefactions spread out from the sound source in the form of a spherical wave. Then, depending on the arrangement of the apparatus with respect to the sound source, those portions of the waves passing through the short-lived beam of light can act prismatically. Even when these circumstances are realized in the most favorable fashion we see that a distinct optical effect will take place only when deviating forces as dissimilar as possible have operated on the individual rays in the field of view. This means, above all, that in the space comprising the field of view there must be the greatest possible variation in density at right angles to the optic axis of the apparatus. Thus, even if a significant variation in density is assumed between the extreme points of a whole wave, even then a large optical variation within the small region of the field of view will not take place, as long as the wavelength is rather considerable, and as long as the density along the whole wavelength goes from one extreme value to the other only very gradually. In short, the air in the direction of the wave propagation must exhibit density variations as large as possible at points as close together as possible. If, therefore, we intend to experiment with the usual acoustic devices, we can expect success only with very high-pitched and very high-intensity tones.

[*Editor's Note:* Toepler then develops this last thought. He says that the sound generator can be either a single impulse (preferably generated by an electric spark) or a regularly repeated train of impulses that produce a musical tone of fixed pitch (preferably caused by a train of very regular electric sparks). It is even better, he continues, to use two spark generators. The first serves as the sound spark; the second, as the illumination spark. The greater the time interval between the two, the larger the expanding spherical sound wave that becomes visible during the momentary illumination. It is therefore desirable to be able to accurately control this time interval.

Toepler then enlarges on this, turning to a discussion of Figure 9, Plate I. We can draw an equivalent circuit of his arrangement whereby we have a d.c. source *Q* in series with the sound spark gap *ed*, which is in series with the illumination spark gap *ab*, which is shunted to ground by a capacitor F (the Leiden jar). But how can this produce a musical tone? Toepler does not bother to discuss this point. It is well understood by him and his contemporaries. Faraday casually mentions these tones in Series XII of his "Experimental Researches in Electricity," articles 1427-1431, published in 1838. If indeed we follow modern practice and let the d.c. source Q be represented by a battery, we will get no tone. If, however, we recall that the source is probably a hand-cranked Wimshurst machine or some similar high-voltage electrostatic generator, we see that the d.c. source Q is a charge generator of nonconstant value (and we have no ready symbol for this, other than a charged capacitor).

The charge generator Q slowly charges up to some high-voltage value. The sound spark gap suddenly breaks down, and the source Q starts discharging into the Leiden jar, which of course starts charging up, at a rate controlled by its RC time constant. Soon this voltage is high enough to discharge the second spark gap (the illuminator). The source Q thereby becomes completely discharged for a moment. The process immediately starts all over again.

Since the generator Q has been accumulating new charges during this whole process, a necessary condition for the generation of a musical tone is that both spark gaps must quickly become nonconducting again (that is, completely extinguished). Otherwise we would get a steady small current and a steady glow discharge. This steady current does not occur, however. We do get complete extinction, and we do get repetition of the cycle in the form of a sawtooth wave. Toepler controlled his time interval between the two spark discharges by changing the level of mercury in his Leiden jar F (Fig. 9), thus changing his capacitance and hence his RC time constant.]

APPEARANCE OF THE OPTICAL IMAGE

With the sensitive arrangement of the analyzer diaphragm we see the waves as sharp, bright curves in the dark field of vision [Fig. 4, Plate II], sometimes near the sound spark and sometimes separated from it. Although in the case in question the mechanical excitation [of the air] is not at all intense, nevertheless the phenomenon is so easily visible that we do not need to make the field of vision fully dark in order to produce sharp shadows. (See the section above on the optical apparatus.) In fact it is desirable to back-off the diaphragm [that is, the shutter] to the extent that the light of the sound spark itself almost disappears in the general brightness of the field of view.

With the lateral position of the spheres, the phenomenon has the appearance shown in Figures 2 and 3 of Plate II. If the path of the spark is straight, as in Figure 2, for a small spark gap, the image is very regular. The spark path *a* is often at a small distance from a shaded type of image *bc*. That is, it is as if *a* were surrounded by a cylindrical envelope. For the most part, however, the latter is farther from the sound path and then takes the form of the dotted lines *b'c'* or *b''c''*. If the spark distance is larger, so that the spark path (Fig. 3) has a zig-zag form, the phenomenon, as the dotted lines *bc* indicate, is in appearance like a broken cylinder. For most sparks, however, the image is that shown by *b'c'* or *b''c''*. We recognize at first glance that the optical image is the projection of a surrounding surface whose points are equidistant from the nearby points of the spark path. The form of this surrounding surface [or envelope] for large distances is that of a spheroid, which approaches the shape of a sphere the larger it becomes. The side view, which will be discussed below for the axial position of the sphere, completely confirms this. For the above spark distance and with the mercury flask half full, as maintained in the following researches, the smaller diameter of the spheroid is in most cases 20 to 40 mm. Larger as well as smaller values are very rare. [*Editor's Note:* By this time Toepler has become as fascinated by the various appearances of electric sparks, as he is by the observation of traveling sound waves momentarily illuminated. He therefore digresses (as far as we are concerned) more and more frequently from discussions of acoustic phenomena.]

REFLECTION OF WAVES

In the lateral arrangement of the discharge spheres in the field of view as indicated in Figures 2 and 3, Plate II, we have the disturbing situation that in addition to the sound spark its mirror images in the surfaces of the principal lens will be seen through the telescope. In case we want to see only the wave in air, we can observe it best by means of the axial arrangement of the discharge spheres, as indicated in Figure 9 of Plate I. In this arrangement the observer finds not only the sound spark but also its envelope masked by the sphere *d*. Moreover, no mirror images can be seen. At the same time, the spark gap of the sound spark is made so small that the path is straight. We then see the projections of the waves as complete, delicately shaded circles, as in Figure 4 of Plate II. In the spark source used by me, the sparks are so numerous that with a 15 mm spark gap they produce an audible tone. In this case we believe that we see many waves simultaneously. However, only a single circle corresponds to each spark, as we recognize at once from the behavior of the electrical apparatus. Furthermore we can convince ourselves that a deception, to the effect that each illuminating spark did not make visible the wave of the corresponding sound spark but rather the previous one, cannot hold.

In Figure 4 of Plate II, *pq* indicates a reflecting glass plate placed at some distance from the discharge spheres. The well-known wave reflection from this surface shows up very well. We may call attention to the fact that the reflected branch originates at precisely the same points *p* and *q* at which the not-yet-reflected part of the wave comes to an end. This is of note since from investigations with the induction spark we can readily exclude a premature reflected branch. This in turn is a consequence of an optical deception (to be mentioned later) in which, with the use of the induction spark, there is at times a masking of the primary wave.

The laws of reflection at curved surfaces can readily be made visible. If instead of the flat plate we place in the neighborhood of the sound spark a section of a large glass cylinder to serve as a concave mirror, in such a way that the path of the spark coincides with the optical focal line of the cylindrical mirror, the reflected branch will be linear. If the spark is displaced some distance beyond the focal line, we see for longer waves the reproduction of a real, acoustical spark image at a certain place in the field of view.

On many occasions the wave spheroid does not project itself as a simple circle, as in Figure 4 of Plate II, but on the contrary, in place of a circle there appear *two* eccentric circular curves, as in Figure 5. This is almost always the case if the spark gap is so large that the spark path is crooked or zig-zag. The explanation is afforded at once from another look at Figure 3, since this is the side view of the same case. If in this figure we represent rays of the principal lens system by the lines *m'm n'n*, these form tangents to the irregular wave surfaces. Accordingly, when looked at from the analyzer, *m'n'* and *mn* correspond to a circular projection. In Figure 5 we see in addition at *r* the rising, heated air, which will always constitute a disturbance as soon as the stream of sparks has lasted for some time.

INVESTIGATIONS ON REFRACTION

If a flame is brought into the neighborhood of the sound spark, or if alternatively a tube is brought from which a light gas is emitted vertically, we see in both cases a smooth, sharply defined column in the field of view if the air in the domain of observation is very quiet. If sound is now excited by a stream of sparks in the neighborhood, we can see that the sections of the spheroid that meet this column are refracted. This result, which I have already described earlier, can, however, be misleading since the sound excitation soon produces a disturbance in the gas column. The laws of refraction can be presented to the eye in a very simple and instructive fashion. We introduce a vessel with parallel side walls and of the form indicated in Figure 10 of Plate I. Two plane schlieren-free glass plates *v* and *w* are so fastened in a metal frame *B* that the whole forms a box open at the top. The upper edges of the box are ground down so as to be as nearly as possible in the same plane. The opening of the box is covered with a thin collodion sheet, which acts as an elastic membrane to permit the propagation of sound from air into various gases placed in the box. For the filling of the box with gas there are two connecting tubes *r* and *s*: one of these is connected by a rubber tube to a gas meter; the other permits the free exit of the displaced air. The whole vessel is exposed to the discharger producing the sound spark in such a way that the spark element *ed* (Fig. 10, Plate I) is at some distance above the middle of the elastic membrane. If we then arrange matters so that the illuminating beam goes through the vessel walls *vw* as nearly at right angles as possible, then at the moment of discharge we see each wave spheroid at the surface of separation split into two branches, viz., a reflected and a refracted one. The form of the latter is dependent on the nature of the gas inside the vessel *B*.

The collodion membrane must be extremely thin if the refracted branch of the wave surface is not to lose its sharpness in comparison with the reflected wave surface. The following scheme was used in order to attach to the open top, in a few minutes, the thinnest skin that can be made artificially. Good collodion is diluted tenfold with ether. The resulting solution is poured uniformly on a clean, flat glass plate, and the excess is at once run off. What is left is then dried so as to leave an invisible coating. In the neighborhood of the edge a cut is made in the coating with a sharp knife and then distilled water is allowed to flow on this cut, while the plate is kept horizontal. The water forces itself under the dried skin that now, splendidly iridescent, swims on the water. The upper edge of the box *B*, lightly greased and inverted, is pressed down onto the middle of the swimming skin. Because of the profuse flow of water from the sides one can readily lift up the skin attached to the greased edge. In this way one can get membranes that because of their small thickness produce the most brilliant interference colors of the first order. With a little care one can assure that the color is uniform over the whole membrane. These membranes are so delicate that by a careless breath or a too rapid movement of the vessel they will tear. A few drops of water that happen to be left on the membrane can be removed by simple evaporation. No mechanical means for removing them is possible without disturbing the fragile structure. In filling the vessel with gas one must also be very careful not to produce a rupture of the membrane through the production of pressure differences.

The investigations have been surprisingly successful. Figures 6 and 7 of Plate II depict two different cases, the first with carbon dioxide as the gas and the second with hydrogen. The bounding membrane is *pq*, above which the wave excitation takes place in air. In the case of hydrogen we note in Figure 7 the action of total reflection. The reflected branch frequently appears much more strongly shaded at the ends *p* and *q* than in the middle at *t* whereby it appears, especially with the larger spheroids, as if the middle piece is wholly absent.*

* This apparent failing of the middle section of the curves suggests still another assumption, namely, that at this place a reversal[2] enters into the shading, which can have optically the same result. One might expect this if we consider the theoretical difference between the reflection at boundaries between more and less dense media. I have not been able to carry my investigations further in this direction. In order to be able to judge shadow relations accurately in instantaneous

The angle of incidence of the impulse at the sides of the spheroid, of course, is so large that total reflection can take place. The refracted branch *x* can therefore be followed sharply to its ends only rarely. It appears to be broader and somewhat more washed out than the wave in the air, due to the increased velocity of propagation. As a matter of fact, refraction in hydrogen corresponds to refraction in the most difficult objects and hence can serve as a test of a good optical apparatus.

[*Editor's Note:* The remainder of this paper, which was spread out over more than one issue of the *Annals,* is concerned primarily with further remarks on the accuracy of the time difference between the spark discharges from two spark gaps in series. The precise control of this time difference is very difficult, the author points out, being a function of geometry of the electrodes, temperature of the air in the two gaps, time history of the train of sparks in a gap, and other factors as well. Toepler here concludes his writings on the schlieren method with the observation that a proper understanding of the discharge process is so interesting that it merits pursuance through other types of investigation as well. He did indeed pursue electrostatic discharge phenomena for years, but he did no more writing on the visualization of sound waves.]

---

illumination, these must show up very strongly. For the reflected branches in the case of gases this was, however, not usually the case, since the original air excitation was not strong enough. In order to repeat the experiments with sparks from a Leiden jar discharge, stronger membranes of exactly proper thickness must be tried. We can understand that the thinner and more delicate membranes work alright for the sparks from the electrostatic machine but tear easily with the use of the Leiden jar. The communications under discussion have had merely the purpose of making clear the connection between the optical phenomena and the sparks. I shall go into the matter just discussed in later observations.

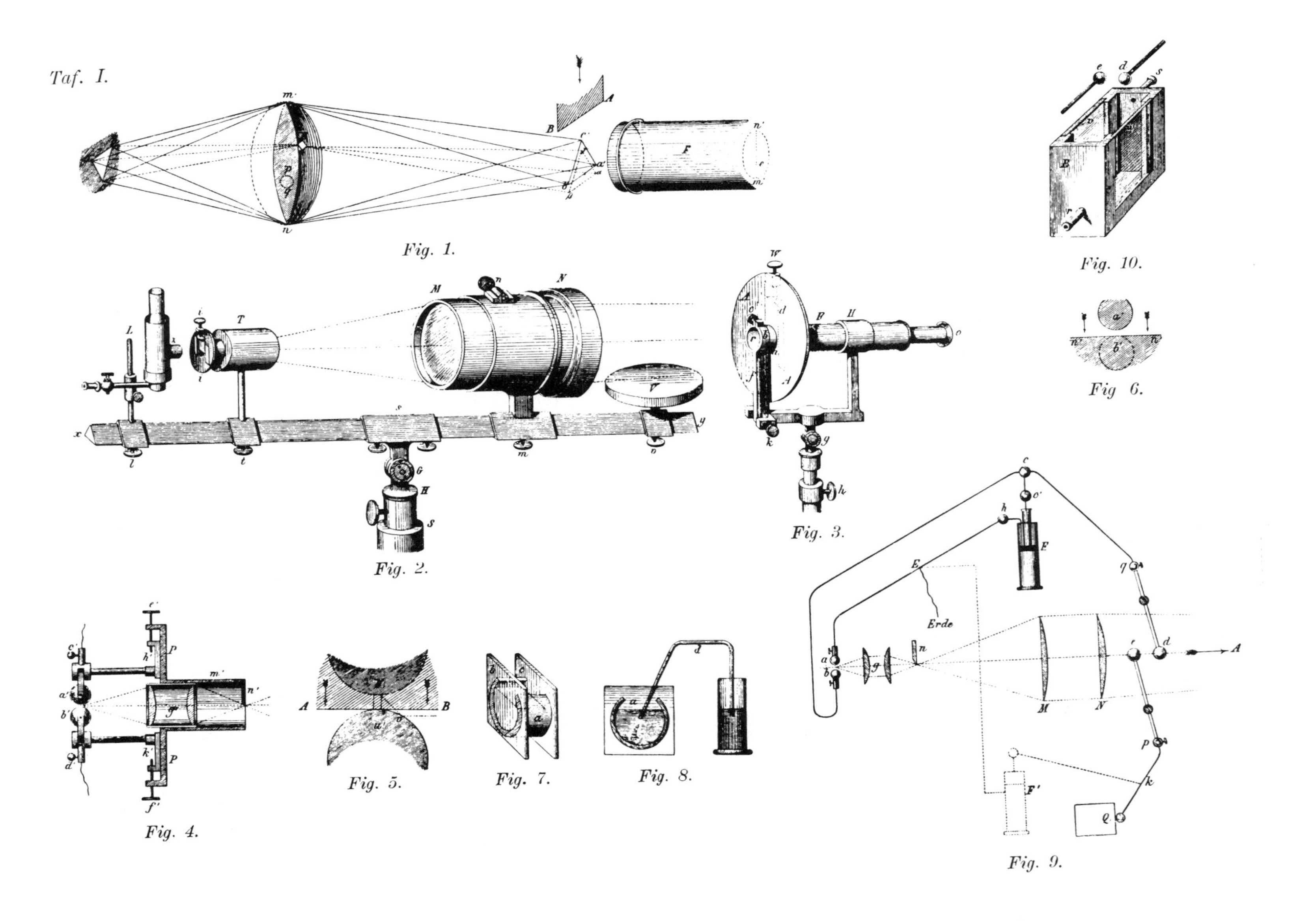

Taf. I.
Fig. 1.
Fig. 2.
Fig. 3.
Fig. 4.
Fig. 5.
Fig. 6.
Fig. 7.
Fig. 8.
Fig. 9.
Fig. 10.
Erde

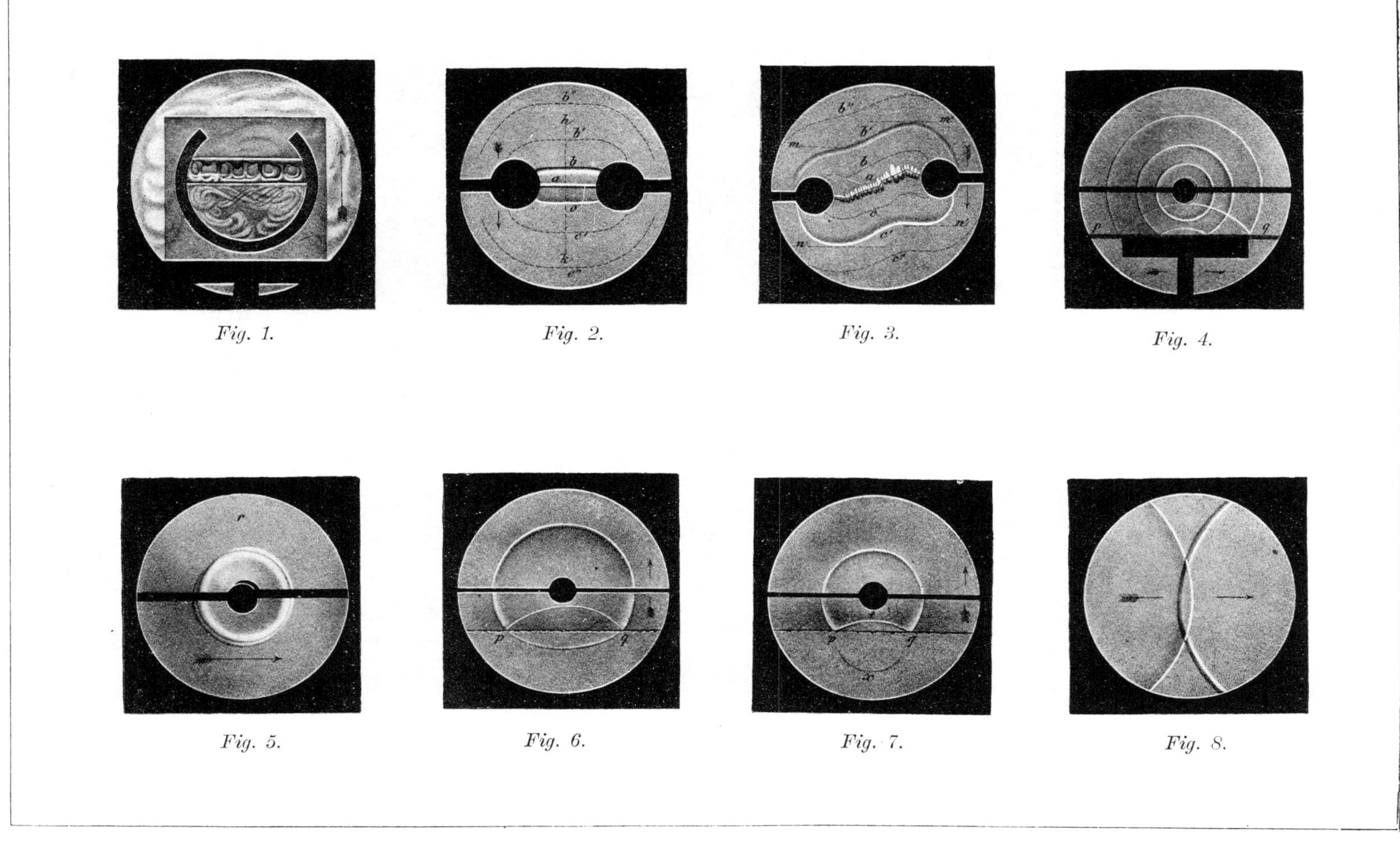

Fig. 1. Fig. 2. Fig. 3. Fig. 4.

Fig. 5. Fig. 6. Fig. 7. Fig. 8.

EDITOR'S NOTES

1. Page **188,** footnote: The reference to Foucault was first given in a long footnote (translated by H. B. Miller) to Toepler's paper in vol. 128 of the *Annalen der Physik* (1866). It reads:

> In my dissertation, "Observations with a New Optical Method" (Bonn, 1864), there will be found the particulars on pp. 23 to 26. Given this opportunity, I cannot leave unmentioned that Prof. Kirchhoff in Heidelberg made me aware of a treatise by L. Foucault, "Memoir on the construction of telescopes using silvered glass" (*Annals of the Imperial Observatory of Paris,* vol. V) [1859, p. 197], in which in the section "Examination of optical surfaces" (p. 203) a principle very analogous to my schlieren method (however without the help of the telescope) is recommended for the examination of spherical mirrors. Foucault says therein (p. 205): "Disregarding the surface itself and considering only the reflected beam, . . . the peculiarities which have been indicated as attributes of a spherical surface, [are associated with the properties] of an exactly conical beam of light. Now, since in optical instruments the sharpness of the image depends expressly on the ultimate convergence of the rays of light, these instruments, whatever they be, make use of the same means of testing."
>
> It can therefore not be doubted that Foucault had already recognized the general utility of the principle for the examination of light-collecting systems. However Foucault seems not to have thought of transferring the method to scientific investigations of an even more general sort, since nothing is mentioned of this in the entire memoir. Furthermore, my observations would probably be repeatable only with difficulty when performed with a reflecting device, since with the use of a concave mirror the conical beam must pass twice through the object under investigation. And a final point: Foucault confirms, in agreement with my observations, the scarcely believable sensitivity of the method, inasmuch as he says (p. 208): ". . . which seems to indicate that this type of examination achieves, with regard to optical surfaces, a kind of testing which is sensitive to an excessive degree."

The question of inventorship is still being argued about. Some people ignore Toepler and credit Foucault with inventing the schlieren method. The editor considers this unjustified. Toepler went far beyond Foucault. Rayleigh uses a good compromise (in *Theory of Sound,* vol. II, Article 279), crediting Toepler by way of Foucault. Zernike, in his own work (Paper 18), builds upon Toepler's work and does not even mention

Foucault. Gabor (*Progress in Optics,* vol. 1, E. Wolf, ed., North-Holland Publishing Co., Amsterdam, 1961) tends to follow Rayleigh by referring to "the Foucault-Toepler schlieren method."

2. Page **194,** footnote: This "reversal of the shading" seems to correlate with the 180° acoustic phase shift that occurs at the air-hydrogen boundary, and may perhaps be explainable by Zernike's phase-contrast work. See Paper 18.

18

Reprinted from pages 119–123 of *Achievements in Optics*, A. Bouwers, ed., Elsevier, Amsterdam, 1946, 135 p.

PHASE CONTRAST, A NEW METHOD FOR THE MICROSCOPIC OBSERVATION OF TRANSPARENT OBJECTS

F. Zernike

[*Editor's Note:* In the original, material precedes and follows this excerpt. The subject matter being discussed here is *The Phase Grating*. Figure 55 has been omitted. It shows a grating in the object plane, illuminated by parallel light; a converging lens; the back focal plane, with the diffraction pattern of the grating "focused" onto it; and the image plane, with an exact image of the object focused onto it.

The problem both Toepler and Zernike had was detecting a transparent phase object in a transparent medium. In Toepler's method, one should be aware of two separate things. First, he blocks out all light from object 1 (the light source) and hence all light from the lens. Second, he now has left the bent radiation from object 2 (the schliere). Now, it would be nice if *all* the diffraction orders from object 2 (appearing in the back focal plane) could be allowed to combine at the image plane, but the schlieren method is unable to allow this. In fact, all the diffraction orders on one side of the central order (including one-half the central order itself) are excluded. These, therefore, cannot assist in forming the image, and so the image is not as sharp as it might be via some other method (for example, Zernike's method.)]

c. *Amplitude grating and phase grating.* The practical bearing of these formulae will become clear from a few particular cases. One extreme case is that of a grating the elements of which only alter the amplitude of the light (e.g. a grating of opaque bars). Let this be called an *amplitude grating*. It is this kind that has always served until now for the discussion of microscopic image formation. In this case the f(x) in our formulae is real, therefore c_m and c_{-m} are conjugate complex numbers and the left and right hand spectra of mth order have equal intensities. The other extreme is that of a grating which only alters the phases, leaving the amplitudes and therefore the intensities of the passing light unchanged. Let this be called a *phase grating*.

It comprises all gratings ruled on transparent plates and in a more general way all structures of unstained microscopic objects (e.g. diatoms). For a phase grating f(x) has unit modulus but this does not correspond to any simple criterion for the c's. It is different, however, in case the changes of phase are only small. The real part of f(x) has then approximately unit value, independent of x, and therefore cancels in c_m, except in c_0, while the imaginary part of f(x) clearly gives rise to values of c_m and c_{-m} with real parts of opposite sign, the imaginary parts being equal. The intensities of both m^{th} order spectra are again equal. In the following we shall restrict the term "phase-grating" to gratings causing small changes of phase. Fig. 56 illustrates these results. The phases are represented by the angles of the resp. vectors with the horizontal direction.

The most obvious difference between an amplitude grating and a phase grating, however, is not to be seen in the plane of the spectra, but in the image plane. Indeed the phase grating must be invisible in that plane, i.e. its image must show uniform intensity, being similar to the object. This means that the interference of the diffraction spectra, though these may be very conspicuous and very similar to those of an amplitude grating, only gives rise to differences of phase in the image.

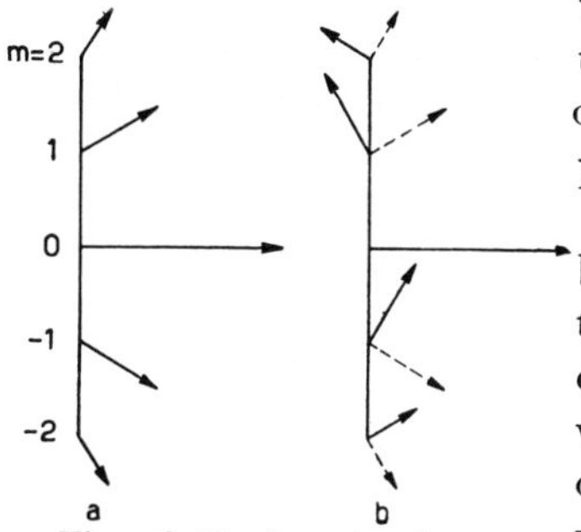

m=2
1
0
-1
-2
a b

Fig. 56. Explanation in text

For different reasons the invisibility is seldom as absolute as theory predicts. Indeed every deviation from the ideal conditions will disturb the exact balance of the different compounding vibrations. Certain larger deviations are used intentionally to get a clearly visible image of a phase grating.

d. *Schlierenmethod.* If the left hand spectra in Fig. 55 are intercepted, the vibration V_s in the image plane will be given by

$$V_s = \sum_{0}^{\infty} c_m e^{-imx}$$

instead of by

$$V = \sum_{-\infty}^{\infty} c_m e^{-imx}$$

In order to take into account that we are observing a phase grating, we put according to the above results

$$c_0 = 1, \qquad c_m = -a_m + ib_m, \qquad c_{-m} = a_m + ib_m$$

and combine the terms with c_m and c_{-m}

$$V = 1 + 2i \sum_{1}^{\infty} (a_m \sin mx + b_m \cos mx)$$

$$V_s = 1 + \sum_{1}^{\infty} (a_m \cos mx - b_m \sin mx) + i \sum_{1}^{\infty} (a_m \sin mx + b_m \cos mx) \tag{124}$$

In squaring these expressions we must remember that they are approximations for small values of the a's and b's, so that squares of these quantities should be neglected compared with first powers. Therefore we find for the intensities

$$|V|^2 = 1 \qquad |V_s|^2 = 1 + 2 \sum^{\infty} (a_m \cos mx - b_m \sin mx)$$

Though it cannot be seen in the image, the imaginary part of V in (124) clearly specifies the structure of the phase grating. The result is therefore that instead of the real structure represented by the series

$$\Sigma_1 = \sum_{1}^{\infty} (a_m \sin mx + b_m \cos mx) \tag{125}$$

the SCHLIEREN method shows us

$$\Sigma_2 = \sum_{1}^{\infty} (a_m \cos mx - b_m \sin mx) \tag{126}$$

Here each term equals the mth part of the derivative of the corresponding term of (125).

The appearance of the schlierenimage V_s may be illustrated in the following way. Suppose we have a plaster relief of the real structure, differences of refractive index being rendered by small differences of height. The relief will become visible when illuminated obliquely, the slopes turned towards the light source showing brighter, those turned away from it showing darker. The differences of intensity will be proportional to the sloping angles, when these are only small. Under ordinary circumstances the structure of the plaster relief will be at once apparent to the observer, which means that he concludes from the observed brightness function to its integral, the height of the relief. In the same way the observer of the SCHLIEREN image (126) will get the impression that he sees an obliquely illuminated relief, the height of which is given by the integral of (126):

$$\sum_{1}^{\infty} \frac{1}{m} (a_m \sin mx + b_m \cos mx) \tag{127}$$

This is indeed rather similar to the real structure (125), with the difference, however, that because of the factor 1/m the higher terms are relatively too small. Especially sharp edges will therefore be seen rounded off, as is clearly shown in Fig. 57a, b in which the calculation according to (126) and (127) of a simple case has been plotted.

The relief effect here described was well known long before ABBE and even before TOEPLER as it appears in the same way when the incident light consists of an extremely oblique pencil. Evidently the impression of an obliquely illuminated relief structure was supposed to be a direct consequence of the obliquely incident light, so that no further explanation was sought for. In 1866 TOEPLER[1] disproved this idea by his arrangement with normally incident light and a shutter inserted from one side above the objective, and pointed out that in both cases only the interception of all light on one side of the direct beam is responsible for the relief effect. He gave an explanation from geometrical optics, which is certainly inadequate near the limit of sensibility of the method. Of the practically found advantages of oblique illumination ABBE in 1873 explained that of the gain in resolving power, with the result that the relief effect appears to have been neglected since then. The gain in resolving power, however, according to our formulae, will be obtained without the relief effect when oblique illumination is made from both sides, so that the left hand spectra remain operative with one

[1] A. TOEPLER, *Pogg. Ann.*, 131 (1867) 180.

[*Editor's Note:* Lines 5 and 6 are unclear. We offer as a substitute: the observer, which means that he infers from the brightness the height of the relief, the brightness being itself the integral of Equation 126.]

beam, the right hand spectra with the other. This is e.g. realised when a hollow cone of light (annular diaphragm in the condenser) is used.

e. *α*. *Oblique Dark Ground Illumination*. In this method only the spectra on one side contribute to the image, either because only these enter the objective, or because the other side and the central image are screened off. From (124) it follows that the intensity of the image is given by

$$[\sum_{1}^{\infty}(a_m \sin mx + b_m \cos mx)]^2 + [\sum_{1}^{\infty}(a_m \cos mx - b_m \sin mx)]^2 \qquad (128)$$

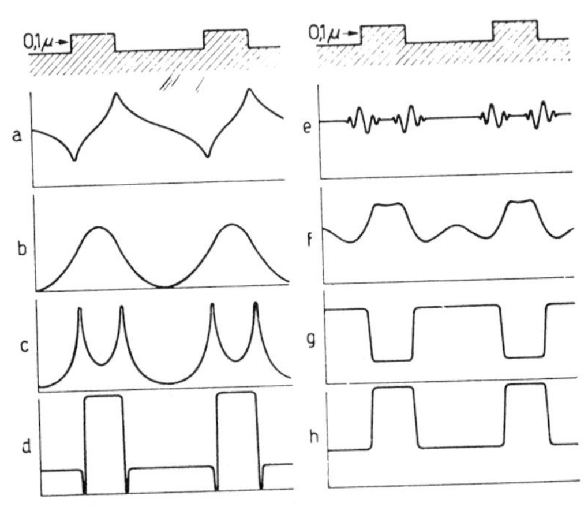

Fig. 57. Explanation in text

Both sums (125) and (126) appear in this result. The first term here is the square of the real structure and does not represent it very clearly as differences of sign disappear in the square. The second term is the square of the "SCHLIEREN"-function (126), which specially emphasises abrupt changes in the structure. This is responsible for the well known dark ground effect in which all contours appear bright and broadened. Fig. 57c gives the calculated result for the same grating.

In this case it makes no difference on which side the spectra are intercepted Therefore obliquely illuminating pencils may as well come from all sides, as in the usual dark ground condensers.

19

ON A NEW, SIMPLE METHOD OF OBSERVING SCHLIEREN

V. Dvorak

This article was translated expressly for this Benchmark volume by R. Bruce Lindsay from "Ueber eine neue einfache Art der Schlierenbeobachtung," in Ann. Phys. *9:502–511 (1880)*

The Toepler method of schlieren observation gives excellent results but is accompanied by circumstances that make its application difficult. In addition to the fact that the schlieren apparatus is expensive and complicated, its field of view is rather small (in Toepler's apparatus the field of view had a diameter of 11.5 cm). Moreover it is not possible for several observers to view the phenomena in question at the same time. In cases in which the Toepler method is not particularly applicable because of the circumstances just mentioned, the method to be described here can be used.

The optical apparatus is very simple. With the use of an ordinary heliostat the sun's rays are allowed to fall on a convex lens *ab* (Plate II, Fig. 7). [*Editor's Note:* The heliostat is an instrument consisting of a mirror moved by clockwork by which a sunbeam is made apparently stationary by being steadily directed to one spot during the whole of its diurnal period.] This lens has a focal length of 22.6 cm and a diameter of 7.1 cm. The rays then pass through a second convex lens *cd* (focal length 19.3 cm, diameter 2.8 cm). The separation of the lenses is a little smaller than the focal length of the lens *ab*. In general I set the lenses 21 cm apart. I have previously used this lens system plus heliostat for the projection of polarization phenomena.

A very important part of the equipment is a metal diaphragm located at *f* (the focal point of the lens system). The diaphragm has a circular opening 1 mm in diameter. The edges of the opening are sharpened and the inside of the diaphragm is well blackened. It is, further, fastened to a short tube (see figure) which is placed on the holder of the lens *cd*. The cone of the sun's radiation emerging from *f* is received on a white screen *mn* in a dark room. The distance *fo* [of the screen] is 4.6 m. The objects to be investigated are placed in the radiation cone at *l*, a distance of 2 to 2.1 meters from *f*. The circular cross section of the cone at *l* has a diameter *ik* equal to 70 cm. That is, therefore, as we readily see, the diameter of the field of view.

In most cases such a large field of view is unnecessary inasmuch as the broad illuminated surface *mn* of the white screen diffuses a rather disturbing brightness into the laboratory. We can cut down the light-circle *mn* by means of a diaphragm *gh* of 4.3 cm diameter, placed before the lens *ab*. Then the diameter of the field of view *ik* becomes

41 cm. In all investigations in which the contrary is not specified I have used the diaphragm *gh*.

The objects placed in the region *ilk* throw a shadow on the screen *mn*, and we investigate actually only the *shadows* of the objects. The edge of the shadow is not sharply bounded but shows diffraction effects.

If it is desired, the lens *cd* may be left out. However, I used it regularly since with its use one can obtain a more uniform illumination of the white screen. Every nonuniformity in the heliostat mirror and in the lens *ab* is clearly shown on the screen *mn* as long as one uses the lens *ab* alone, but [this "noise"] disappears almost completely as soon as the lens *cd* is employed. One can readily be convinced of this by moistening the mirror or lens at one point. Concerning the metal diaphragm at *f*: it increases considerably the sensitivity of the apparatus.

Naturally we need not adhere precisely to the apparatus dimensions set forth here. By trial one can readily determine what distances of separation of the two lenses and the screen give the best results.

In the following I shall first describe some experiments that can be carried out with our apparatus. At the end I shall add a few comments on the theory of the apparatus.

TEST OF SENSITIVITY

The sensitivity of the apparatus is surprising and possibly approaches the remarkable sensitivity of the Toepler equipment.

If one brings the warm hand into the cone cross section at *k*, one sees clearly on the screen *mn* the air heated by the hand streaming up. Our apparatus also satisfies the sensitivity test Toepler used for his. We take two similar vessels or cells made of first-rate plate glass, such as are used in light-absorption investigations. The vertical walls of each cell must be as nearly parallel as possible. The cells are then filled with pure water and are joined with a siphon formed by a bent glass tube of 1.8 mm inner diameter. The siphon tube has been previously filled with water. If we now raise the one cell *A* (see Fig. 8 in Plate II), the water in the siphon tube flows into the cell *B*. The stream of water emerging from the siphon tube is clearly visible in cell *B* with our apparatus. [*Editor's Note:* The shadows appearing in Figure 8, due to a change in refractive index, are very similar to the shadows of ripples on the bottom of a bathtub.] The difference between the index of refraction of the water stream itself and that of the surrounding water [probably due to temperature difference] is certainly too small for accurate measurement. Hence we may conclude that the sensitivity of the apparatus is practically unlimited. [The reader may not concur.]

FURTHER EXPERIMENTS

I introduce here merely a few lecture demonstrations that can readily be carried out with our apparatus. A piece of ordinary window glass is shown to be very nonuniform in character, as might be expected. Even good plate glass shows spots and streaks as a rule, especially if the glass surface is held inclined to the direction of the sun's radiation.

For liquids we use a vessel or cell of rather large size (around 3 cm in width). If we fill this with water and then blow on the surface with a small bellows (following Toepler), the water is cooled by evaporation and some of the surface water sinks. We can see this very clearly on the screen *mn*. If next we heat a soldering iron and hold the point in contact with the bottom of the cell for a few seconds (it is understood that the bottom is made of metal and not too thick), at once we observe a rather sharply defined dark column rising from the heated place. It has a mushroom-shaped head, spreading out over the water surface.

If we use a pair of forceps to hold a pinch of salt at the water surface for a few seconds, dark streaks of salt solution are observed to fall through the water.

We can also easily do experiments with gases and vapors. If we open a flask of ether or carbon disulphide and tilt this so that the liquid is almost on the verge of flowing out, we clearly see (over a large extent) the vapor sinking down. For this it is important that the surrounding air be very quiet. We can also readily show the well-known experiment of the flow of carbon dioxide from one vessel into another, as well as the flow of carbon dioxide over the mouth of a vessel, etc. The following is a particularly easy experiment to perform. We place some water and sodium bicarbonate in a tall vessel open at the top. Then we add some sulphuric acid. Immediately there takes place an eruption of carbon dioxide, which produces a startling appearance on the screen *mn*.

A beautiful picture is provided by a Bunsen burner with a rather small flame. We see masses of heated air rising from the burner, with characteristic forms and sharp boundaries. For this it is well to remove the diaphragm *gh*, so as to provide a larger field of view.

Our method can be readily applied to the investigation of sensitive gas jets. Tyndall* made his jets visible by mixing them with ammonium chloride smoke. However, he believed that the sensitivity of the jet (to small disturbances in the surrounding air) would be greater if one could see it without the use of the ammonium chloride smoke. At another place he mentions that "the smallest amount of impurity mechanically communicated to the flame exerts the greatest influence on the flame" [he had this time replaced the air jet by a sensitive flame]. The ammonium chloride smoke is itself such an impurity and it often nearly blocks the aperture of the vessel from which the jet or flame is emerging. Since we can make a pure gas jet readily visible with our method we do not need the inconvenient ammonium chloride. In one of my experiments I used a Koenig bellows, which worked pretty uniformly at an average pressure of 65 mm of water. The air went from the bellows into a rather large flask *A* (Plate II, Fig. 9) and then from this to a smaller flask *B* through a tube reaching nearly to the bottom of *B*. The air could then stream out through tube *C*, with an opening from 0.8 to 2.0 mm in diameter. The exit velocity was regulated by the stopcock *D*. If the stopcock is opened too wide or is not wide enough the jet turns out to be short and insensitive. We put in flask *B* some pieces of cotton soaked in ether. The bellows is filled and then left to itself until it is completely empty. In the process of supplying air flow it must neither hiss nor squeak. Naturally we could to advantage replace the bellows with a gasometer kept at constant pressure. If we have adjusted the stopcock properly

* Tyndall, John, *Sound,* 2nd ed., Longmans, Green, & Co., New York, 1875, p. 289.

a beautiful gas jet is visible on the screen *mn*, which sometimes occupies the full field of view and is very sensitive. [He obviously means sensitive to sound, as in Tyndall's experiments with sensitive jets and flames.] For high tones as well as for hisses, the jet is not sensitive, as indeed Tyndall observed with his smoke jets. But with low tones, the sensitivity is extraordinary. If one speaks or sings at it with a voice of only moderate intensity, the jet shows very great disturbance. Just walking around or scraping the feet on the floor produces a great effect. If one taps inaudibly with the finger on the table on which *A* and *B* rest, the jet suddenly shrinks to a third of its original height. At times it becomes divided in two in the shape of a fork.

We can also form a jet with illuminating gas if we are not afraid of the smell. Then we need neither bellows nor flasks. The main thing again to look out for is the control of the flow by a stopcock. However, ether vapor is more easily visible with our apparatus than illuminating gas.

In addition to the above I carried out several experiments with heated and cooled air. For these we use the bellows and the large flask *A*. We insert a rubber tube beyond the stopcock *D*, and to this tube we attach a copper tube *abcd* of helical form (Plate II, Fig. 12). The helical part is heated to red heat by a Bunsen burner. Naturally the straight part *cd* must be long enough so that the rubber tube attached at *d* is not damaged by the heating. After the helical part of the copper tube has been sufficiently heated we remove the burner, fill the bellows and let the air stream out at *a*, an opening of 1 to 1.5 mm. The sensitivity of the air jet is again regulated by stopcock *D*. We can then see the jet very clearly if we touch with the hand the helical part of the copper tube (assuming its temperature has dropped to about 35° C).

The heated air rising from the upper surface of the copper tube can disturb and partially mask the heated air jet. We can either incline the whole tube *abcd* so that the air rising from the surface of the tube does not meet the jet, or we can leave the tube vertical and place a vessel *fg* around the tube with the opening *h* made as nearly airtight as possible.

In case one wishes to do experiments with cooled air jets we place the helical part of the copper tube in a freezing mixture. Jets obtained in this way are readily visible and sensitive.

EXPERIMENTS WITH THE ELECTRICAL SPARK

Toepler has made many experiments in this field with his schlieren apparatus. I shall add here only a few, using sparks from an induction coil. From different electrical phenomena it has been judged that the air flowing away from the positive pole has greater velocity than that driven away in the opposite direction from the negative pole. Using the schlieren method we observe phenomena that appear to confirm this. [*Editor's Note:* Dvorak then discusses the details of his own experiments, using a Ruhmkorff induction coil and a Foucault interrupter. He describes the expanding cloud of heated air shown in Figure 10 and tells how he set up a stroboscopic disc that could synchronize with the Foucault interrupter and make the expanding cloud stand still or move slowly, and so on. One important point: "One can sharpen up the shadow picture on the screen if one makes the opening of the diaphragm at *f* very small—for example, about 0.5 mm in diameter." This is the second time Dvorak has referred to the use of a very small hole at *f*, the first occasion being when he described "the cone of rays" in Figure 7, coming from the point source *f*. He needs a point source in order to throw the darkest possible shadow onto the screen. If he used an extended source he would partially illuminate the edges of the shadow, thereby degrading the contrast. This is similar to the relation between the penumbra and the umbra when, during a solar eclipse, the moon throws its shadow onto the earth.]

THEORY OF THE APPARATUS

We insert a round glass rod *qp* in the cone of rays at *l* (Plate II, Fig. 7). In spite of its transparency it throws a dark shadow *rs* on the screen, just as if it were a completely opaque object. The reason is that the light rays that go through *qp* are scattered in all possible directions before they reach the screen *mn*. Very few rays manage to get into the geometrical shadow *rs*, and hence this region appears dark. A somewhat similar situation arises if we insert a column of gas in the light ray cone. However, because of the small difference in the index of refraction of the gas and the air, the rays through the gas column are only slightly deviated. Moreover, the rays that do go through the gas column interfere with those that pass by it on either side. We see that the complete theory of the shadows formed by transparent bodies is complicated; we shall not go into this further. I content myself with mentioning some striking interference phenomena observed in my experiments. Thus, consider again flask *B* (Fig. 9 in Plate II) containing the cotton soaked in ether. The exit tube *C* is now to be rather wide (e.g., about 1 cm in diameter). Its edge is now to be covered with a thin metal plate having a square hole in it about 5 mm on a side. We glue on to the metal plate a small piece of heavy tin foil having a small round opening in it. If then we blow through flask *B*, a jet of ether vapor rises upward from this opening. The interference phenomena in the shadow of the jet are dependent on the diameter of the opening. If the latter is 0.56 mm and if [in Fig. 7] *lf* = 2.15 m, *fo* = 4.6 m, and the opening of the diaphragm at *f* is 0.5 mm, we see in the middle of the shadow a bright band and on both sides colored bright and dark bands. Again if the diameter of the opening in the foil in 3.4 mm [i.e., 6 times larger], there are two narrow dark bands in the middle of the shadow. And again, we see a beautiful phenomenon if we make the opening in the tin foil rectangular with 4 mm length and 0.25 mm width. The jet gets much narrower above the opening (vena contracta effect) and shows brilliantly colored interference bands. Interesting interference phenomena are also observed if the tin foil piece is not fastened airtight to the metal plate so that at various points between the foil and plate narrow gas jets escape. These always display brilliant colored bands.

Agram (Zagreb), October 25, 1879

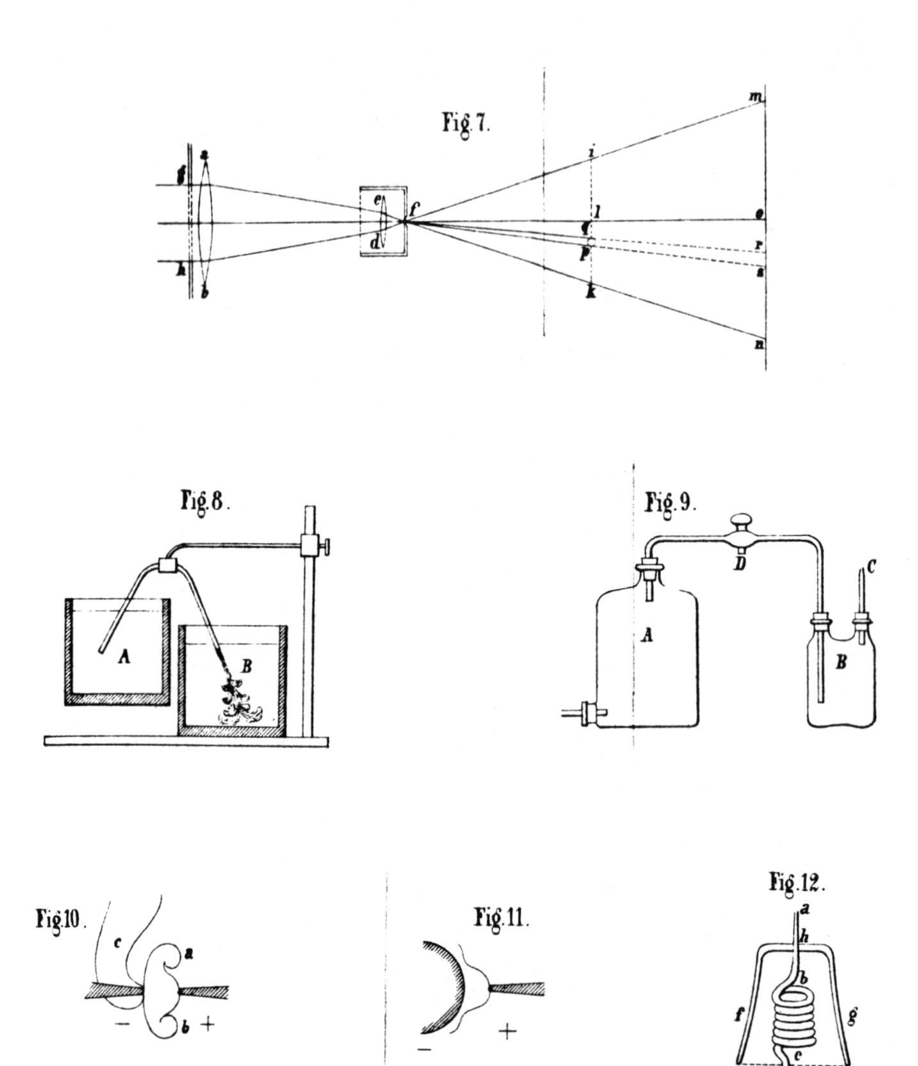
Fig. 7.
Fig. 8.
A
B
Fig. 9.
A
B
C
D
Fig. 10.
Fig. 11.
Fig. 12.

20

Reprinted from pages 373, 381–382, and 386 and Plates II and V of *Phys. Rev.* **35**:373–386 (1912)

A NEW METHOD OF PHOTOGRAPHING SOUND WAVES.

By Arthur L. Foley and Wilmer H. Souder.

AT the Chicago meeting of the American Physical Society in 1905, the senior author of this paper exhibited photographs of interference and diffraction fringes about electric discharges and fluid streams, taken by what might be called the point source shadow method.[1] The light from a point source is allowed to fall directly upon a photographic dry plate several meters distant from the source. About half way between the source and the plate is placed the fluid stream or whatever is to produce a shadow on the plate. Say the object is a stream of warm air issuing from a glass tube. The air stream maintains for a time a more or less definite lateral surface and produces diffraction fringes on the plate. The light is refracted on passing through the stream of air of density differing from the air about it, and so the stream casts a sort of shadow on the plate. The method is so sensitive that it gives a shadow of a stream of water in water. Inasmuch as a sound wave in air consists of a compression and rarefaction, the method should enable one to photograph the shadow of a wave, provided a sufficiently strong instantaneous point source of light is obtainable. Mr. Wilmer H. Souder, while teaching fellow at Indiana University, undertook the problem and succeeded in obtaining sound wave photographs, but discontinued the work to accept a position elsewhere. The writer continued and extended the work, and succeeded in obtaining a light source many times as intense as the source previously used—so intense that, in photographing sound waves, no further increase is desirable.

[*Editor's Note:* Material has been omitted at this point.]

[1] "Diffraction Fringes from Electric Discharges and from Fluid Streams," A. L. Foley and J. H. Haseman, Physical Review, Vol. 20, 1905, page 399; Proceedings Indiana Academy of Science, 1904, p. 206.

So far as the writer knows all previous work on the visualization or photography of sound waves has been by the so-called "Schlieren Methode" devised by A. Toepler,[1] and used by Mach,[2] M. Toepler[3] and Wood.[4] The Toepler Schlieren method is exceedingly ingenious and has very wide application. However, compared with it, the method described in this paper has some obvious advantages.

In the first place no high grade lenses or concave mirrors are required. All optical part are dispensed with.

In the second place the size of the sound wave is not limited by the diameter of a lens or concave mirror, but only by the size of the photographic plate or screen on which the wave shadow is to fall. Since the sound wave can be large, so can the sound lens, the sound gratings, etc.

In the third place, the wave shadow itself may be seen and photographed full size. By the Schlieren method the wave images are small, usually but a few millimeters in diameter. This makes it difficult to study the waves visually and impossible to exhibit them to a class. This defect cannot be overcome by changing the lens system, for the size of the images is limited by the lack of light, most of the light being cut off by the diaphragms which the Schlieren method requires.

In the fourth place the method in this investigation gives a field of uniform illumination and distinct sound wave shadows in all points of the field. The Schlieren method gives a field of varying and uncertain illumination and wave images much more definite at some than at other points of the field.

[*Editor's Note:* The descriptions and discussions of photographs 1 through 24 have been omitted. Photographs 9 and 10 (Plate II) show the temperature wave, which travels much more slowly than the sound wave and so appears only long after the sound wave has streaked past. The temperature wave, being a diffusion wave, has high attenuation in addition to its slow velocity, whereas the sound wave, obeying the wave equation, is attenuated only by geometrical spreading. (Compare Paper 14 by Arnold and Crandall).]

[1] A. Toepler, Pogg. Am., 131, p. 33, N. 180, 1867.

[2] E. Mach, Sitzungsber. d. k. Akad. d. Wissensch. zu Wien, 98, p. 1333, 1889.

[3] M. Toepler, Ann. d. Phys., 14, p. 838, 1904. Also Ann. d. Phys., 27, p. 1043, 1908.

[4] R. W. Wood, Phil. Mag., 48, p. 218, 1899. Also Phil. Mag., 50, p. 148, 1900.

Photographs 25, 26, 27 and 28 show four positions of a sound wave reflected from and transmitted by a diffraction grating. The grating was made by cutting four equal and equally spaced rectangular slits (*O*, Fig. 5) in a strip of sheet tin 6 cm. wide and 18 cm. long. The slits were 7 mm. wide and 3.5 cm. long, with a strip of tin 7 mm. wide between the openings. The tin was tacked to a wooden block which served as a supporting base. The grating is placed with its apertures parallel with the spark gap. In this position the shadow on the photographic plate is edge on and the location of the apertures is not shown on the plate. To correct this fault pieces of heavy wire (*W*, Fig. 5) were soldered to the upper, lower and intermediate strips of tin, so that an edge on shadow of an aperture is narrow, of a reflecting surface it is broad.

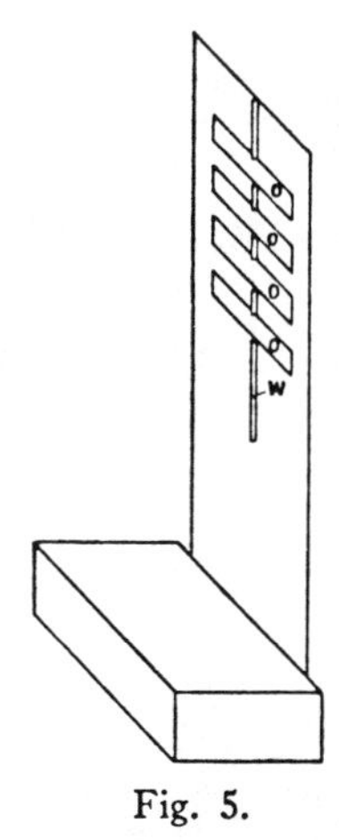

Fig. 5.

Both the reflected and transmitted system of waves are in complete accord with Huygens's principle. It seems to the writer that such photographs as these, or better, a chance to see the waves themselves, might be used by a teacher to give a concrete and definite idea of Huygens's principle and deductions from it, to students who have difficulty in forming mental images. Reality would be given to such terms as secondary waves, center of disturbance, pole of wave, wave front, common tangent, diffraction, etc.

Photographs 29 and 30. These pictures were made by substituting a curved grating for the plane grating of the last four photographs. The grating is similar to the plane grating already described except that it has eight slits instead of four, and is cylindrical—having been curved over a piece of gas pipe. It was intended that the spark gap should be on the axis of the cylinder but it was accidentally displaced slightly before the photographs were taken. This explains why the transmitted and reflected wave systems are not symmetrical with respect to the grating and the spark gap.

All the photographs having a reflected wave show that a condensation is reflected as a condensation. This is true in the case of the hydrogen lens. While the film separating the rare hydrogen and the denser air is exceedingly thin, nevertheless it is a dense body so far as sound wave reflection is concerned. In a later paper the author will show waves reflected from the boundary between two gases of different density—when there is no film or other object separating them.

Physics Laboratory,
Indiana University,
June, 1912.

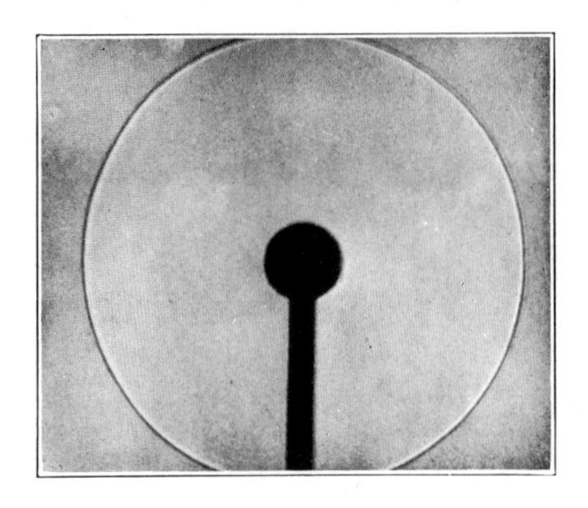

Photograph 7.

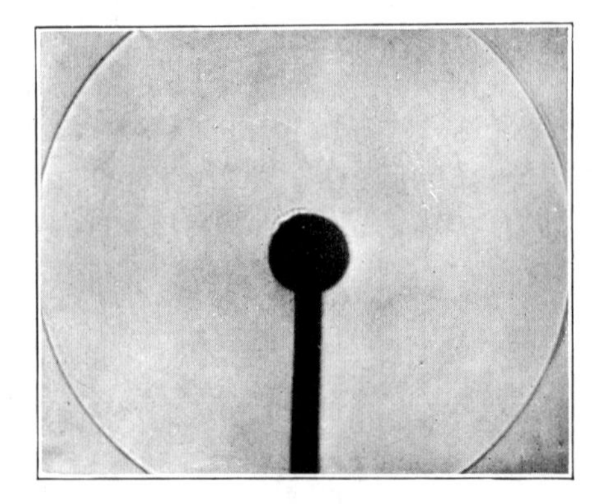

Photograph 8.

Photograph 9.

Photograph 10.

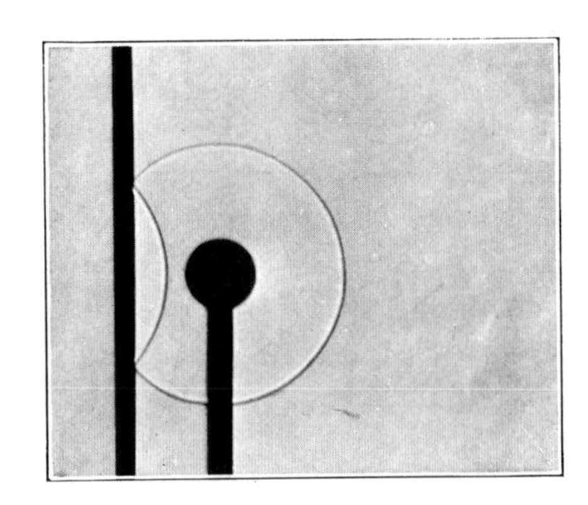

Photograph 11.

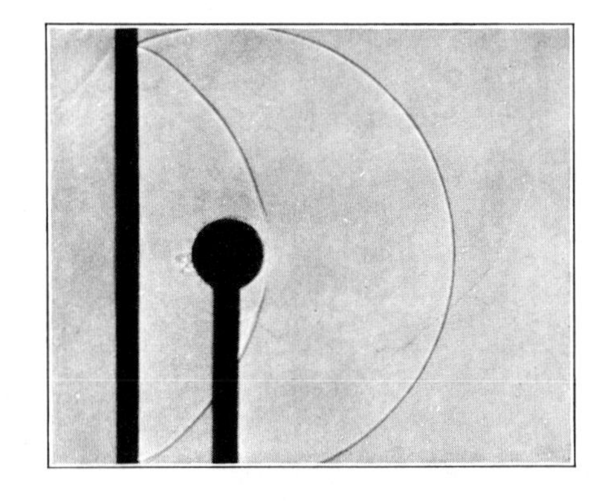

Photograph 12.

A. L. FOLEY AND W. H. SOUDER.

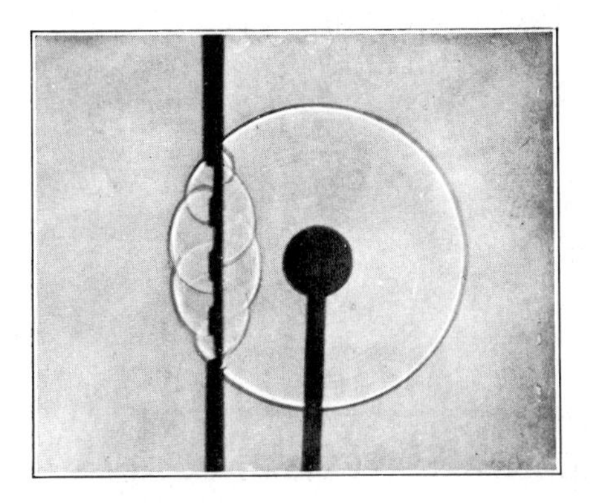

Photograph 25.

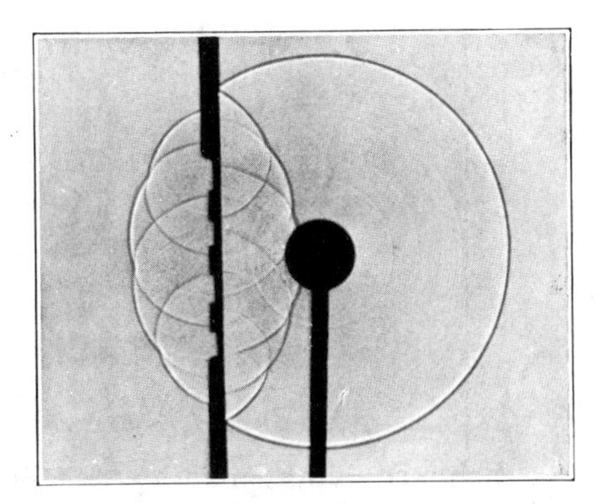

Photograph 26.

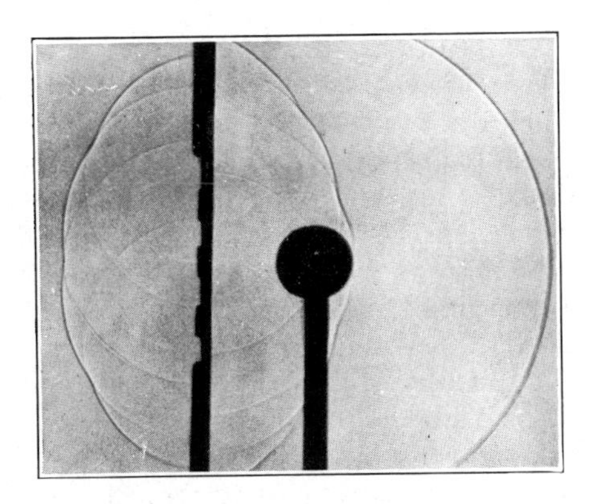

Photograph 27.

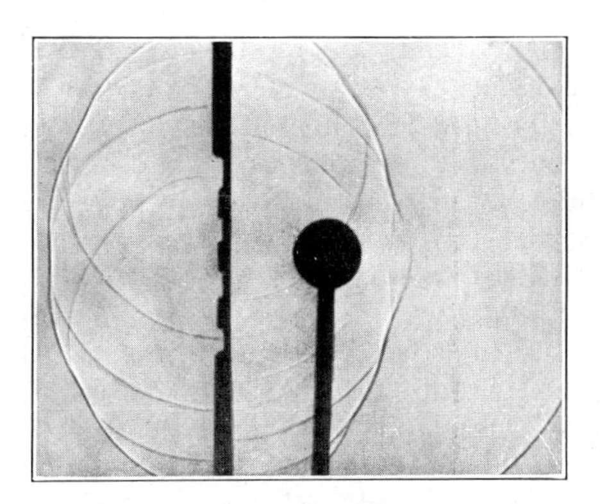

Photograph 28.

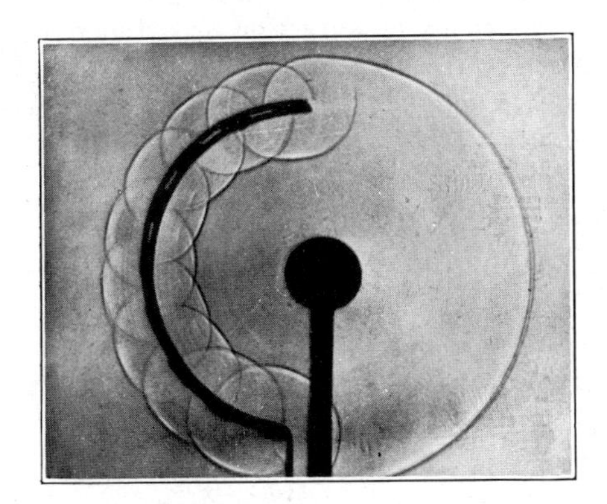

Photograph 29.

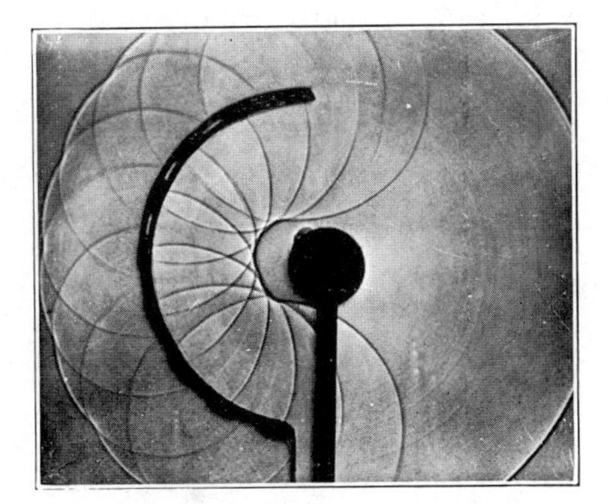

Photograph 30.

A. L. FOLEY AND W. H. SOUDER

21

Reprinted from pages 497–498 and 505–509 of *Opt. Soc. Am. J.* **35**:497–509 (1945)

Schlieren and Shadowgraph Equipment for Air Flow Analysis

NORMAN F. BARNES AND S. LAWRENCE BELLINGER
General Engineering and Consulting Laboratory, General Electric Company, Schenectady, New York

(Received February 15, 1945)

THOUGH the science of making the flow of air visible to the human eye is still in its infancy, the fundamental principle dates back about 86 years. At that time Foucault devised a test, which now bears his name, for observing the surface configuration of a concave telescope mirror. Though in his tests Foucault probably observed what were apparently heat waves passing in front of his mirror surface, it remained for Toepler to realize the real significance of this additional phenomenon. Since the latter's publication in 1864, the method has been often referred to as the Toepler Schlieren Method.

The operation of the schlieren method can be best described with reference to Fig. 1. Light from an illuminated pinhole S is allowed to fall upon lens L_1 and be converged to the image point at P. If the eye is placed slightly behind the image point, as in the Foucault test, the lens L_1 will appear as a uniformly illuminated field. In usual practice, however, a lens L_2 is substituted for the eye, this lens focusing the striation upon a viewing screen or upon a photographic plate. If the knife-edge E is then moved laterally across the image point until all the rays passing through that image are obscured, the field as viewed upon the screen will be uniformly dark.

Next to be considered is the effect of introducing air into a very small element of volume dv (Fig. 1). The index of refraction n of this air is different from that of the surrounding air n_0 for which the knife-edge has been adjusted to produce a uniformly dark field upon the viewing screen. A light ray passing through the volume dv will be refracted. Therefore, it will no longer proceed along its original path, except for the singular instance for which the gradient or change in the index of refraction as a function of the space position is parallel to that light ray.

If the gradient of n through the refraction is so directionally constituted that the light ray is refracted in a counterclockwise manner, this ray

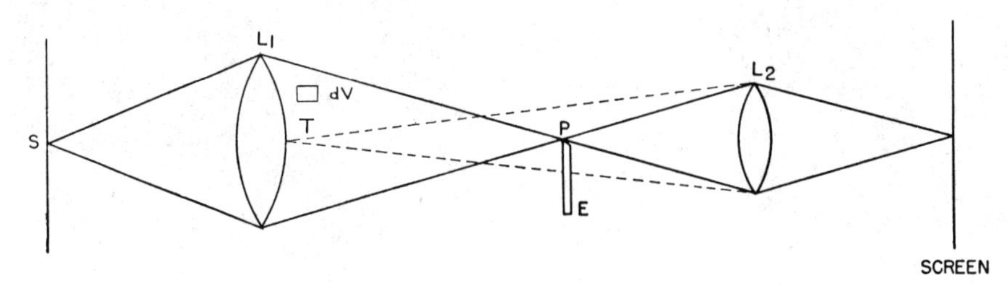

FIG. 1. Schematic diagram of the basic elements of a lens schlieren system.

will no longer pass through the image point P but will travel above it. Hence, this ray will not be obscured by the knife-edge but will pass on to the screen. There it will illuminate a point corresponding to the location of the elementary volume dv. Thus for every point in the region T for which a similar refraction takes place, there will be a corresponding point illuminated on the viewing screen. The composite of all such points forms the image of the phenomenon being investigated.

If the refraction in the elementary volume dv is in a clockwise direction, the rays will be bent down toward the base of the knife-edge, and the corresponding points on the viewing screen will be dark.

In making the lateral adjustment of the knife-edge when no air flow or striation is present in the region T, it is usually desirable to allow some of the rays to pass over the knife-edge in order to produce a uniformly low background illumination. The presence of this background makes it possible to see more clearly in silhouette or outline the objects or models used in producing the airflow phenomena. Thus, referring to the earlier paragraphs about the refraction produced by the elementary volume dv, a counterclockwise bending of a light ray will produce an increase in illumination at the corresponding screen point; a clockwise bending will produce a decrease in screen illumination.

It is apparent from the foregoing that, if the knife-edge is allowed to cut the light beam laterally from the top rather than from the bottom, the resultant screen pattern will be the negative of the former one. In other words, where the picture was previously light, that same area will now be dark.

The refraction of the light rays in the region T can be produced in four ways:

(a) by a change in temperature of the fluid being investigated—air, for example;
(b) by a change in the pressure of the fluid;
(c) by the introduction of a transparent object into the light path; and
(d) by the use of a gas having an index of refraction different from that of air.

Since this method is extremely sensitive to light refraction produced either as a result of a continuous index of refraction gradient or as a result of surface refraction, it is often useful for checking the optical homogeneity and surface flatness of transparent materials, such as glass.

For an effectively circular light source, such as an illuminated pinhole, the rays which contribute to the greatest sensitivity of the system pass through the region of the light-source image near the knife-edge. Thus rays initially passing through this point have to be refracted upward only slightly in order for them to pass over the knife-edge and contribute to the illumination on the viewing screen. On the other hand, those rays which pass through the lower part of the light-source image away from the knife-edge must undergo a large refraction in the region T of Fig. 1 before they can contribute to the screen illumination. For the highest sensitivity, therefore, the maximum number of rays in the light-source image must lie close to the knife-edge. This requirement is most easily satisfied by the use of a slit source of light. Here, no rays lie at a large distance from the edge itself; consequently, more of the rays can be bent over the knife-edge for a given amount of refraction in the striation.

An additional requirement for maximum sensitivity is that the light-source image adjacent to the knife-edge be as straight and as sharply defined as possible. Then it will conform to the sharp edge of the knife-edge so that accurate positioning of the knife-edge can be made with respect to the image.

If the index of refraction gradient lies only in a horizontal plane, thereby producing a corresponding refraction in that plane, the effect is to produce a lateral shift of an affected ray parallel to the knife-edge. Such a displacement, if strictly parallel to the edge, cannot affect the viewing-screen illumination. It can be seen, therefore, that a further requirement for maximum sensitivity of the system is that the effective light source, its image, and the knife-edge must be perpendicular to the index of refraction gradient being considered. An optical method for adjusting the light source and image orientation will be described later.

[*Editor's Notes:* Material has been omitted at this point.]

If a horizontal slit source is desired, the lamp house is so rotated in a horizontal plane that the axis of the lamp is perpendicular to the axis of the cone of light through which it illuminates the first mirror. The prism is again adjusted so that it directs the light to the first mirror; thus an effectively horizontal light source is obtained. This type of manipulation is necessary because the H6 lamps cannot be burned in a vertical position. If this is done, all of the mercury tends to collect in the bottom end of the tube, and the electrode in the upper end disintegrates. (This statement applies only to a.c. operation of the lamp. If d.c. is used, the lamp may be burned vertically provided the lower electrode is always positive.)

The effective distance of the light source from the first mirror is very critical. As has already been pointed out, in order to have the beam through the field made up of light as nearly parallel as possible with this type of system, the effective distance of the light source from the first mirror must be exactly the focal length of the mirror. This adjustment is obtained by slight fore and aft movement of the lamp house until the light which falls on the second mirror just reaches its outer edges all the way around. Since the two mirrors are of the same diameter, this condition shows that the source is at the focal point of the first mirror. After this has been done, the knife-edge is then positioned at the image focus at the other end of the system.

An investigation has been made of the effect upon the resolving power of the system produced as a result of using various distances between the first mirror, the object, and the second mirror. From the standpoint of classical theory, an object placed in any part of the parallel-light region between the two mirrors will be in focus upon the viewing screen at all times. However, when refraction of the rays takes place in the striae, these rays are no longer parallel to the optical axis between the two mirrors, and they must be considered as skew rays. For this reason the striation region must be placed at a distance greater than the focal length of the second mirror away from that mirror. Otherwise the projector lens or camera lens will not resolve a sharply defined image of the striation.

Theoretically, there should be no upper limit to the possible distance between the first and second mirrors. However, there is a very definite lower limit. It has been shown that an object in the region where the parallel beam and either cone of light overlap will produce a double image in the optical system. It is immediately apparent, therefore, that the object must be sufficiently far from the first mirror to avoid any interference in the cone of light which illuminates that mirror, and, as just shown, it must be farther from the second mirror than the knife-edge is.

In practice, using this system in which the matched mirrors are of eight-foot focal length, a spacing of eighteen feet between mirrors has been found to be convenient for most applications. However, there is one factor which may make it desirable at times to try to work with a somewhat shorter distance. That factor is the background disturbance which will appear in the picture as a result of the circulation of normal air currents around the room.

It must be remembered that any and all disturbances of the air in the entire region between the light source and the knife-edge will produce their own schlieren images on the screen, especially where the system is being used at peak sensitivity. If maximum sensitivity must be used, it then becomes important to control, and to reduce to an absolute minimum, turbulent air conditions in the vicinity of the apparatus. Often this will mean that, once the set-up is completed and ready to be photographed, all ventilators must be stopped, doors closed, and the operator seated quietly at his control box for several minutes before actually taking the picture. However, where strong disturbances, such as high pressure jets are being studied, the effect of background can be practically eliminated by somewhat increasing the amount of cut-off at the knife-edge, and thus reducing the sensitivity of the system.

If one places a viewing screen between the jet, or other object being studied, and the second mirror, the image of the jet as seen on the screen will show what is commonly referred to as a shadowgraph. Its formation is also due to changes in the refraction of light by conditions within

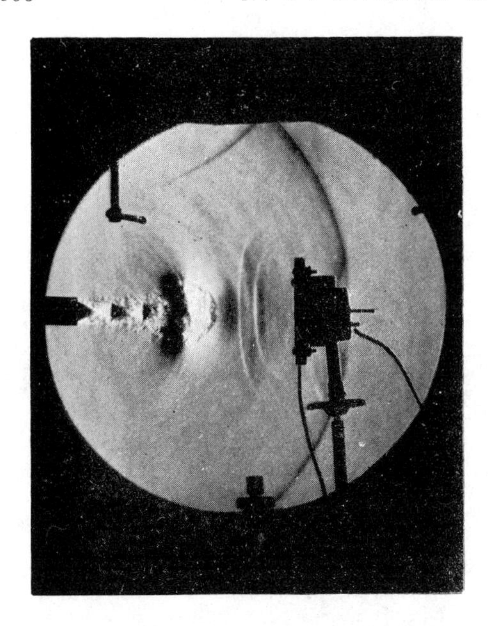

FIG. 6. Schlieren photograph of a jet, showing a sound wave and its reflection from the flash triggering microphone.

the jet. However, there is a distinct difference between a shadowgraph image and a schlieren image.

A shadowgraph shows only the boundaries of sharp wave fronts in the medium, while a schlieren picture shows small gradients as well as wave fronts. If the viewing screen is moved up close to the object, the shadowgraph practically disappears. As the screen is moved farther and farther away, the image appears with ever-increasing contrast upon the screen. The light, as it passes through the boundary layers of the sharp wave fronts, is subjected to refraction very much as it would be if it were to pass through the edge of a very thin-walled glass cylinder, at an angle perpendicular to the cylinder axis and tangent to the cylinder. As it goes through, it is slightly deflected, so that, directly beyond the cylinder, the image shows a decrease in light intensity immediately adjacent to an increase in light intensity. The longer the optical lever arm, the wider this contrasting band becomes, and accordingly, the greater the sensitivity of the system. However, a point is soon reached beyond which the resolution of the image rapidly deteriorates. Thus, a compromise position is chosen in the making of shadowgraphs. The position is generally the point nearest to the object which will give adequate sensitivity.

While work of this type may be carried out in the parallel beam between the two schlieren mirrors, it can be accomplished much more simply and with somewhat better results in another way. All that is required is the object, a spark-gap unit, and a film holder. No lenses are involved. When making the shadowgraphs, some experimentation was required in order to arrive at the distance ratios which would give the most satisfactory results. In most instances, the spark gap was placed about fifteen feet from the jet on one side of, and perpendicular to its axis, while the film was held at a distance of eighteen inches to the other side of, and parallel to, the jet axis.

There is an important reason for using a spark gap in the place of the H6 lamp; namely, circuit characteristics of present equipment make possible a considerably shorter exposure with the gap than is possible with the lamp. The duration of the spark gap exposure is about one-fifth of a microsecond, while the duration of the lamp's flash is about twenty times as long. If the lamp were masked down to the size of the spark gap, there would be ample light to secure a good shadowgraphic exposure, but it has developed in practice that shadowgraphs made with the spark gap produce more favorable results.

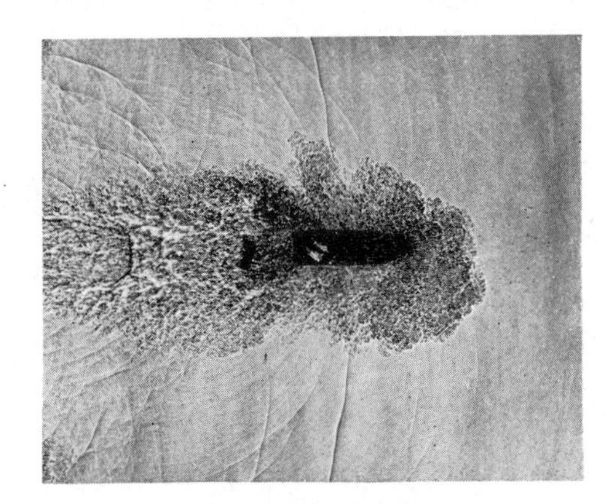

FIG. 7. Shadowgraph of a bullet being discharged from a gun. The many curved lines in the picture are sound waves generated when the compressed gases expand from the muzzle. The bullet is centered in the air that it pushes out of the barrel by its piston action. Expanding turbulent gas behind the bullet gives it its acceleration.

If the study being made with this equipment involves transient conditions, it becomes necessary to synchronize the tripping of the spark gap with the particular point in the transient condition which it is desired to photograph. If the rate of change of the transient condition is very rapid, then this synchronization becomes a problem in electronic controls. Properly worked out, however, the equipment may be made very reliable, and can be made to show conditions at almost any previously chosen time in a transient condition lasting only a thousandth of a second or even less. The picture in Fig. 6 illustrates the method of allowing a sound wave to fall upon a microphone, thereby producing the impulse which triggers the flash.

Figure 7 is a shadowgraph of a bullet being discharged from a gun. Ordinary process films and contrast developers were used in making the shadowgraphs. Even if the film is placed at a distance of 20 feet from the spark gap, very good photographic exposures can be obtained, although the duration of the exposure time, as in this case, is only a fraction of a millionth of a second.

In many types of studies, shadowgraphs will make available practically all of the required information. If this is so, then the simplicity of the technique makes it highly preferable to the use of schlieren equipment. However, if the last possible detail of information regarding conditions and gradients is needed, then the shadowgraph is greatly lacking, and the use of schlieren equipment becomes almost mandatory.

The following bibliography covering the schlieren and shadowgraph methods is probably not entirely complete, some references being necessarily omitted because of military restrictions. However, it is hoped that this bibliography will afford a means of tying together the many loose ends of this rapidly expanding field. The authors wish to take this opportunity to thank Henriette Davidge for her excellent work in preparing this bibliography.

BIBLIOGRAPHY

Schlieren and Shadowgraph Method of Airflow Analysis

(1) J. Ackert, *Handbuch der Physik* (Verlagsbuchhandlung, Julius Springer, Berlin, 1927), Vol. VII, Pp. 308–342.

(2) H. S. Allen, Proc. Roy. Phil. Soc., Glasgow **33**, **71** (1902).

(3) *Annual Report of the National Physical Laboratory for 1933* (H. M. Stationery Office, London).

(4) F. Auerback and W. Hort, *Handbuch d. physik. u. techn. Mechanike* (Johann Ambrosius Barth, Leipzig, 1928), Vol. 5, p. 486.

(5) L. Bairstow, *Applied Aerodynamics* (Longmans Green, London, 1939).

(6) H. Bénard, Comptes rendus **147**, 970 (1908).

(7) G. B. Brown, Phil. Mag. **13**, 161 (1932).

(8) G. Burniston Brown, Proc. Phys. Soc. **47**, 703 (1935).

(9) W. Bucerius, Betriebsführung **11**, 33 (1932).

(10) W. Bucerius, Schornsteinfeger und Technik **57**, 65 (1932).

(11) A. Busemann, Luftwissen **9**, 173 (1942).

(12) A. Busemann, Z.V.D.I. **84**, 857 (1940).

(13) A. Busemann, Verh. d. III. Intern. Kongr. f. techn. Mech., Stockholm **I**, 282–86 (1930).

(14) C. Cranz, *Anwendung der elektrischen Momentphotographie auf die Untersuchung von Schusswaffen.* (Knapp, Halle, 1901).

(15) C. Cranz, *Lehrbuch der Ballistik* (Verlagsbuchhandlung, J. Springer; Berlin, 1927), second edition, Vol. 3, pp. 259, 264.

(16) C. Cranz and E. Bames, Zeits. f. angew. Chemie **36**, 76 (1923).

(17) C. Cranz and B. Glatzel, Ann. d. Physik **43**, 1186 (1914).

(18) C. Cranz, P. A. Günther, and F. Külp, Zeits. f. d. ges. Schiess- u. Sprengstoffwesen **9**, 61 (1914).

(19) C. Cranz, and P. A. Günther, Zeits. f. d. ges. Schiess- u. Sprengstoffwesen **7**, 317 (1912).

(20) C. Cranz and R. Kock, Ann. d. Physik **3**, 247 (1900).

(21) C. Cranz and F. Külp, Artillerist. M. **88**, 251 (1914).

(22) C. Cranz and H. Schardin, Zeits. f. Physik **56**, 147 (1929).

(23) S. Czapski, Zeits. f. Instrumentenk. **6**, 139 (1886).

(24) V. Dvorak, Ann. d. Physik **9**, 502 (1880).

(25) V. Dvorak, Zeits. f. physik. chem. Unterricht **21**, 17 (1908).

(26) E. Eckert and W. Weise, V.D.I. Forschung (Nov.-Dec. 1942).

(27) R. Emden, Ann. d. Physik **69**, 264 (1899).

(28) R. Emden, Wiedemanns Ann. **69**, 264, 426 (1899).

(29) S. Erk, Z.V.D.I. **77**, 1119 (1933).

(30) S. Exner, Repert. d. Physik **21**, 555 (1886).

(31) A. Ferri, Tech. Memo. No. 901, National Advisory Comm. for Aeronautics (July 1939).

(32) Léon Foucault, Ann. de l'observatoire imp. de Paris **5**, 197 (1859).

(33) Léon Foucault, *Recueil des travaux scientifiques* (Paris, 1878).

(34) Garside, Hall, and Townend, Nature **152**, 748 (1943).

(35) D. B. Gawthrop, Rev. Sci. Inst. **2**, 522 (1931).

(36) D. B. Gawthrop, J. Frank. Inst. **214**, 647 (1932).

(37) D. B. Gawthrop, W. C. F. Shepherd, and G. St. Perrolt, J. Frank. Inst. **211**, 67 (1931).

(38) C. W. M. Gregor, Metals and Alloys **4**, 19 (1933).

(39) C. W. M. Gregor, World Power **19**, 75 (1933).
(40) R. Hermann, Luffahrtforschung **19**, 201 (1942).
(41) R. H. Humphrey and R. S. Jane, Trans. Faraday Soc. **2**, 420 (1926).
(42) K. Kieser, Phot. Ind. **28**, 729 (1930).
(43) K. Kieser, Zeits. f. angew. Chemie **43**, 587 (1930).
(44) H. Klug, Ann. d. Physik **11**, 53 (1931).
(45) F. Kohlrausch, *Lehrbuch der praktischen Physik* (B. G. Teubner; Leipzig and Berlin, 1930), sixth edition, p. 304.
(46) A. Lafay, Comptes rendus **CL11**, 694 (1911).
(47) A. Lafay, Tech. Aéron **3**, 169 (1911).
(48) A. Lafay, Tech. Aéron **4**, 91 (1911).
(49) B. Lewis and G. VonElbe, *Combustion Flames and Explosions of Gases* (The University Press, Cambridge, England, 1938), pp. 149–155.
(50) W. Lindner, V.D.I. Forschungsheft No. 326 (V.D.I. Verlag, Berlin, 1930).
(51) F. Luft, Photo-Woche **2**, 24 (1933).
(52) E. Mach, Ann. d. Physik u. Chemie **41**, 140 (1890).
(53) E. Mach, Pogg. Ann. **159**, 330 (1876).
(54) E. Mach, Wiener Berichte **95**, 764 (1887).
(55) E. Mach, Wiener Berichte **98**, 41, 1303, 1310, 1327, and 1333 (1889).
(56) E. Mach and G. Gruss, Wiener Berichte **78**, 467 (1878).
(57) H. Mach, Forschung auf dem Gebiete des Ingen. **14**, 77 (1943).
(58) L. Mach, Zeits. f. Luftschiffahrt u. Physik der Atmosphäre **15**, 129 (1896).
(59) L. Mach, Wiener Berichte **105**, 605 (1896).
(60) L. Mach, Wiener Berichte **106**, 1025 (1897).
(61) H. Malz and Conrady, Autog. Metallbearb. **25**, 117 (1932).
(62) Hermann Malz, *Die Grenzen der Schneidegeschwindigkeit beim Brennschneiden. Diss.* (Berlin, 1932).
(63) A. Nadai, Proc. Am. Soc. Test. Mat. **31**, 2 (1931).
(64) *National Physical Laboratory Report for the Year 1935*, p. 179–180.
(65) R. A. Nelson, Phys. Rev. **23**, 94 (1924).
(66) A. V. Obermayer, Zeits. f. Luftschiffahrt u. Physik der Atmosphäre **15**, 120 (1896).
(67) A. V. Obermayer, Mitt. Gegenstände des Artillerie und Genie-Wesens 815 (1897).
(68) W. Oschatz, V.D.I. Forschung **11**, 296 (1940).
(69) W. Payman and H. Robenson, Safety in Mines Research Board Paper No. 18 and 29 (1926).
(70) W. Payman and Titman, Proc. Roy. Soc. **A152**, 418 (1935).
(71) W. Payman and D. B. Woodhead, Proc. Roy. Soc. Inst. **2**, 522 (1931).
(72) J. St. L. Philpot, Nature **141**, 283 (1938).
(73) S. Ch. Pramanik, Proc. Indian. Ass. Cult. Sci. **7**, 115 (1922).
(74) L. Prandtl, Physik. Zeits. **8**, 23 (1907).
(75) Proc. Roy. Soc. **A145** (1934).
(76) B. Ray, Proc. Ind. Assoc. Cult. Sci. **6**, 95 (1920).
(77) Reports and Memos of British Aero Research Committee No. 1349.
(78) Reports and Memos of British Aero Research Committee No. 1454.
(79) Reports and Memos of British Aero Research Committee No. 1767.
(80) Reports and Memos of British Aero Research Committee No. 1803.
(81) Rieckeherr, Kriegstechn. Zeits. **3**, 383, 439, 513 (1900).
(82) A. M. Rothrock and R. C. Spencer, National Advisory Committee of Aeronautics 24th Annual Report 213 (1938).
(83) A. M. Rothrock and R. C. Spencer, National Advisory Committee of Aeronautics 25th Annual Report 313 (1939).
(84) H. Schardin, V.D.I. Forschungsheft No. 367 (V.D.I. Verlag, Berlin, 1934).
(85) H. Schardin, Heizung und Lüftung 167 (1932).
(86) E. Schmidt, Forsch. Gebiete des Ingen **3**, 181 (1932).
(87) E. Schmidt, Z.V.D.I. **74**, 1487 (1930).
(88) P. Schrott, Kinotechn. **12**, 40 (1930).
(89) W. C. F. Shepherd, Bur. Mines Bull. No. 345 (1932).
(90) A. Stodola, *Dampf-und Gas-Turbinen* (Verlagsbuchhandlung, J. Springer, Berlin, 1922), 98–112.
(91) R. Straubel, in Winkelmann *Handbuch der Physik* (Johann Ambrosius Barth, Leipzig, 1906), Vol. VI, p. 485.
(92) T. Suhara *et al.*, Report of the Aero Research Institute Tokyo Imperial University **50**, 187 (1930).
(93) Harry Svensson, Kolloid Zeits. **87**, 181 (1939). Harry Svensson, Kolloid Zeits. **90**, 141 (1940).
(94) A. Tanakadate, Comptes rendus **151**, 211 (1910).
(95) E. P. Tawil, Comptes rendus **191**, 92, 998 (1930).
(96) H. G. Taylor and J. M. Waldram, J. Sci. Inst. **10**, 378 (1933).
(97) H. G. Taylor and J. M. Waldram, J. Sci. Inst. **11**, 31 (1934).
(98) Technical Memo 1066 N.A.C.A. (June, 1944).
(99) Terazawa, Yamazaki and Akishino, Tokyo Report of Aeronautical Research Institute **1**, 213 (1924).
(100) O. Tietjens, in W. Wien and F. Harms *Handbuch der Experimentalphysik* (Leipzig, 1931), Vol. I, p. 695.
(101) Arne Tiselius, Trans. Faraday Soc. **33**, 524 (1937).
(102) A. Toepler, Ann. d. Physik u. Chemie **131**, 33–35 (1867).
(103) A. Toepler, Arch. d. Artillerie- und Ingenieurwesens 485 (1887).
(104) A. Toepler, *Beobachtungen nach einer neuen optischen Methode* (Bonn, 1864).
(105) A. Toepler, "Beobachtungen nach einer neuen optischen Methode," *Ostwalds Klassiker der exakten Wissenschaften* No. 157 (Leipzig, 1906).
(106) A. Toepler, "Beobachtungen nach der Schlierenmethode," *Ostwalds Klassiker der exakten Wissenschaften* No. 158 (Leipzig, 1906).
(107) A. Toepler, Pogg. Ann. **127**, 556 (1866).
(108) A. Toepler, Pogg. Ann. **128**, 126 (1866).
(109) A. Toepler, Pogg. Ann. **131**, 33 and 180 (1867).
(110) A. Toepler, Pogg. Ann. **134**, 194 (1868).

(111) M. Toepler, Ann. d. Physik **27**, 1043 (1908).
(112) Townend, Aeronautical Research Committee, Reports and Memoranda No. 1349.
(113) H. C. H. Townend, J. Ae. Sci. **3**, 343 (1936).
(114) H. C. H. Townend, J. Sci. Inst. **11**, 184 (1934).
(115) H. C. H. Townend, Phil. Mag. **14**, 700 (1932).
(116) Townend, Reports and Memos of British Aero. Research Committee No. 1614 (1933).
(117) Townend, Reports and Memos of British Aero. Research Committee No. 1634 (1934).
(118) Vielle, Mém. des Poudres et Salpêtres **10**, 177 (1899).
(119) R. Wachsmuth, Ann. d. Physik **14**, 469 (1904).
(120) Weinhold, Zeits. f. physik. chemie Unterr. **21**, 281 (1908).
(121) O. Wiener, Wiedemanns Ann. **49**, 105 (1893).
(122) Melchior Wierz, *Beiträge zur Theorie der Lichtbahnen und Wellenflächen in heterogenen isotropen Medien. Diss.* (Rostock, 1901).
(123) R. W. Wood, Nature 62 (August 9, 1900).
(124) R. W. Wood, Phil. Mag. **48**, 218 (1899).
(125) R. W. Wood, Phil. Mag. **50**, 148 (1900).
(126) R. W. Wood, Phil. Mag. **52**, 589 (1901).
(127) R. W. Wood, *Physical Optics* (The Macmillan Company, London, 1905), p. 94.

22

Reprinted from page 180 of *Collected Papers on Acoustics,* Dover Publications, New York, 1964, 299 p.

THEATRE ACOUSTICS

W. C. Sabine

[*Editor's Note:* In the original, material precedes and follows this excerpt.]

The immediate problem is the discussion of the reflections from the ceiling, from the side walls near the stage, from the screen and parapet in front of the first row of boxes and from the wall at the rear of these boxes. To illustrate this I have taken photographs of the actual sound and its echoes passing through a model of the

FIG. 14. Photograph of a sound-wave, *WW*, entering a model of the New Theatre, and of the echoes a_1, produced by the orchestra screen, a_2 from the main floor, a_3, from the floor of the orchestra pit, a_4, the reflection from the orchestra screen of the wave a_3, a_5 the wave originating at the edge of the stage.

theatre by a modification of what may be called the Toeppler-Boys-Foley method of photographing air disturbances. The details of the adaptation of the method to the present investigation will be explained in another paper. It is sufficient here to say that the method consists essentially of taking off the sides of the model, and, as the sound is passing through it, illuminating it instantaneously by the light from a very fine and somewhat distant electric spark. After passing through the model the light falls on a photographic plate placed at a little distance on the other side. The light is refracted by the sound-waves, which thus act practically as their own lens in producing the photograph.

23

Reprinted by permission of Doubleday & Company, Inc., from pages 49 and 50 of *Sound Waves and Light Waves,* Doubleday, New York, 1965

WAVE PROPAGATION

W. E. Kock

[*Editor's Note:* In the original, material precedes this excerpt.]

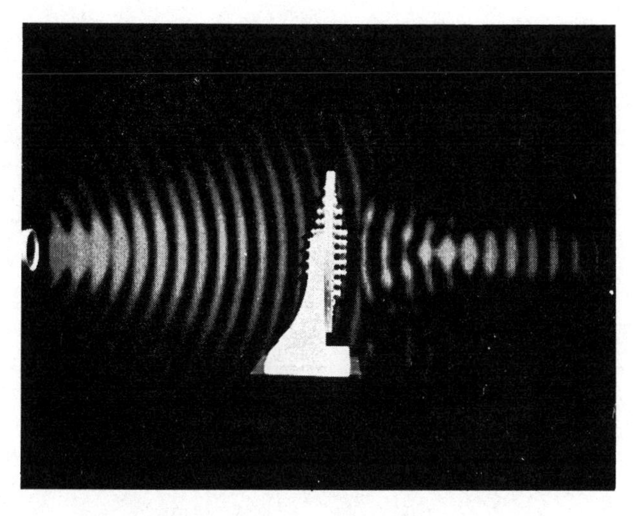

FIG. 34. Circularly diverging sound waves issuing from the horn at the left are converted, by the acoustic lens, into circularly converging waves at the right.

A lens can be employed also to cause spreading energy to concentrate again into a second focal area. The lens then has two focal points. Figure 34 shows this situation with the wave field sampled and portrayed on both sides of the lens. The energy is seen spreading out from the horn at the left; the circular lines now visible in this figure (they were not visible in Figure 33) are evidence of the circular spreading waves. After the waves pass through the lens the wave fronts become concave inward, and a concentration of wave energy is seen to the right of the lens. This particular lens is called *double convex* because both its front and back surfaces are curved. The lens in Figure 33 is called plano-convex because one surface is plane and the other is curved (convex).

24

Reprinted from pages 271 and 273 of *Sound,* 3rd ed., Appleton, New York, 1896, 480 p.

SENSITIVE SMOKE-JETS

J. Tyndall

[*Editor's Note:* In the original material precedes this excerpt.]

It is not to the flame, as such, that we owe the extraordinary phenomena which have been just described. Effects substantially the same are obtained when a jet of unignited gas, of carbonic acid, hydrogen, or even air itself, issues from an orifice under proper pressure. None of these gases, however, can be seen in its passage through air, and, therefore, we must associate with them some substance which, while sharing their motions, will reveal them to the eye. The method employed from time to time in this place of rendering aërial vortices visible is well known to many of you. By tapping a membrane which closes the mouth of a large funnel filled with smoke, we obtain beautiful smoke-rings, which reveal the motion of the air. By associating smoke with our gas-jets, in the present instance, we can also trace their course, and, when this is done, the unignited gas proves as sensitive as the flames. The smoke-jets jump, shorten, split into forks, or lengthen into columns, when the proper notes are sounded.

[*Editor's Note:* Material has been omitted at this point.]

In a perfectly still atmosphere these slender smoke-colums rise sometimes to a height of nearly two feet, apparently vanishing into air at the summit. When this is the case, our most sensitive flames fall far behind them in delicacy; and though less striking than the flames, the smoke-wreaths are often more graceful. Not only special words, but every word, and even every syllable, of the foregoing stanza from Spenser, tumbles a really sensitive smoke-jet into confusion. To produce such effects, a perfectly tranquil atmosphere is necessary.

Fig. 139.

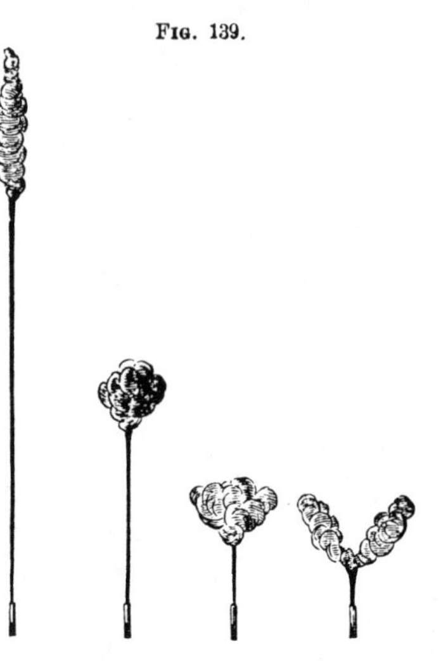

Flame-experiments, in fact, are possible in an atmosphere where smoke-jets are utterly unmanageable.[1]

[1] Referring to these effects, Helmholtz says: "Die erstaunliche Empfindlichkeit eines mit Rauch imprägnirten cylindrischen Luftstrahls gegen Schall ist von Herrn Tyndall beschrieben worden; ich habe dieselbe bestätigt gefunden. Es ist dies offenbar eine Eigenschaft der Trennungsflächen die für das Anblasen der Pfeifen von grösster Wichtigkeit ist."—("Discontinuirliche Luftbewegung," Monatsbericht, April 1868.)

Part IV

FREE-FIELD CALIBRATION METHODS AND THE DEVELOPMENT OF THE RECIPROCITY METHOD

Editor's Comments on Papers 25 Through 33

Paper 25, by S. Ballantine, taking up where Wente et al. left off, starts with the closed-cavity calibration of microphones and proceeds to free-field calibration methods. In his forty-two page article published in 1932, of which we reproduce only seven pages, Ballantine not only gives his own detailed analysis of the thermophone, agreeing nicely with Wente's formula, but also he presents a substitute for a standard sound source—namely, an electrostatically charged grille pulling and pushing on the diaphragm of a Wente-type condenser microphone. Space constraints force the omission of this closed-cavity material. We therefore begin Paper 25 with Ballantine's discussion of free-field measurement techniques and problems. These include the use of the Rayleigh disc, the advantage of embedding the microphone in a rigid sphere, and cavity-resonance effects. (See also Annotated Bibliography.)

Paper 26 by F. M. Wiener goes beyond the calculations of Ballantine and others on the diffraction of sound by various bodies. Wiener presents experimental measurements of the complex sound pressure not only at a point on a sphere but also at a point on a *finite* cylinder, a problem that his predecessors could not solve by calculation. We reproduce six of the original eight pages.

Paper 27, written in 1882, is by Lord Rayleigh on the workings of the Rayleigh disc. This is an instrument for measuring the particle-velocity squared in a wave. Indeed it has been for years the standard instrument for the absolute measurement of the sound intensity. Recall that Rayleigh himself had earlier (Strutt, 1876) used such devices as Tyndall's sensitive flames to detect the diffraction of sound around an obstacle. Hence the Rayleigh disc was a great leap forward in acoustical measurements, carrying the art from a qualitative probe to an unsurpassed quantitative probe. Rayleigh mentions that the invention was born accidentally in 1880 during his work on the absolute measurement of the ohm. Paper 28 is an extract from Rayleigh and Schuster's paper of 1881, "The Determination of the Ohm in Absolute Measure," containing the description by Rayleigh of the accident that led to the birth of the Rayleigh disc. (See also Lamb in the Annotated Bibliography.)

Around the year 1924, the growing radio receiver industry had a great need for a high-power wide-range loudspeaker. In America the most intensive work was being done at General Electric (for example, by Rice and Kellogg), at Westinghouse (for example, by Hanna and Slepian), and at AT&T (for example, by Flanders and Harrison). In Germany the most intensive work was being done at the Siemens Company (for example, by Schottky and Gerlach). One design approach was to use a telephone-type earphone and couple it to a long horn similar to the horns shown in Paper 3 by D. C. Miller, in order to maximize the acoustic power radiated. This was the approach taken by Hanna and Slepian at Westinghouse. In order to measure the frequency response of various horns to be used in acoustic radiation, Slepian invoked the Helmholtz-Rayleigh reciprocity theorem and measured the frequency response of his horns in reception, with the aid of D. C. Miller's phonodeik. This was one of the first applications of acoustic reciprocity since Rayleigh introduced it fifty years earlier.

In Paper 29, from the discussion section of the 1924 paper by Hanna and Slepian, Slepian discusses his reciprocity measurement of horns (see also H. B. Miller, 1977). Paper 29 is a one-page extract from the 1924 paper. The full paper is given in *Acoustic Transducers* (Groves, 1981).

Meanwhile, at the Siemens Company in Germany, W. Schottky, who was making measurements on the frequency responses of microphones and loudspeakers, discovered an interesting relationship between the two responses. An English translation of his forty-eight-page 1926 paper exists, but the translation is not satisfactory. It is not included in this book but can be obtained from the Crerar Library in Chicago. We will refer to his discovery as the "Schottky reciprocity law."

Paper 30 consists of a two-page extract from Ballantine's twenty-three-page 1929 paper. In this paper, which discusses first antennas and then electroacoustic transducers, Ballantine points out that the overall transmission curve of a loudspeaker working into a microphone can be made to yield the absolute frequency response of either one alone. The extract, Paper 30, is so similar to Schottky's work and so foreshadowing that of MacLean, that the wonder now is that Ballantine missed being the inventor of the absolute reciprocity calibration method. (See H. B. Miller, 1977.) Moreover, in his 1932 paper (Paper 25), Ballantine omitted all reference to the method, thereby presumably implying that he felt it had no future.

Paper 31 consists of a one-page extract from W. R. MacLean's seven-page reciprocity paper published in July 1940. The extract contains MacLean's famous formula as Equation 20. The complete paper is reprinted in Albers (1972). MacLean dealt with both free-field reciprocity and closed-cavity reciprocity.

Paper 32 is the complete reciprocity paper of R. K. Cook, originally delivered orally before the Acoustical Society of America in April 1940. Cook points out that he made use of Lippmann's reciprocity principle, which in turn indirectly derives from Rayleigh's reciprocity principle. Cook makes no mention of Schottky, who was not concerned with closed-cavity measurements, just as Cook was not concerned with free-field measurements. Cook's Equation 7 is seen to be the counterpart to MacLean's Equation 20. All the greater credit, surely, belongs to Cook and to MacLean for having the courage to challenge the great prestige of Ballantine.

To complete this selection and correlate the material contained, an original paper by the volume editor is included (Paper 33). Here, he discusses various reciprocities and their relationship.

REFERENCES

Albers, V. M., 1972, *Underwater Sound,* Dowden, Hutchinson & Ross, Stroudsburg, Pa., pp. 366–372.

Groves, I., ed., 1981, *Acoustic Transducers,* Hutchinson Ross Publishing Company, Stroudsburg, Pa., pp. 155–166.

Miller, H. B., 1977, Acoustical Measurements and Instrumentation, *Acoust. Soc. Am. J.* **61**:274–282.

Strutt, J. W. (Lord Rayleigh), 1876, On the Application of the Principle of Reciprocity to Acoustics, *R. Soc. London Proc.* **25**:118–122. (Paper 36 in *Acoustics: Historical and Philosophical Developments,* R. B. Lindsay, ed., Dowden, Hutchinson & Ross, Stroudsburg, Pa., pp. 402–406. See especially page 404.)

ANNOTATED BIBLIOGRAPHY

Ballantine, S., 1928, Note on the Effect of Reflection by the Microphone in Sound Measurements, *Inst. Radio Eng. Proc.* **16**:1639–1644.

In this paper Ballantine anticipated the age of impulse-response measurements by stressing the importance of compensating for the phase distortion created by the diffraction of sound around a microphone. In addition, he pointed out that by virtue of the (Helmholtz-Rayleigh) reciprocity theorem, the radiation patterns from a small loudspeaker embedded in a rigid sphere will exhibit the same perturbation from diffraction as the patterns for a microphone embedded in the same sphere. This point is further discussed by Stewart and Lindsay, *Acoustics*, D. Van Nostrand, New York, 1930, pp. 15–20.

Lamb, H., 1945, *Hydrodynamics*, 6th ed., Dover Publications, New York.

See especially Chapter IV, article 71, page 86 for a good discussion of a flat body setting itself broadside to a stream. Rayleigh realized that the behavior of the flat body toward a d.c. flow would be unchanged for an a.c. flow. For further discussion of this point, see W. West, *Acoustical Engineering*, Pitman, London, 1932, pp. 198–214. See especially page 199 where it is explicitly pointed out that the direction of the torque is the same if the flow is reversed.

25

Reprinted from pages 319–320, 341–342, 343–345, and 347 of *Acoust. Soc. Am. J.* 3:319–360 (1932)

TECHNIQUE OF MICROPHONE CALIBRATION

By STUART BALLANTINE
Boonton Research Corporation

The usual purpose of microphone calibration is to determine the relation between the open-circuit voltage generated by the microphone and the pressure at the surface of its diaphragm or the pressure in a sound field. For technical purposes and with microphones having inaccessible diaphragms it is usually sufficient to compare the unknown microphone with a standard microphone; in other cases an absolute calibration may be desired. A variety of methods of absolute calibration are available but no attempt shall be made here to consider all of them. A detailed discussion of the technique of several methods which have been found convenient in practice may be presumed to be of more value to practicing engineers in view of the scarcity of publications dealing with the manipulative side of the subject.

Particular reference shall be made to the condenser microphone of the stretched diaphragm type.

1. Calibration in Terms of Diaphragm Pressure and Pressure in a Progressive Sound Wave.—Until recently, microphone calibrations have been exclusively performed by applying known alternating pressures to the diaphragm. A suitable technique, employing the thermophone and pistonphone, was worked out by Arnold, Crandall, and Wente and this early work made possible and gave impetus to a number of important quantitative investigations in acoustics and hearing which have since been made.

For the majority of uses to which the microphone is put, however, this type of calibration is of questionable significance because the pressure at the diaphragm may not be the same as the undisturbed pressure in the sound field under observation. This was discovered experimentally a few years ago as the result of an attempt to check the thermophone calibration by means of the Rayleigh disk and the discrepancy was attributed in part to the effect of diffraction around the microphone whereby the pressure on the diaphragm was increased over that existing in the undisturbed sound wave.[1] Employing a spherical microphone

[1]Stuart Ballantine: "Effect of Diffraction Around the Microphone in Sound Measurements"; *Phys. Rev., 32,* 988, 1928; *Proc. Inst. Radio Engs., 16,* 1639, 1928.

mounting for which this effect could be computed, more careful comparisons with the Rayleigh disk revealed a residual discrepancy which is ascribed to acoustical resonance in the recess in front of the diaphragm.[2] The possibility of the diffractive effect was recognized by Crandall, who attempted unsuccessfully to evaluate it; both effects were suspected by Aldridge[3] and Barnes,[4] and the cavity resonance was independently discovered by West,[5] Oliver[6] and Hartmann.[7]

These effects are of considerable importance; for example, in the measurement of the pressure in a plane progressive sound wave their combined action leads to errors as high as 300 per cent at frequencies of the order of 3,000 cycles. The extent to which important older investigations in acoustics and hearing have been prejudiced by these effects in condenser microphones calibrated by means of the thermophone can best be judged by those who have been responsible for them and who are familiar with the details of the technique employed. At least some review of those results would seem desirable for the purpose of applying the proper corrections where they are found to be necessary.

An instructive physical picture of the diffractive effect may be obtained by considering the simple case of a microphone of spherical form for which the pressure increase can be calculated theoretically. The pressure ratio varies from unity at low frequencies to 2 at high frequencies. The calculated variation for spheres of various diameters are reproduced in Fig. 12.

Resonance in the cavity in front of the diaphragm produces an additional increase of pressure which also varies with the frequency. In a microphone having a $1\frac{1}{2}'' \times \frac{1}{2}''$ cavity the pressure increase amounts to about $2\times$ at 3,000 cycles.

In view of these effects it seems desirable to supplement the usual calibration of the microphone in terms of pressures applied to the diaphragm by such additional tests as will establish the relation of the response to the actual pressure in an undisturbed wave field.

[2] Stuart Ballantine: "Effect of Cavity Resonance on the Frequency Response Characteristic of the Condenser Microphone" *Contributions from Radio Frequency Laboratories, No. 18*, April 15, 1930; *Proc. I.R.E.*, *18*, 1206, 1930.

[3] A. J. Aldridge: *Jour. Post Office Elec. Engs.*, *21*, 223, 1928.

[4] E. J. Barnes: *Proc. Wireless Section, I.E.E.*, *3*, No. 7, 59, 1928.

[5] W. West: "Pressure on the Diaphragm of a Condenser Transmitter": *Proc. Inst. Elec. Engs.*, **5**, 145, June 1930.

[6] D. A. Oliver: "An Improved Condenser Microphone for Sound Pressure Measurements"; *Jour. Scientific Inst.*, **7**, 113, 1930.

[7] C. A. Hartmann: *Elek. Nach. Tech.*, **7**, 104, 1930.

[*Editor's Note:* Material has been omitted at this point.]

The rotation of the disk is observed optically by means of a lamp and scale. Observations are made outside the chamber by means of a telescope mounted in the door.

To avoid the necessity for opening the door during a frequency run the source shown is sometimes replaced by a small horn 2″ in diameter, driven by a moving-coil and specially damped so as to give a

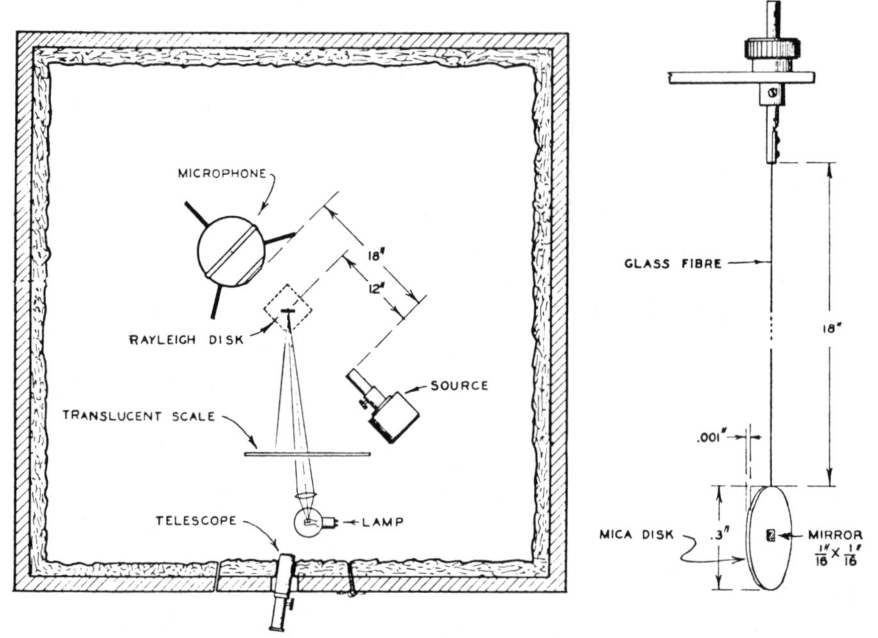

FIG. 11. (*a*) *Arrangement of apparatus for wave calibration by means of the Rayleigh disk;* (*b*) *dimensions of typical Rayleigh disk system.*

substantially uniform output over the frequency range employed. Errors arising from the proximity of the Rayleigh disk to a source of finite size have been discussed by West.[13]

18. Rayleigh Disk.—This is a light circular disk suspended by means of a delicate fibre so as to rotate about a vertical axis against the restoring torque of the fibre. If a flat object of this sort be suspended in the path of a current of air it tends to turn itself so that its plane is normal to the direction of the air stream. This tendency persists if the direction of air flow be reversed so that, as first suggested by Lord

[13] W. West: *Jour. Inst. Elec. Eng.*, **67**, 1140, September 1929.

Rayleigh,[14] a steady deflection will be obtained in a periodic air current and the device may be used to measure the particle velocity in a sound wave.

The steady torque so produced by a circular disk in an inviscid fluid of sinusoidally varying velocity has been calculated by Koenig. For a disk of negligible thickness:

$$\text{Torque} = \frac{1}{6}\rho d^3 v^2 \sin 2\theta \tag{20}$$

where ρ = density of the gas
d = diameter of disk
v = rms velocity of gas
θ = angle between the direction of velocity and the normal to the disk.

Maximum sensitivity is obtained when $\theta = 45$ and the disk is generally so adjusted. The steady deflection against the torque of the fibre is thus roughly proportional to v^2.

References on the Rayleigh disk technique are given below.[15] An exceptionally valuable article on the manipulative aspects of the subject has been published by E. J. Barnes and W. West (*Jour. Inst. Elec. Engs.*, 65, 871, September, 1927). The following practical suggestions on the construction and use of the disk may be of interest.

[*Editor's Note:* Material has been omitted at this point.]

19. Increase of Pressure by Diffraction.—If the microphone diaphragm were of infinite extent, or part of an infinite wall, the pressure at the diaphragm due to an incident sound wave would be doubled at all frequencies. If, on the other hand, the dimensions of the microphone were small in comparison with the wavelength of the sound, the pressure at any point of its surface would tend to approach that in the undisturbed

[14] Rayleigh: *Phil. Mag.*, **14,** 186, 1882.

[15] Mallett and Dutton: *Jour. Inst. Elec. Eng.*, **63,** 502, 1925. Geiger and Scheel: "Handbuch der Physik," Vol. 8, Akustik, p. 573 (Berlin: 1927).

sound field. In the actual case, in which neither of these conditions obtains, the pressure increase due to diffraction varies from unity at low frequencies to 2 at high frequencies.

The nature of this variation may be brought out by considering the case of a spherical obstacle for which the increase of pressure can be calculated theoretically. The pressure ratio for a sphere of diameter a and a plane progressive wave of the type $Exp\ i\omega(t-r/c)$ may be expressed in terms of Bessel functions whose order is half an odd integer as follows:

$$\frac{P}{P_0}=\sqrt{\frac{2}{\pi ka}}e^{-ika}\sum_{n=0}^{\infty}\frac{-(2n+1)i^nP_n(\cos\phi)}{(-)^n(nJ_{-n-1/2}+kaJ_{-n-3/2})-i(nJ_{n+1/2}-kaJ_{n+3/2})} \qquad (23)$$

where the arguments of the Bessel functions are ka, $k=2\pi f/c$, c is the velocity of sound, and ϕ is the angle between the wave-normal and the axis of the sphere. Numerical calculations for $\phi=0$ over a large range of ka have been performed and tabulated in an earlier paper (Reference 1). The pressure ratios for spheres of various diameters are shown in Fig. 12, as a function of the frequency.

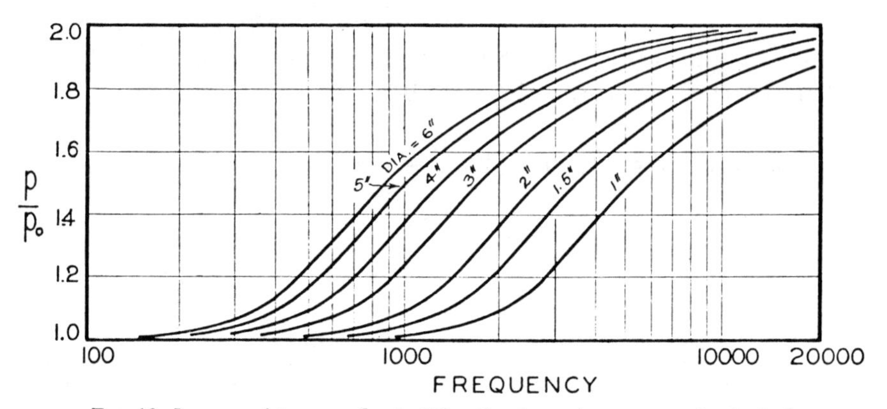

FIG. 12. *Increase of pressure due to diffraction for a plane wave and spherical microphone housings of various diameters.*

In order to render the diffraction effect definite and calculable the microphone may be built into a spherical housing. The microphone may be designed so that the diaphragm is substantially a part of the spherical surface. In this way cavity resonance is eliminated and the pressure ratio is due only to diffraction. A standard microphone of this type is shown in Fig. 13.

The pressure-ratios as experimentally measured by means of the

Rayleigh disk for a group of microphones (Fig. 13) are shown in Fig. 14. The curve for the spherical microphone agrees pretty well with the

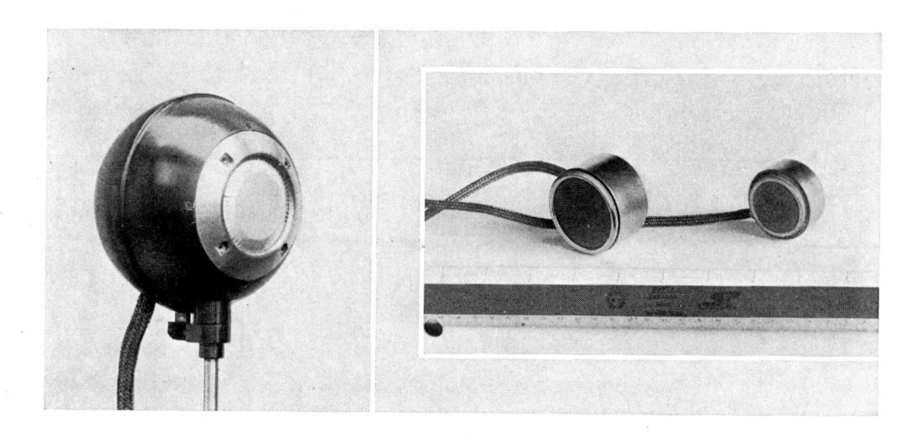

Fig. 13. *Group of microphones with flush diaphragms; (a) standard spherical condenser microphone; (b) and (c) electromagnetic microphones of small size.*

theoretical curve (dotted). The solid angle subtended by the diaphragm was approximately 35° and the diameter of the spherical mounting was

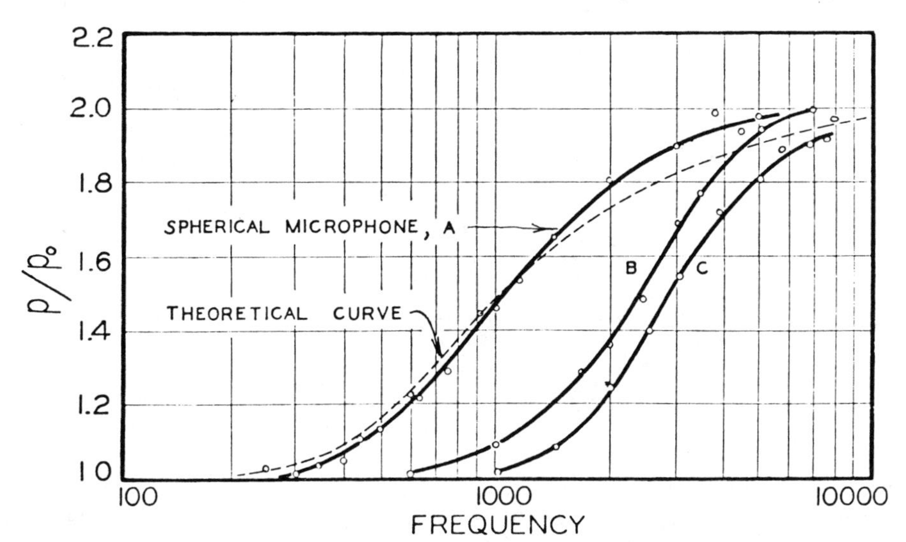

Fig. 14. *Experimental values of pressure ratios for group of microphones illustrated in Fig. 13.*

5 inches. The microphones shown at the right of Fig. 14 (B and C) are of the electromagnetic type and are comparatively small in size, ...

[*Editor's Note:* Material has been omitted at this point.]

Pressure-ratio curves as determined from the pressure and wave calibrations for two specimen microphones are shown in Fig. 15. The contours and dimensions of the microphones are given in the lower part of the figure. Microphone A was of the spherical type with recessed di-

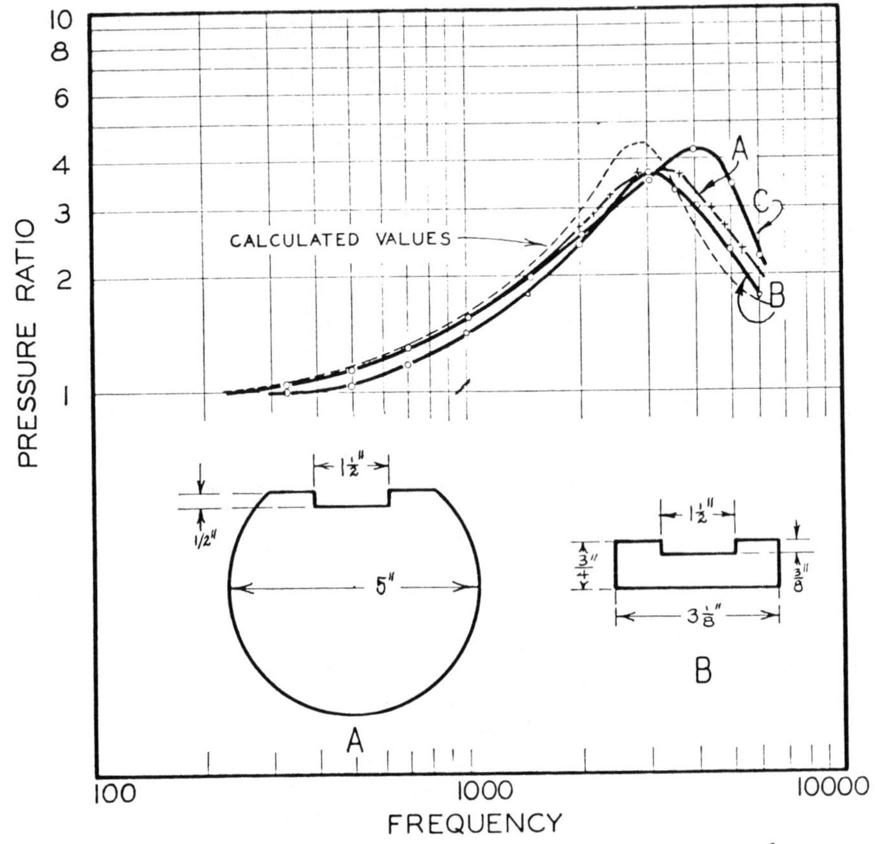

FIG. 15. *Pressure ratios for several condenser microphones having recessed diaphragms showing effects of cavity resonance.*

aphragm. The dotted curve represents the theoretical values calculated from (23) and (24). The rise of pressure is seen to amount to about 4×, or 300 percent. Microphone B was of more conventional form and was measured without the diaphragm protecting grille and amplifier housing. The dimensions of microphone B were about the same as those of several types sold commercially, and the pressure curve shown at B, Fig. 15, is in good agreement with that published by West . . . .

[*Editor's Note:* In the original, material follows this excerpt.]

26

Reprinted from pages 444–449 of *Acoust. Soc. Am. J.* **19**:444–451 (1947)

Sound Diffraction by Rigid Spheres and Circular Cylinders*

FRANCIS M. WIENER†

The results of calculations of the pressure distribution on the surface of a stationary rigid sphere and a stationary rigid circular cylinder of infinite length, when exposed to a plane progressive sound wave, are compared with experiment. A small probe microphone was used to measure the sound pressures on the surface of the obstacles in a room essentially free from acoustic wall reflections under a variety of experimental conditions. The sound pressures p on the surface are conveniently expressed relative to the free-field pressure p_0 in the undisturbed incident wave.

In the case of the sphere, reasonably good agreement was obtained between theory and experiment in the range of $\frac{1}{3} < ka < 10$, where k is the wave number of the incident wave and a the radius of the obstacle. In particular, the existence of the "bright" spot diametrically opposite the point nearest the sound source was verified experimentally. This comparison of experiment with theory affords a valuable means of estimating the validity of the experimental procedure.

In the case of the cylinder whose axis was oriented parallel to the wave front of the incident wave and whose length was chosen to equal its diameter, the pressures were measured for the most part in the median plane. There is a marked similarity between the results obtained for this finite cylinder and the results obtained for a sphere. The pressure at the point in the median plane farthest away from the source of sound ($\theta = 180°$) is substantially equal to the free-field pressure only for $ka < 5$. As the frequency is increased further, the sound pressure near $\theta = 180°$ drops rapidly up to and most likely beyond $ka = 10$ at an approximately constant rate, in decibels per frequency octave. A small microphone located on the surface of such a cylinder at $\theta = 180°$ will exhibit a low pass filter action with a cut-off at $k \approx 5$.

Tables are furnished of the ratio p/p_0 computed for a rigid circular cylinder of infinite length. This ratio is tabulated in magnitude and phase in the range of $\frac{1}{2} \leq ka \leq 10$, for azimuths θ varying from 0 to 180° in 10-degree steps. The results are very similar to the spherical case for θ not exceeding about 150 degrees. There is only an insignificant trace of the bright spot at $\theta = 180°$.

I. INTRODUCTION

THE disturbance of a plane sound wave due to obstacles whose dimensions are comparable with the wave-length has long been of interest to workers in both theoretical and applied acoustics. Corresponding problems have been under investigation in electromagnetic theory. Rigorous solutions for obstacles whose dimensions are all finite, and which are valid for all frequencies, have been obtained only for the sphere.[1] Attempts have been made to calculate the diffraction effect produced by finite obstacles of various shapes[2,3] at certain points on the surface by approximate methods. Rigorous solutions are also available for certain cases which involve obstacles of semi-infinite or infinite extent.[2,4–6] A common characteristic of all available solutions is the fact that numerical calculations become lengthy and tedious, because of slow convergence of the series involved, in the frequency range which is of greatest interest in acoustics.

Recently tables[7] have been prepared dealing with diffraction and radiation by spheres and circular cylinders of infinite length with a variety of boundary conditions admissible at the surface of the obstacles. They greatly facilitate the numerical computations.

In view of the difficulties attending the rigorous theoretical treatment of obstacles of finite dimensions other than spheres, an essentially experimental approach was adopted. The results of measurements on a sphere and a circular cylinder are compared with the values obtained from theoretical treatment of the sphere and the

* This research was carried out under contract with the Office of Naval Research (Contract N5ori-76, Report PNR-25).

† Now at Bell Telephone Laboratories, Murray Hill New Jersey.

[1] Lord Rayleigh, *Theory of Sound* (MacMillan and Company, London, 1929). Section 328; also *Scientific Papers*, Vol. 5, p. 149.

[2] L. J. Sivian and H. T. O'Neil, J. Acous. Soc. Am. **3**, 483 (1932).

[3] G. G. Muller, R. Black, and T. E. Davis, J. Acous. Soc. Am. **10**, 6 (1938).

[4] A. Sommerfeld, Math. Ann. **47**, 317 (1895).

[5] H. Lamb, Proc. London Math. Soc. **4**, 190 (1906).

[6] H. Lamb, *Hydrodynamics* (Dover Publications, New York, 1945), p. 529.

[7] Scattering and Radiation from Circular Cylinders and Spheres, prepared by A. N. Cowan, P. M. Morse, H. Feshbach, and M. Lax, Massachusetts Institute of Technology, 1945. (Applied Mathematics Panel, Report No. 62.1R.)

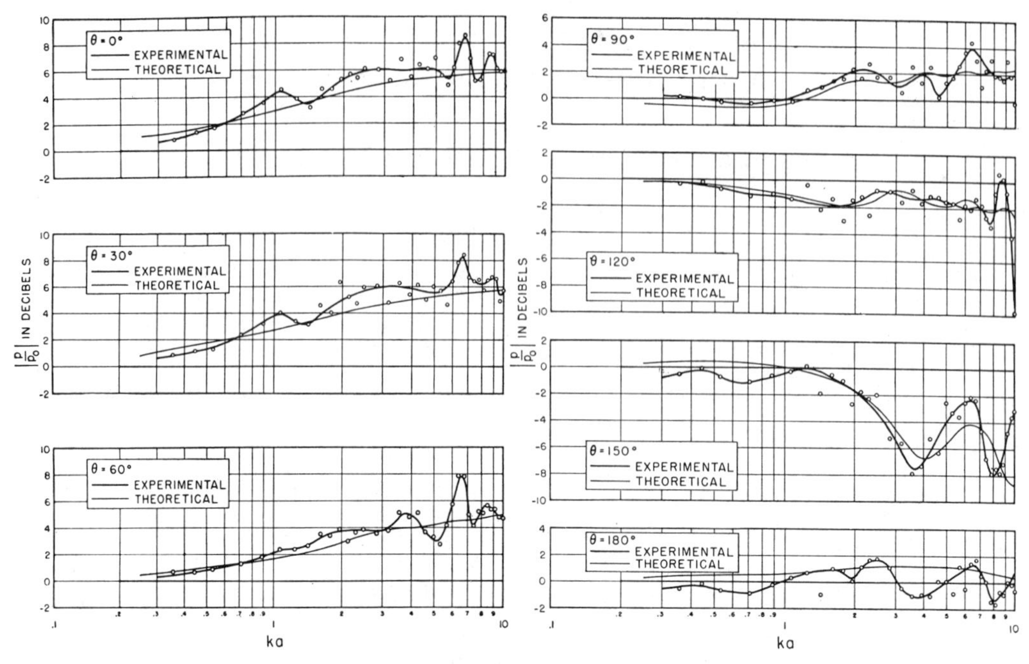

FIG. 1. Showing the ratio of the sound pressure p to the free-field pressure p_0 on the surface of a rigid sphere in a plane wave of wave number k. The magnitude of this ratio is plotted, in decibels, as a function of ka ($a=9.7$ cm). For comparison, the theoretical values are also shown.

circular cylinder of infinite length, a rigorous solution for a cylinder of finite length not being available at present. Tables are furnished for the infinite cylinder showing the numerical solution in magnitude and phase.

II. EXPERIMENTAL TECHNIQUE

The obstacles, embodied by carefully machined and smoothly finished wooden models, were placed in front of a sound source in a anechoic (echo-free) chamber. The experiments were, in part, carried out in the large chamber described by Beranek and Sleeper,[8] where W. E. types 750-A and 751-B loudspeakers were used. The free sound field in the vicinity of the location of the obstacle (6 feet distant from the source), was approximately that of a plane progressive wave. The variations of sound pressure inside a spherical region somewhat larger than the obstacle did not exceed ±2 db at any frequency in the range of 200 to 6000 c.p.s. The obstacle size was chosen so that a range of ka between about $\frac{1}{3}$ and 10 was covered. Measurements were also made in a smaller chamber at the Psycho-Acoustic Laboratory.[8] The distance between obstacle and sound source was reduced to about $4\frac{1}{2}$ feet and the variation of the free-field pressure with position near the location of the obstacle was somewhat less favorable (±3 db).

The sound pressures near the surface of the bodies and in the free field were determined, in magnitude, by means of a small probe microphone. The probe consisted of a small tube, about 4 inches in length, which was coupled to a W. E. type 640-AA condenser microphone. The effective area of the instrument was that of a circle less than 0.1 cm in diameter.[9] Symmetry about the line joining the points $\theta=0$ (point nearest the sound source) and $\theta=180$ degrees was assumed throughout and the measurements were confined to one side of the obstacle. Measurements for a number of corresponding points on the other side of the obstacle justified the

[8] L. L. Beranek and H. P. Sleeper, Jr., J. Acous. Soc. Am. **18**, 140 (1946).

[9] F. M. Wiener and D. A. Ross, J. Acous. Soc. Am. **18**, 401 (1946).

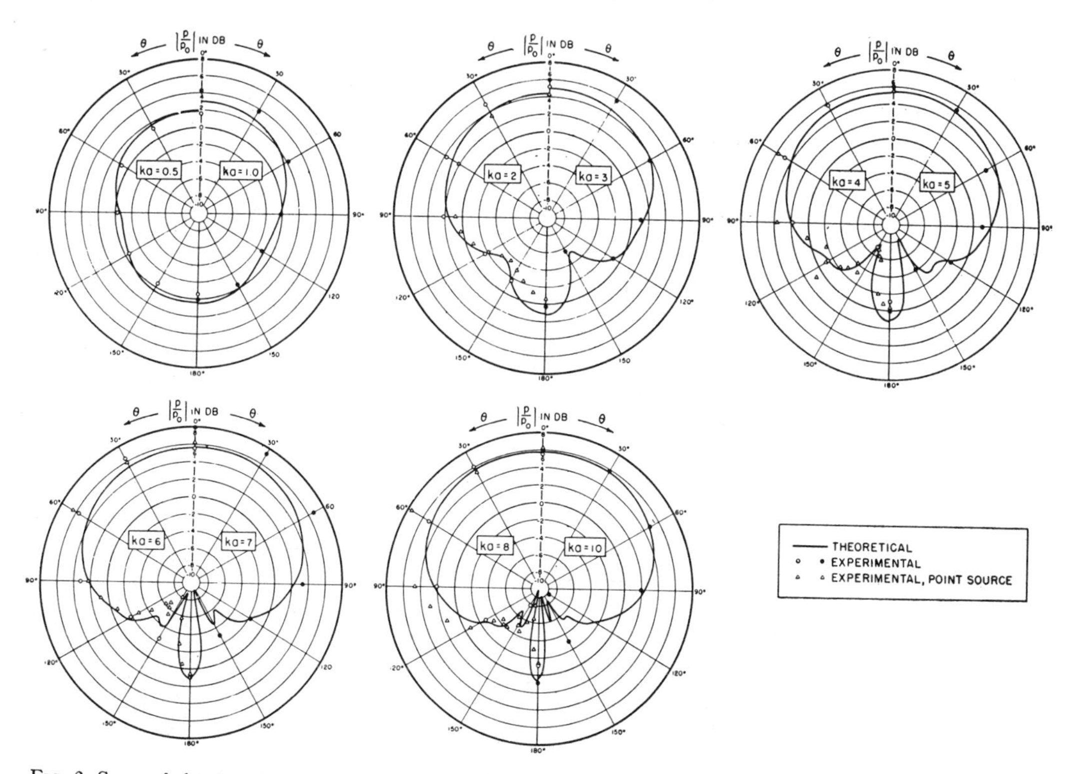

FIG. 2. Some of the data from Fig. 1 are re-plotted in polar form and data obtained with a point source of sound are added. Note that p near $\theta=180°$ is substantially equal to p_0.

assumption of symmetry within the accuracy of the experiments.

Appropriate coordinate systems were ruled on the surface and the microphone was placed by means of adjustable clamps close to the point under observation. The opening of the probe tube was adjusted to be within 0.1 cm of the surface. The sound pressure was then measured as a function of frequency and the procedure repeated for values of θ between 0 and 180 degrees in steps of 30 degrees.

It soon became evident that this procedure was not entirely satisfactory in the region of sound "shadow," where the pressure on the surface varies rapidly with position. In order to trace out the pressure distribution in those regions more satisfactorily, a small sound source whose maximum dimensions were equal to one wave-length at about 6000 c.p.s. ("point" source) was used at selected values of ka lying between 2 and 10. Data were then obtained by keeping the frequency fixed and by measuring the sound pressure for small increments of θ.

Preliminary experiments have shown that the probe microphone is non-directional in the frequency region considered here and is activated only by the sound pressure existing in the immediate vicinity of the opening of the tube. The signal-to-noise ratio is adequate except in certain regions of exceptionally deep pressure minima at high frequencies. In tracing out such minima it was found that in some cases the pressures measured there were, to a certain extent, dependent on the orientation of the probe. This points to the possibility of inter-action between the obstacle and the housing for the condenser microphone and its supporting clamps.

In all cases, the sound pressures p obtained on the surface of the obstacle were expressed relative to the free-field sound pressure p_0 in the incident wave at $\theta=0$. This magnitude is taken to be a measure of the diffraction effect.

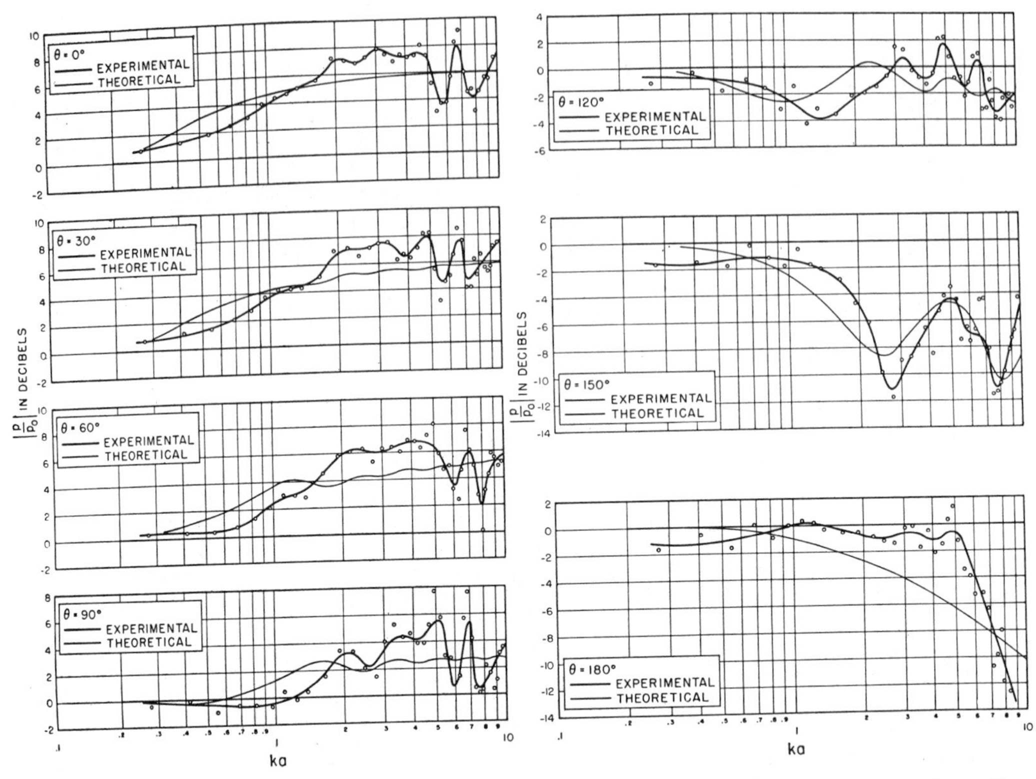

FIG. 3. Showing the pressure distribution on the surface of a rigid circular cylinder for points lying in the median plane perpendicular to the axis. The wave front of the incident plane wave and the axis of the cylinder are parallel. The cylinder has a length equal to its diameter $2a$ ($a=7.5$ cm). Note low pass filter characteristic for $\theta=180°$. For comparison, the theoretical values for a cylinder of infinite length are also shown.

III. RESULTS

1. Rigid Sphere

Figure 1 shows the experimental results for the sphere, compared with the theoretical curves obtained from tables given by Schwarz.[10] In view of the relative experimental difficulties, the agreement is, on the whole, satisfactory. The greater spread of the experimental points at the higher frequencies is, at least in part, attributable to the fact that both loudspeaker and microphone exhibited local irregularities in their respective response characteristics which tend to magnify the effect of even small errors in resetting the oscillator dial. The scatter of the experimental points for a given frequency is generally higher for large values of the angle θ. Residual room patterns and the spurious reflections mentioned above are probably responsible. The curve for the point on the sphere farthest away from the source ($\theta=180°$) shows that the pressures there are substantially equal to the free-field sound pressure over the whole frequency range considered here. This corresponds to the well-known bright spot in optical diffraction. At the point on the sphere closest to the source ($\theta=0$) the pressure at the high frequencies is doubled, whereas a region of sound shadow exists over a large part of the hemisphere which is pointed away from the source of sound.

Figure 2 shows the pressure distribution on the sphere plotted in polar form for selected values of ka. The points designated by circles and dots are taken from the smoothed data of Fig. 1. The data obtained with the point sound source described above are shown by triangles. The two sets of data are fairly consistent and

[10] L. Schwarz, Akust. Zeits. 8, 91 (1943).

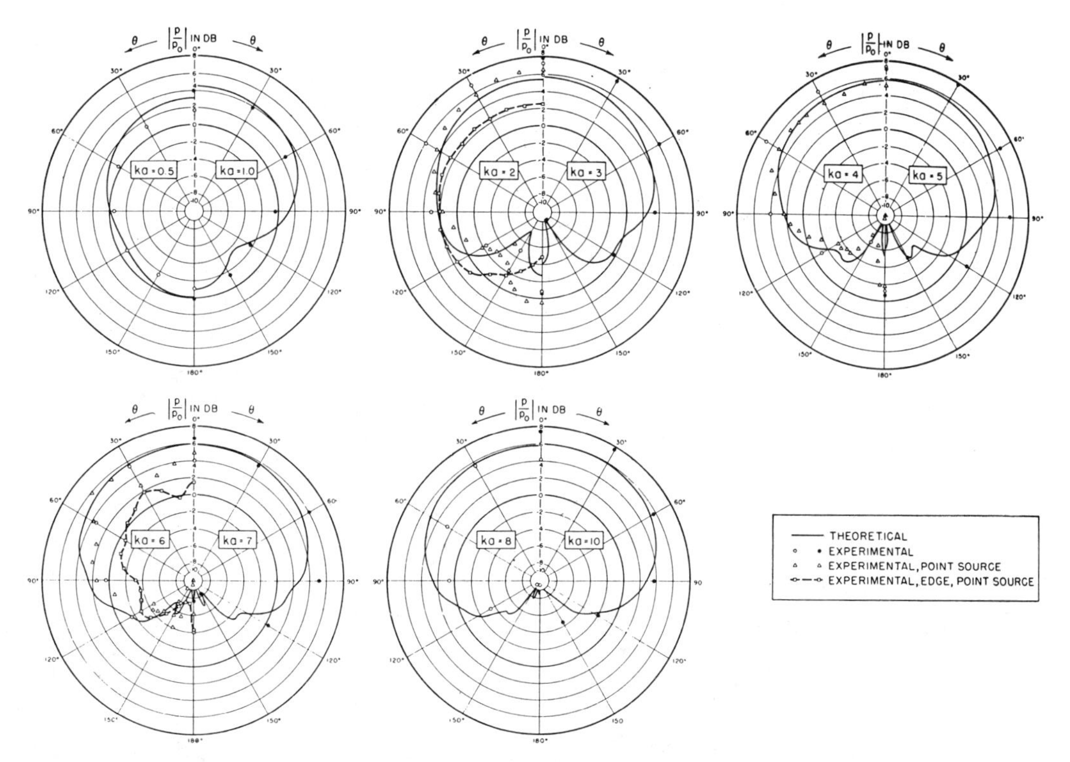

FIG. 4. Some of the data from Fig. 3 are re-plotted in polar form and data obtained with a point source of sound are added. Circles and triangles denote the data for the median plane while squares show the pressures near the edge of the cylinder.

the agreement with theory is shown to be reasonably good. It can be concluded that the results obtained with this experimental arrangement for other obstacles of similar dimensions can also be viewed with confidence as to their validity.

The complete pressure distribution on the sphere due to a plane progressive wave of wave number k is obtained by rotating the directional diagrams of Fig. 2 about the line joining the points $\theta=0$ and $\theta=180$ degrees. By reciprocity, this surface shows the distribution-in-angle of the sound pressure at large distances from the sphere, when a point source of sound is located on the sphere at $\theta=0$.

2. Rigid Circular Cylinder of Finite Length

Figure 3 shows the results of measurements in the median plane of a cylinder whose length was chosen equal to its diameter. Attention should be called to the fact that p/p_0 for $\theta=180$ degrees is close to unity only for values of ka less than about 5. For frequencies beyond this "cut-off," the pressure drops rapidly as the frequency is increased, up to (and probably beyond) $ka=10$ in the manner of a low pass filter. There is a general similarity with the spherical case for $ka<5$. At higher frequencies the differences are most marked in the region of $\theta>90$ degrees.

In Fig. 4 there are plotted the experimental values in polar form for the cylinder, which correspond to Fig. 2 for the sphere. From the smoothed data of Fig. 3 the points designated by circles and dots have been taken. Further data for the median plane obtained with the point source are shown by triangles. As a matter of interest, a limited number of measurements of the pressures near the edge of the cylinder using the point source are included. Squares have been used to designate them. Particularly at high frequencies, the positioning of the probe near the edge becomes critical, because of the high pressure gradients likely to exist in that region.

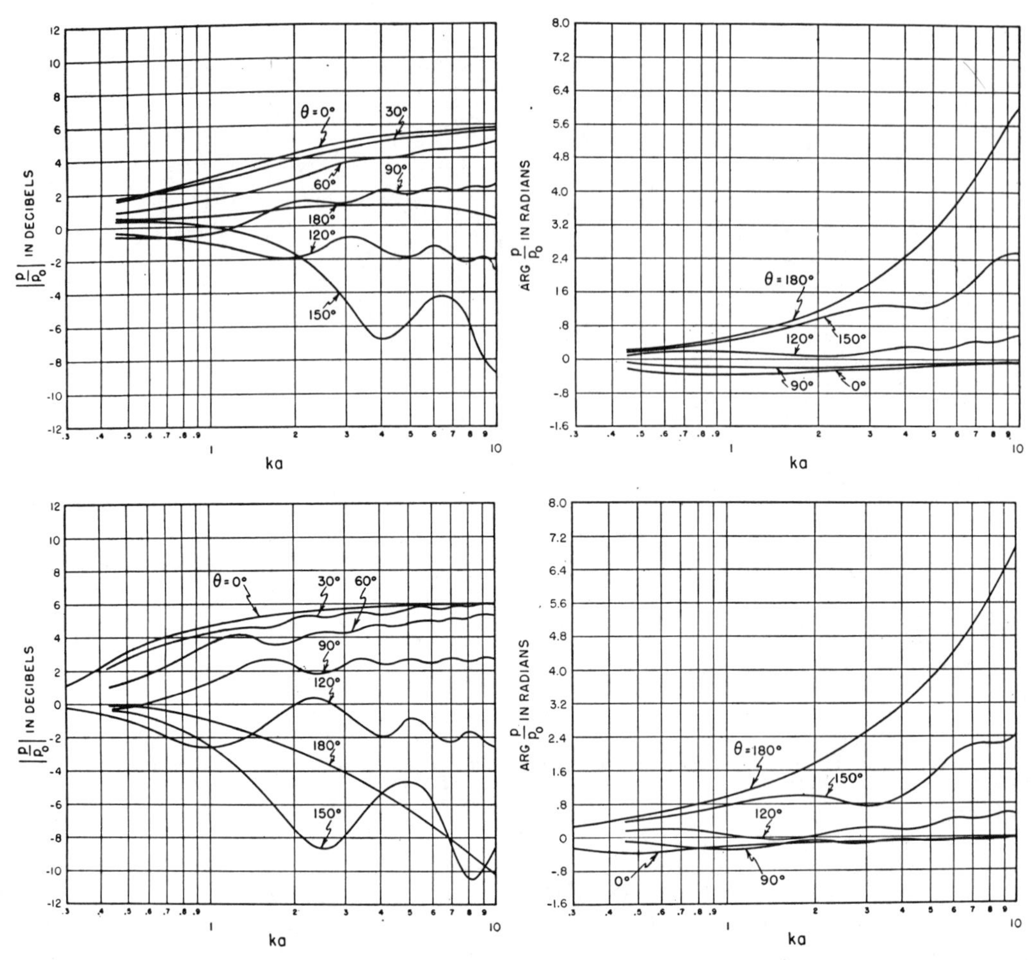

FIG. 5. Comparison of the pressure distributions calculated from theory for a rigid sphere (top) and a rigid circular cylinder of infinite length (bottom). A positive phase angle denotes a time lag of the pressure at the obstacle with respect to the free-field pressure.

It is hoped that the case of a cylinder of finite length will eventually prove amenable to theoretical treatment. The best that can be done at present is to use the values obtained for a cylinder of infinite length. Curves for such a case taken from the tables given below, are shown in Figs. 3 and 4 for comparison. The similarity with the spherical case is surprisingly close with the exception of the region $150<\theta<180$ degrees. There are only insignificant traces of the bright spot formation at $\theta=180$ degrees.

The complete pressure distribution on the surface of the cylinder of infinite length in the field of a plane progressive wave for a given value of ka is obtained by mirroring the computed polar diagram of Fig. 4 about the line joining the points $\theta=0$ and $\theta=180$ degrees and moving this diagram parallel to the cylinder axis. By reciprocity, this surface yields the distribution-in-angle of the sound pressure at large distances from the cylinder due to a line source of sound located on the cylinder at $\theta=0$.

To permit a better comparison between sphere and infinite cylinder the theoretical pressure distributions are plotted in Fig. 5, in magnitude and phase, for selected values of the angle θ.

[*Editor's Note:* In the original, material follows this excerpt.]

27

Reprinted from *Philos. Mag.* **14**:186–187 (1882)

ON AN INSTRUMENT CAPABLE OF MEASURING THE INTENSITY OF AERIAL VIBRATIONS

By LORD RAYLEIGH, *F.R.S.**

THIS instrument arose out of an experiment described in the 'Proceedings of the Cambridge Philosophical Society'†, Nov. 1880, from which it appeared that a light disk, capable of rotation about a vertical diameter, tends with some decision to set itself at right angles to the direction of alternating aerial currents. In fig. 1, A is a brass tube closed at one end with a glass plate B, behind which is a slit C backed by a lamp. D is a light mirror with attached magnets, such as are used for reflecting-galvanometers, and is suspended by a silk fibre. The light from the slit is incident

* Communicated by the Author.
† See also Proc. Roy. Soc. May 5, 1881, p. 110.

upon the mirror at an angle of 45°, and, after reflection, escapes from the tube through a glass window at E. It then

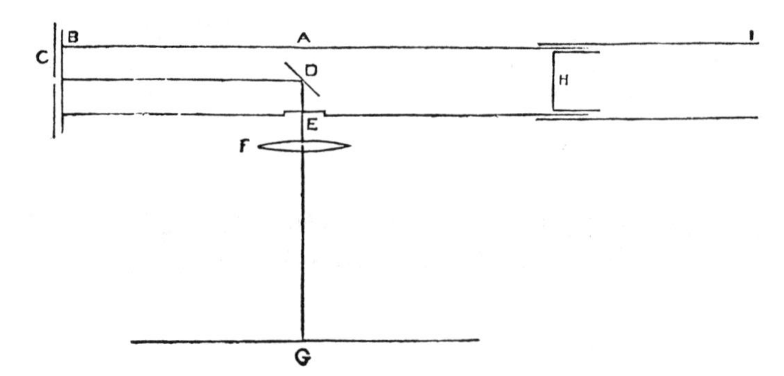

falls upon a lens F, and throws an image of the slit upon a scale G. At a distance DH, equal to DC, the tube is closed by a diaphragm of tissue paper, beyond which it is acoustically prolonged by a sliding tube I.

When the instrument is exposed to sounds whose half wave-length is equal to C H, H becomes a node of the stationary vibrations, and the paper diaphragm offers but little impediment. Its office is to screen the suspended parts from accidental currents of air. At D there is a loop; and the mirror tends to set itself at right angles to the tube under the influence of the vibratory motion. This tendency is opposed by the magnetic forces; but the image upon the scale shifts its position through a distance proportional to the intensity of the action.

As in galvanometers, increased sensitiveness may be obtained by compensating the earth's magnetic force with an external magnet. Inasmuch, however, as the effect to be measured is not magnetic, it is better to obtain a small force of restitution by diminishing the moment of the suspended magnet, rather than by diminishing the intensity of the field in which it works. In this way the zero will be less liable to be affected by accidental magnetic disturbances.

So far as I have tested it hitherto, the performance of the instrument is satisfactory. What strikes one most in its use is the enormous disproportion that it reveals between sounds which, when heard consecutively, appear to be of the same order of magnitude.

June, 1882.

28

Reprinted from pages 109–110 of *R. Soc. London Proc.* **32**:104–141 (1881)

ON THE DETERMINATION OF THE OHM (B. A. UNIT) IN ABSOLUTE MEASURE

Lord Rayleigh and A. Schuster

[*Editor's Note:* In the original, material precedes and follows this excerpt.]

The first matter for examination was the behaviour of the magnet and mirror when the coil was spinning with circuit open. At low speeds the result was fairly satisfactory, but at six or more revolutions per second a violent disturbance set in. This could not be attributed to the direct action of wind, as the case surrounding the suspended parts was nearly air-tight, except at the top. It was noticed by Mr. Darwin that even at low speeds a disturbance was caused at every stroke of the bell. This observation pointed to mechanical tremor, communicated through the frame, as the cause of the difficulty, and the next step was to support the case surrounding the suspended parts independently. A rough trial indicated some improvement, but at this point the experiments had to be laid aside for a time.

From the fact that the disturbance in question was produced by the slightest touch (as by a tap of the finger nail), upon the box, while the upper parts of the tube could be shaken with impunity, it appeared that it must depend upon a reaction between the air included in the box and the mirror. It is known that a flat body tends to set itself across the direction of any steady current of the fluid in which it is immersed, and we may fairly suppose that an effect of the same character will follow from an alternating current. At the moment of the tap upon the box the air inside is made to move past the mirror, and probably executes several vibrations. While these vibrations last, the mirror is subject to a twisting force tending to set it at right angles to the direction of vibration. The whole action being over in a time very small compared with that of the free vibrations of the magnet and mirror, the observed effect is as if an impulse had been given to the suspended parts.

In order to illustrate this effect I contrived the following experiment.* A small disk of paper, about the size of a sixpence, was hung by a fine silk fibre across the mouth of a resonator of pitch 128. When a sound of this pitch is excited in the neighbourhood, there is a powerful rush of air in and out of the resonator, and the disk sets itself promptly across the passage. A fork of pitch 128 may be held near the resonator, but it is better to use a second resonator at a little distance in order to avoid any possible disturbance due to the neighbourhood of the vibrating prongs. The experiment, though rather less striking, was also successful with forks and resonators of pitch 256.

* "Proc. Camb. Phil. Soc.," Nov. 8, 1880.

29

Reprinted with permission from page 410 of *Am. Inst. Electr. Eng. Trans.* **43**:393–411 (1924)

THE FUNCTION AND DESIGN OF HORNS FOR LOUD SPEAKERS

C. R. Hanna and J. Slepian

[*Editor's Note:* In the original, material precedes and follows this excerpt.]

This being a gathering of electrical men, I shall give the electrical form for this relation; first, this is known as the reciprocal relation for electrical networks. Let A and B, Fig. 1A, be two pairs of terminals joined by a network built up out of simple impedances in any manner. Then if one ampere is made to flow into the terminals A, the same voltage will be produced at B as would be produced at A if one ampere is caused to flow at B. It is very interesting to check this relation for a few simple cases.

The acoustic analog of this relation is then as follows: In a given space containing bounding walls of any description, a source of sound of given intensity at a point A, Fig. 1B, will produce the same pressure on a diaphragm at a second point B, as would be produced on a diaphragm at A if the source of sound were moved to B. A more precise statement of this proposition may be found in Rayleigh's Theory of Sound, Vol. II, p. 131.

It follows then that on sounding organ pipes of different frequencies, but with the same intensitybefore a horn, the curve of pressures developed on an attached diaphragm will be the same as the curve of acoustic loading which the horn would impose on the diaphragm if used as a loud speaker. For an exponential horn, we should then expect to reproduce the curve, Fig. 8, of Hanna and Slepian's paper.

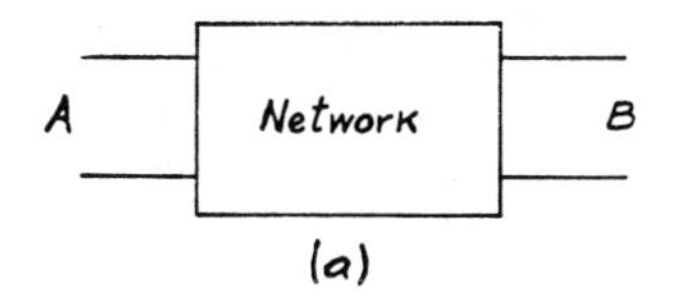

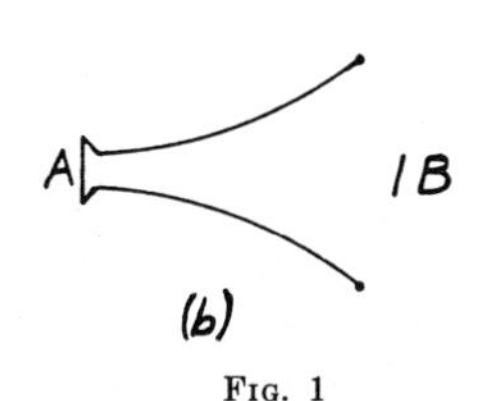

FIG. 1

However, Prof. Miller did not measure the pressure developed at the diaphragm, but the motion resulting from this pressure. We should therefore consider what motion would be produced in the diaphragm by a constant force of varying frequency. The motion of the diaphragm will be determined by its mass, stiffness and damping. The influence of the horn, when it is effective, will be to increase the damping. An important point brought out in Hanna and Slepian's paper is that the loading of a horn for the frequencies for which it is effective, is the same as that which would be produced by an infinite straight tube having the same section as the throat of the horn.

Mr. Hanna has calculated what the motion of the diaphragm would be under a constant force when loaded by an infinite straight tube of the same section as the throat of the horn. His results are shown by the smooth curves in Figs. 2 and 3.

Before discussing these curves, I would like to call attention for comparison to Fig. 10 of Martin and Fletcher's paper. Curve A of that figure is said to be the output curve of one of the best commercial types of loud speakers. The response begins to fall off at about 800 cycles and is one tenth of its maximum at 600 cycles. Now the falling off in response of this type of loud speaker at low frequencies is entirely due to the failure of its horn to sufficiently load the diaphragm, for the electromagnetic force for a given current is just as great or greater at low frequencies as for high. That is, when the acoustic loading of the horn is too small, such large amplitudes of vibration of the diaphragm are necessay to produce appreciable sound that most of the magnetic force is used in overcoming the stiffness of the diaphragm, and little is left for doing work on the air. We should expect then that if a better horn were used, one which would load down to lower frequencies, a much better response curve would be obtained.

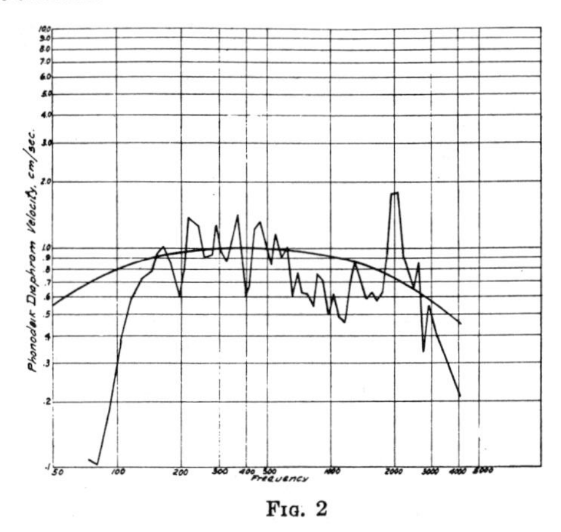

FIG. 2

The curves obtained by Prof. Miller bear this out. Fig. 2 is for a horn whose throat area was 0.81 cm^2. and whose section increased at the rate of 4.5 per cent per cm. of length. We see that it gives the full loading corresponding to an infinite straight tube of the same section down to about 150 cycles, and we must go down to about 100 cycles before its loading effect is reduced to one tenth. This is not only very much better than the curve for the commercial type of loud speaker shown by Messrs. Martin and Fletcher, but even compares well with their laboratory models.

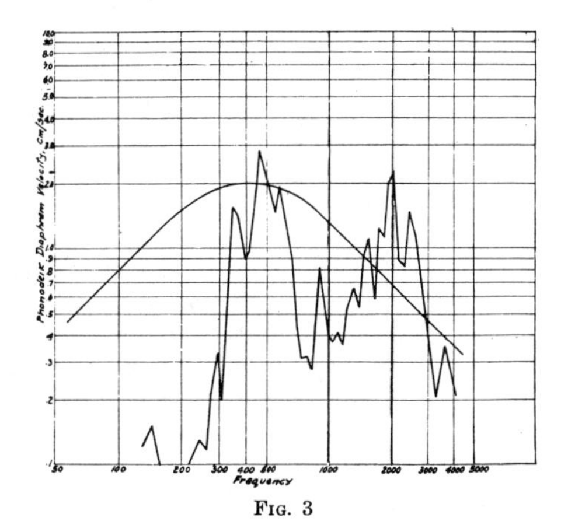

FIG. 3

Fig. 3 is for an exponential horn having three times the throat area, and two times the rate of increase of section as the horn of Fig. 2. According to the theory, the loading of this horn with larger throat should be only one third that of the other horn, and so the resonances of the diaphragm should be more marked. This is seen to actually be the case.

30

Reprinted with permission from pages 929 and 947–948 of *Inst. Radio Eng. Proc.* **17**:929–951 (1929)

RECIPROCITY IN ELECTROMAGNETIC, MECHANICAL, ACOUSTICAL, AND INTERCONNECTED SYSTEMS*

BY

STUART BALLANTINE

(Radio Frequency Laboratories, Inc., Boonton, New Jersey)

Summary—*Recent criticism by Carson of the statement of the reciprocal relations in a radio communication system given by Sommerfeld is supported by a simple example showing the incorrectness of this statement.*

Carson's proof of the extention of Rayleigh's reciprocity theorem to a general electromagnetic system was limited to $\mu = 1$ and to sources consisting of ponderomotive forces on the electricity. A new proof is given under more general conditions, ϵ, μ, σ being merely restricted to be scalars and the impressed forces are of the electric type introduced by Heaviside and Abraham. These may be regarded as impressed charges and currents, including ether displacement current. The theorem is finally stated in terms of volume and surface integrals, and thus combines the viewpoints of both Lorentz and Carson.

The reciprocity relations in a mechanical system are reviewed.

The interconnection of electrical and mechanical systems is next considered and a "transduction coefficient" is defined. This concept is useful in formulating mechanical problems in electric-circuit form. An example of symmetrical transductance (copper coil in steady magnetic field) is given. The subject of units is taken up and it is proposed that in order to bring the mechanical quantities into agreement with the electrical ones when the latter are expressed in "practical" units, the mechanical quantities be expressed in "mechanical-volts," amperes, etc., i.e., in "practical-electric units of the mechanical quantities." A table for converting the principal mechanical quantities from c.g.s. to practical-electrical units is given.

Reciprocity is shown to exist in interconnected electro-mechanical systems with reversible transduction.

The equations of sound propagation in a gas are developed, regarding the velocity and excess pressure as the fundamental quantities. An acoustical reciprocity theorem involving these quantities is then proved.

Interconnections of mechanical and acoustical systems are then discussed and reciprocal relations are shown to exist in the composite system. Such relations are also valid for a system comprising electrical, mechanical and acoustical systems in series connection.

* Dewey decimal classification: R510. Original manuscript received by the Institute, February 26, 1929. Presented before joint meeting of the Institute and the International Union of Scientific Radiotelegraphy, American Section, at Fourth Annual Convention, Washington, D. C., May 15, 1929.

[*Editor's Note:* Material has been omitted at this point.]

10. Example of the Application of the Extended Reciprocity Relations. Frequency Calibration of an Electrophone. The above extended reciprocity relations between two reversible electrophones has an interesting and useful application in a method for determining the frequency response characteristics of an emitting electrophone or a receiving microphone. This may be called the "method of three electrophones" and is analogous to the method of measuring the resistance of a ground by the method of three grounds, or of the effective height of an antenna by using two additional antennas. The frequency characteristic of the electrophone usually desired is that in a certain direction, say normal to the sound-emitting area. (Characteristics in other directions may obviously be obtained in the same way.)

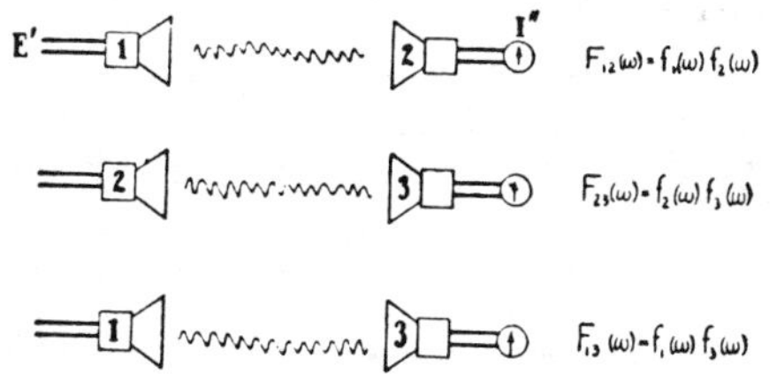

Fig. 9

The electrophones are used two at a time (Fig. 9) and are situated so far apart that the wave-front is substantially plane and the interaction between them is negligible. The purpose of this is to obtain data which will be fundamentally related to plane waves; investigations based on divergent waves may be carried out by proper adjustments of separation. One unit is excited by a generator E' of variable frequency and the current (or equivalent) in the other is measured. Let the frequency characteristics of the individual electrophones with respect to plane waves be denoted by $f_1(\omega)$, $f_2(\omega)$ and $f_3(\omega)$, respectively. Then for transmission from No. 1 to No. 2, for example, the measurement will give $F_{12}(\omega) = f_1(\omega)f_2(\omega)$. The transmission characteristic of No. 2 to No. 3 (F_{23}) is next obtained, and then that from No. 1 to No. 3 (F_{13}). This experimental information suffices to determine f_1, f_2, or f_3. For

$$
\begin{aligned}
F_{12}(\omega) &= f_1(\omega)f_2(\omega) \\
F_{23}(\omega) &= f_2(\omega)f_3(\omega) \qquad (51) \\
F_{13}(\omega) &= f_1(\omega)f_3(\omega),
\end{aligned}
$$

which is a set of simultaneous functional equations to be solved for the f's. The solutions are

$$\begin{aligned} f_1(\omega) &= (F_{12}F_{13}/F_{23})^{1/2} \\ f_2(\omega) &= (F_{12}F_{23}/F_{13})^{1/2} \\ f_3(\omega) &= (F_{13}F_{23}/F_{12})^{1/2}. \end{aligned} \qquad (52)$$

If absolute values for the f's are desired care must be taken to measure properly the F's for each pair. For most purposes, however, all that is required is a curve of the relative variations of $f(\omega)$ with frequency; in this case absolute values of the F's are of no importance.

It may be noted that in the special case of two electrophones having identical frequency characteristics it is merely necessary to measure the frequency-transmission F for this pair since $F_{12}(\omega) = f_{12}(\omega) f_{21}(\omega)$ and $f_1 = (F_{12})^{1/2}$. This match will rarely be sufficiently exact in practical apparatus, except perhaps with precision microphones of the air-damped variety which have been accurately made and adjusted as to membrane tension.

It is also worth noticing that this method of calibration is equivalent to a Rayleigh-disk calibration in that it also takes into account the effect of reflection by the microphone.[11]

As will be seen from Fig. 9 this application of reciprocity requires only one of the electrophones (No. 2) to be *reversible*. Also note that one of them, No. 1, is always used as a source, and that the other, No. 3, is always used as a receiver. Thus a thermophone (irreversible) might be used at No. 1 as a source, with a reversible electrophone No. 2 as a comparator, to calibrate a carbon microphone, No. 3 (irreversible). It is needless to say that in order to obtain fundamental data the work should be performed in the open air and at a sufficient height to avoid disturbance due to reflection from the earth.

[*Editor's Note:* In the original, material follows this excerpt.]

[11] For an account of this effect and of another method for evaluating it, see Stuart Ballantine, *Contributions from the Radio Frequency Laboratories*, No. 9, December 1928; *Phys. Rev.*, 32, 988, 1928; Proc. I. R. E., 16, 1639; December, 1928.

31

Reprinted from pages 143–144 of *Acoust. Soc. Am. J.* **12**:140–146 (1940)

ABSOLUTE MEASUREMENT OF SOUND WITHOUT A PRIMARY STANDARD

W. R. MacLean

[*Editor's Note:* In the original, material precedes this excerpt.]

In the experiment leading to the f.f. calibration of the microphone there enter: (a) the microphone to be calibrated, U, (b) a reversible unit, U_x, and (c) a sound generator, U_o. Since it is desired ultimately to obtain the plane wave f.f. calibration in terms of pressure, and since the curvature of the wave affects the calibration of a velocity microphone at low frequencies, a somewhat more strictly specified experiment is needed in this case than for a pressure microphone. This more restricted experiment will be described since it includes both cases.

The sound generator U_o should be a zero order radiator, i.e., one which behaves as a point source at sufficiently low frequencies. On, in, or near U, U_x, and U_o arbitrary centers are chosen. Similarly, three axes are chosen. Let l be the greatest physical dimension of any of the three devices (the centers are included as part of physical configuration).

U and U_x are now placed one after the other at the same spot in the sound field of U_o. More specifically, U is placed co-axially with U_o, and their centers are separated by a distance d. This distance is chosen so that $l \ll d$. Without changing U_o in any way, U_x is substituted for U. The voltages generated by each, E and E_x, are recorded. If M_o and M_o^x are the o.c. free field calibrations of U and U_x for waves of radius d, then since the sound field was the same in both cases:

$$\frac{E}{M_o}=\frac{E_x}{M_o^x} \quad \text{whence:} \quad \frac{M_o}{M_o^x}=\frac{E}{E_x}. \tag{17}$$

The use in (17) of M (which is for waves of radius d) must be justified, since U_o does not necessarily emit spherical waves at all frequencies. For high frequencies with $\lambda \ll d$, the values for response of U and U_x in plane, spherical, or actual field would be nearly identical since $l \ll d$. For a velocity microphone, the curvature effect becomes apparent for $d \sim \lambda$ or $d \ll \lambda$. But then we have also $l \ll \lambda$ and therefore U_o behaves as a spherical radiator. Consequently the use of M is justified.

U_o is now removed, and the two units placed in juxtaposition with their axes coincident but opposed, and their centers separated distance d. U_x is driven with a current I'. The f.f. pressure at U's center is $I'S_o^x$, where, as before, S_o^x is the calibration of U_x as a speaker. If U now puts out the o.c. voltage E', then:

$$E' = I'S_o^x M_o, \tag{18}$$

where the use of M is similarly justified. But now from (12) we have the value for the ratio of M_o^x to S_o^x. So from (18) and (12):

$$M_o M_o^x = \frac{E'}{I'} \frac{2d\lambda}{r}, \tag{19}$$

which together with (17) gives:

$$M_o = \left(\frac{E'E}{I'E_x} \frac{2d\lambda}{r}\right). \tag{20}$$

(20) is the absolute calibration of the given microphone in o.c. abvolts per f.f. bar for waves of radius d. Quite similar results are obtained for s.c. abamperes per f.f. bar, and analogous results for the calibration of a unit as a speaker, by applying (12) to the above, for instance.

If the microphone being tested is one of the pressure type, M is practically equal to the plane wave value; if the microphone is a velocity type, and the plane wave value is desired, the correction factor $\{1+(c/\omega d)^2\}^{-\frac{1}{2}}$ must be applied.

The Absolute Measurement of Sound—Chamber Pressure

To apply the method to obtain the pressure calibration of a given microphone, there enter the same three devices as before, except that this time they would want to be of such physical size and otherwise so chosen as to be suitable for work in a chamber. Coupling all three to a small chamber and driving U_o, the analog of (17) is obtained:

$$\frac{T_o}{T_o^x} = \frac{E}{E_x}. \tag{17'}$$

[*Editor's Note:* In the original, material follows this excerpt.]

32

Reprinted from *Nat. Bur. Stand. J. Res.* **25**:489–505 (1940)

ABSOLUTE PRESSURE CALIBRATIONS OF MICROPHONES

By Richard K. Cook

ABSTRACT

A tourmaline-crystal disk was used both as a microphone (direct piezoelectric effect) and as a sound source (converse piezoelectric effect). Application of a principle of reciprocity to the acoustic measurements gave an absolute determination of the piezoelectric modulus $d_{33}+2d_{31}$ of tourmaline under hydrostatic pressure. A condenser microphone was calibrated by the tourmaline disk. The same principle was applied to data obtained by using a condenser microphone as both source and microphone to secure an absolute calibration of another condenser microphone. It was proved experimentally that the tourmaline disk and the condenser microphones satisfied the principle of reciprocity. The absolute acoustic determination of the piezoelectric modulus gave $d_{33}+2d_{31}=2.22\times 10^{-17}$ coul/dyne. The "reciprocity" calibrations agreed with the results of electrostatic actuator, pistonphone, and "smoke particle" calibrations, but disagreed with thermophone calibrations of the condenser microphones.

I. INTRODUCTION

Absolute measurements of sound intensity are of basic importance in acoustics. Such measurements are most conveniently made by means of calibrated microphones, and it is important to have accurate methods of making both absolute pressure and absolute free-field calibrations of microphones. Absolute pressure calibrations are needed at low frequencies to supplement free-field Rayleigh-disk measurements at higher frequencies, and are needed to check Rayleigh-disk measurements in the frequency region in which both pressure and free-field calibrations should be the same. Absolute pressure calibrations are also needed in audiometric work.

First calibrations of a condenser microphone made with an electrostatic actuator and with a thermophone, both of conventional design, showed a difference much greater than the probable error of the determination. The difference between the thermophone and the free-field Rayleigh-disk calibrations of a condenser microphone, in the frequency region in which both calibrations should be the same, was also greater than the probable error. It was decided to use piezoelectric crystals to throw some light on the discrepancies between the results obtained with the thermophone, the electrostatic actuator, and the Rayleigh disk.

II. PRINCIPLE OF RECIPROCITY

1. GENERAL

Shortly after the discovery of the direct piezoelectric effect by the brothers P. and J. Curie, M. G. Lippmann predicted the existence of a converse piezoelectric effect, and showed quantitatively how large it should be.

Lippmann's proof, generalized by W. Voigt [1][1], is in reality a special case of a principle of reciprocity. Other special cases of the principle are well known in applied mechanics, electrical engineering, and acoustics. This principle was discussed by Lord Rayleigh [2] for mechanical systems and electrical systems, and was discussed by S. Ballantine [3] for certain mixed mechanical and electrical systems. The same principle of reciprocity can be applied not only to piezoelectric crystals but also to condenser microphones, moving-coil microphones, and to the thermophone. By means of this principle, the output of a transducer (i. e., a device which is used as both sound source and microphone) as a sound source can be computed from its response as a microphone, and vice versa.

Suppose that a pressure sinusoidal in time

$$p=p_0e^{i\omega t}$$

is applied to the diaphragm of a transducer (e. g., a condenser microphone); suppose electric charge

$$q=q_0e^{i\omega t}$$

is released from the transducer. The release of this charge, q, by the pressure, p, is called the direct effect (see fig. 1). It is assumed that the charge released is directly proportional to the pressure. This assumption of linearity is sufficient for the application of the reciprocity principle.

If the proportionality between p and q is given by

$$q=\tau p, \tag{1}$$

the principle of reciprocity asserts that

$$v=\tau u, \tag{2}$$

where τ is numerically the same in both eq 1 and 2, if q, p, v, and u are all in the same system of units, and where

$$v=v_0e^{i\omega t}$$

[1] Figures in brackets indicate the literature references at the end of this paper.

is the change of volume caused by the motion of the diaphragm due to an applied voltage

$$u = u_0 e^{i\omega t}.$$

The volume change, v, due to the voltage, u, is called the converse effect (see fig. 2). There might be differences in phase among the quantities q, p, v, and u. To account for phase differences, the

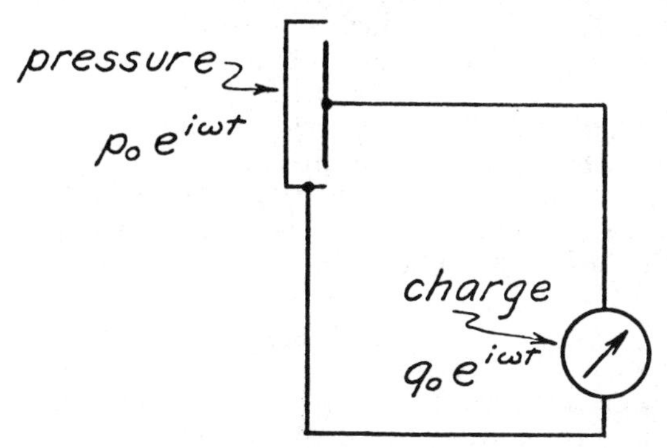

FIGURE 1.—*Direct effect.*
Transducer used as microphone.

coefficients q_0, p_0, v_0, and u_0, and the transduction coefficient, τ, are in general complex numbers, and τ might also be a function of frequency.

The principle of reciprocity applies without regard to the mechanism of operation of the transducer, provided only it is linear. The principle applies even if there are frictional forces or thermal actions in the

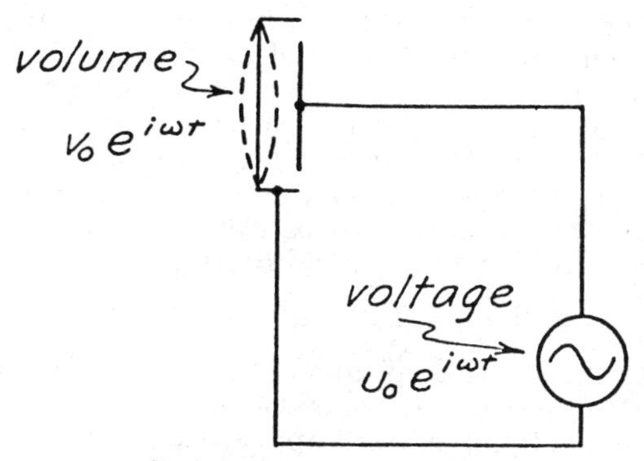

FIGURE 2.—*Converse effect.*
Transducer used as sound source.

transducer, provided these are linear (i. e., $\tau = q/p$ is independent of p). Thus a thermophone without a polarizing current is not linear when used as a sound source and the principle does not apply. Indeed, the thermophone used as a microphone under such circumstances has no response. But with a polarizing current the thermophone becomes a linear sound source (within certain limits), and the reciprocity theorem can be applied within these limits to compute the response as a microphone from the output of the thermophone as a source of sound. Similarly, the principle can be applied to a condenser microphone having a polarizing voltage, but the principle is not applicable to an unpolarized condenser microphone.

Besides the assumption of linearity, it is also assumed that the only force applied in the direct effect is the pressure on the diaphragm (i. e., the electric terminals of the transducer are shorted, so that no electric forces are applied), and that the only force applied in the converse effect is the voltage on the electric terminals (i. e., the diaphragm volume change takes place with no reaction from the gas in front of the diaphragm). It presently will be shown how closely these conditions are realized in the laboratory, and how to take into account departures from the specified conditions.

If p and v are, respectively, in dynes per square centimeter and cubic centimeter and Q and E in coulombs and volts, then eq 1 and 2 become, in this mixed system of units,

$$\left.\begin{aligned} Q &= \tau p \\ v &= 10^7 \tau E \end{aligned}\right\} \tag{3}$$

and τ is in square centimeters coulomb per dyne. These units will be used throughout the remainder of this article. The quantities representing voltage, charge, pressure, and volume changes will henceforth have root-mean-square (rms) values.

2. APPLICATION TO PIEZOELECTRIC CRYSTALS

Tourmaline, a polar crystal of the trigonal system, acquires a uniform electric moment per unit volume under hydrostatic pressure. The magnitude of the moment per cubic centimeter is

$$M = (d_{33} + 2d_{31})p \text{ coul/cm}^2,$$

and its direction is parallel to the principal (optic and piezoelectric) crystal axis. The quantities d_{33} and d_{31} (Voigt's [1] notation) are, respectively, the piezoelectric modulus for normal pressures on faces perpendicular to the principal axis, and a piezoelectric modulus for normal pressures on any two faces which are parallel to the principal axis and to each other. The quantity $d_{33}+2d_{31}$ is the piezoelectric modulus of the crystal under hydrostatic pressure.

Voigt's measurements on Brazilian tourmaline gave

$$d_{33} + 2d_{31} = 2.42 \times 10^{-17} \text{ coul/dyne.}$$

For a crystal cut in the shape of a circular disk of area A and having its parallel flat faces perpendicular to the principal axis, the electric moment per unit volume, M, will appear as a uniform electric charge per unit area, M, on the flat faces, or the total charge on each flat face will be

$$Q = AM = A(d_{33} + 2d_{31})p \text{ coul.}$$

Then, by the principle of reciprocity, a potential difference of e'_s (volts) applied between the flat faces will cause a volume change of the crystal of amount

$$v = 10^7 A(d_{33} + 2d_{31})e'_s \text{ cm}^3. \tag{4}$$

A tourmaline crystal is then used in the following way as both sound source and microphone to secure an absolute pressure calibration of (for example) a condenser microphone, and to secure also the hydrostatic piezoelectric modulus of the crystal.

Let the response of the condenser microphone be

$$\rho=\frac{e_m}{p}\text{cm}^2 \text{ volts/dyne},$$

where e_m is the voltage output of the microphone due to the sound pressure, p. The response of the crystal used as a microphone (direct effect) is

$$\frac{A(d_{33}+2d_{31})}{C}=\frac{Q}{Cp} \text{ cm}^2 \text{ volts/dyne},$$

where C (farads) is the capacity of the crystal microphone. The crystal response is expressed in this way because Q/C (the voltage output of the crystal microphone) can be readily measured, whereas it is not feasible to measure Q directly.

The first step is to measure the ratio of the condenser-microphone response to the crystal-microphone response. The ratio is

$$\frac{\rho C}{A(d_{33}+2d_{31})}=\frac{e_m}{e_s}, \tag{5}$$

and is got experimentally by applying the same pressure (or pressures in a known ratio) to both microphones and by measuring the ratio of the output voltages, e_m/e_s, by means of an attenuation box or known resistances. The voltage output of the condenser microphone is e_m, and that of the crystal for the same sound pressure is e_s.

The second step is to use the crystal as a source, acting on the condenser microphone as a microphone through a gas-filled cavity of volume, V (cm^3). If B (dynes/cm^2) is the barometric pressure in the gas and γ is the ratio of specific heats of the gas, then the piezoelectric volume change of the crystal given by eq 4 will give rise to a pressure

$$p=\left(\frac{10^7\gamma B}{V}\right)A(d_{33}+2d_{31})e'_s \text{ dynes/cm}^2,$$

provided v is small in comparison with V. The condenser microphone voltage output due to this pressure will be

$$e'_m=\left(\frac{10^7\gamma B}{V}\right)A(d_{33}+2d_{31})\rho e'_s \text{ volts}. \tag{6}$$

Simultaneous solution of eq 5 and 6 yields

$$\left.\begin{aligned} \rho&=\sqrt{\frac{10^{-7}V}{\gamma BC}\left(\frac{e_m}{e_s}\right)\left(\frac{e'_m}{e'_s}\right)} \text{ cm}^2 \text{ volts/dyne} \\ A(d_{33}+2d_{31})&=\sqrt{\frac{10^{-7}CV}{\gamma B}\left(\frac{e_s}{e_m}\right)\left(\frac{e'_m}{e'_s}\right)} \text{ cm}^2 \text{ coul/dyne} \end{aligned}\right\}. \tag{7}$$

Thus an absolute pressure calibration of a condenser microphone is obtained by absolute measurements of a volume, the barometric pressure, and a capacitance, and by measurements of two voltage ratios. An absolute determination of the adiabatic piezoelectric modulus of the tourmaline crystal under hydrostatic pressure is obtained by measurement of the crystal area in addition to the quantities enumerated above. The only quantity which is not measured directly is γ, whose values for air, hydrogen, and helium are accurately known.

Superimposed on the piezoelectric action in the direct effect will be the pyroelectric effect (since temperature changes in the gas which accompany pressure changes will penetrate into the crystal), and superimposed on the piezoelectric action in the converse effect will be the electrocaloric effect (since temperature changes in the crystal which accompany voltage changes will penetrate into the gas). But the principle of reciprocity will hold nevertheless, and eq 7 will give a quantity which represents a mixed piezoelectric and thermoelectric modulus at low frequencies, and which will approach asymptotically the adiabatic piezoelectric modulus at high frequencies.

In cavities of small volume, the elastic volume changes of the crystal (used as a microphone) due to an applied pressure, will in themselves give rise to a pressure superimposed on the applied pressure calculated on the assumption of no yielding of the crystal. The crystal suffers a volume change also if the faces are not at the same potential, because the potential difference, Q/C, acquired by the crystal causes a piezoelectric volume change by virtue of the converse effect, and likewise will give rise to a pressure superimposed on the calculated applied pressure. In effect, the piezoelectric modulus will appear to be a function of cavity volume, if the crystal volume changes are appreciable and due allowance is not made for them. This will be called the volume effect. For tourmaline the adiabatic compressibility is about 10^{-12} cm^2/dyne. Therefore, the ratio of the pressure caused by elastic volume change to the applied pressure is less than 10^{-6} for cavity volumes greater than 10 cm^3 and crystal volumes less than 4 cm^3. For tourmaline the adiabatic piezoelectric modulus is about 2.4×10^{-10} cm/volt. Therefore, the ratio of the pressure caused by piezoelectric volume change to the applied pressure is less than 10^{-8} for cavity volumes greater than 10 cm^3, crystal areas less than 15 cm^2, and crystal-microphone capacitance greater than 30×10^{-12} farad. With these restrictions, the volume effect for a tourmaline crystal used as a microphone is completely negligible, and by the principle of reciprocity the volume effect for a tourmaline crystal used as a source is also negligible, with the same restrictions. The volume effect for condenser microphones will be discussed in section 3.

Quartz and Rochelle salt have no electric moment under hydrostatic pressure. This conclusion is reached from considerations of the crystal symmetry of these materials. Consequently, such crystals cannot be used in the same way as tourmaline is used to obtain an absolute calibration of a microphone.

3. APPLICATION TO CONDENSER MICROPHONES

Let ρ_1(cm^2 volts/dyne) be the response and C_1 (farads) the capacitance of a condenser microphone, *I*, which is to be used as both source and microphone to calibrate a second condenser microphone, *II*, whose response is ρ_2(cm^2 volts/dyne). The equations of the direct and converse effects for the condenser microphone are

$$\left.\begin{aligned} Q&=\rho_1 C_1 p \\ v&=10^7\rho_1 C_1 E \end{aligned}\right\} \tag{8}$$

These correspond to eq 3. Microphone *I* is then used in exactly the same way as was the tourmaline crystal to secure an absolute pressure calibration of microphone *II*. That is to say, the first step is to obtain the ratio of the responses, ρ_1/ρ_2, by applying the same pressure to both microphones, and the second step is to couple microphone *I* (used as a sound source) to microphone *II* by means of a gas-filled cavity of known volume, from which is obtained the product of the responses, $\rho_1 \rho_2$. The final equations for the responses, ρ_1 and ρ_2, are

$$\left.\begin{aligned}\rho_1&=\sqrt{\frac{10^{-7}V}{\gamma BC_1}\left(\frac{e_1}{e_2}\right)\left(\frac{e'_2}{e'_1}\right)}\text{ cm}^2\text{ volts/dyne}\\ \rho_2&=\sqrt{\frac{10^{-7}V}{\gamma BC_1}\left(\frac{e_2}{e_1}\right)\left(\frac{e'_2}{e'_1}\right)}\text{ cm}^2\text{ volts/dyne}\end{aligned}\right\} \qquad (9)$$

These correspond to eq 7. In these equations V (cm^3) is the volume of the cavity in which microphone *I* is used as a sound source, B (dynes/cm^2) is the barometric pressure in the cavity, γ is the ratio of the specific heats of the gas in the cavity, e_1 and e_2 are, respectively, the voltage outputs of microphones *I* and *II* for the same applied pressure, and e'_2 is the voltage output of microphone *II* for a voltage e'_1 applied to microphone *I* when used as a sound source (the two microphones being coupled by the cavity of volume V).

The condenser microphone diaphragm suffers a volume change if the potential difference between the diaphragm and back plate is not steady, because the alternating potential difference, Q/C, causes a volume change by virtue of the converse effect, and will therefore give rise to a pressure superimposed on the applied pressure calculated on the assumption of no volume change of the microphone. The ratio of the pressure caused by this volume change to the applied pressure is less than 5×10^{-3} for a condenser microphone having a response less than 3×10^{-3} cm^2 volt/dyne and a capacitance less than 3.5×10^{-10} farad, and for cavity volumes greater than 10 cm^3. This part of the volume effect was therefore neglected.

The compressibility (admittance) of the diaphragm gives rise to a volume effect which is not negligible. An absolute calculation of the amount would require an exact knowledge of the construction of the microphone. In the absence of such knowledge, the amount of the volume effect must be determined experimentally by making response measurements at several different cavity volumes. Unfortunately, at high-frequencies volume effects due to cavity resonances swamp those due to diaphragm compressibility, and there is no way of separating the two effects. This means that a precise determination of the volume effect can be made only at low frequencies. Let the compressibility of the diaphragm *II* be c_2(cm^5/dyne); that is, a pressure of p(dynes/cm^2) on the diaphragm causes a volume change of c_2p(cm^3). Then at frequencies less than the lowest natural frequency of diaphragm *II* the compressibility will have the effect of increasing the cavity volume which appears in eq 9 from V to $V+\gamma Bc_2$. It is clear from the principle of reciprocity that the same volume effect will be present when the microphone is used as a sound source. If c_1 is the compressibility of diaphragm *I*, then the total effect will be to increase the volume which appears in eq (9) from V to $V+\gamma Bc_1+\gamma Bc_2$. The quantity γBc(cm³) will hereafter be denoted by κ(cm³), and will be called the volume constant of the microphone.

III. EXPERIMENTAL RESULTS

1. REVERSIBILITY OF TRANSDUCERS

A "reversible" transducer is one which satisfies the principle of reciprocity. A simple experimental way of ascertaining whether or not two transducers are reversible is the following. The transducers are coupled by a cavity of fixed volume. A voltage e_1 is applied to transducer *I* (source), and the corresponding voltage output e'_2 of transducer *II* (microphone) is measured. Then a voltage e_2 is applied to transducer *II* (source), and the corresponding voltage output e'_1, of transducer *I* (microphone) is measured. Then the over-all coupled system is reversible, if the equation

$$\frac{C_2}{C_1}=\left(\frac{e_1}{e'_2}\right)\left(\frac{e'_1}{e_2}\right) \tag{10}$$

is satisfied, where C_2 and C_1 are the capacitances of transducers *II* and *I*, respectively. Thus eq 10 is a necessary condition for the reversibility of each transducer. If a third transducer is used and eq 10 holds for each pair of transducers, this will be a sufficient condition for the reversibility of each of the three. Actually, it will be entirely safe to assume each of two transducers is reversible if eq 10 holds when the two transducers are coupled. Thus the reversibility of a transducer is determined experimentally by purely electrical measurements of two voltage ratios and a capacitance ratio.

Such measurements were made on the following: A Western Electric condenser microphone type 394, No. 4829 (denoted hereafter as microphone 2); a Western Electric condenser microphone type D96436, No. 70 (denoted hereafter as microphone 70); a tourmaline crystal transducer (denoted hereafter as transducer *CD*). Transducer *CD* was made of black California tourmaline, 3.8 cm in diameter and 0.3 cm in thickness. The front face was coated with tin foil affixed with wax, and the tourmaline was wrung on to a brass block. The block was insulated from the outside brass block (at ground potential) which was connected to the tin foil.

The voltage ratios appearing in eq 10 were measured at various frequencies by means of an attenuation box to the nearest decibel and by means of an output meter to the nearest one-tenth decibel. The capacitances were separately measured at 1,000 c/s (by means of a strict substitution method) with a variable capacitance calibrated in the capacitance section of this Bureau. The results are shown in figure 3. The gases which were used with each pair of transducers are indicated in the legend. The sound pressure was about 2×10^{-2} dyne/cm² with the tourmaline transducer *CD* as source, and about 10 dynes/cm² with transducer *CD* as microphone. In the other two cases plotted in figure 3, the pressure was about 1 dyne/cm².

The estimated probable error in 20 $\log_{10}$ (C_2/C_1) is 0.1 db, and in 20 $\log_{10}$ (e_1/e'_2) (e'_1/e_2) is less than 0.2 db. The conclusion from the results plotted in figure 3 is, that within the experimental uncertainty eq10 is satisfied. Consequently, condenser microphones 2 and 70 and tourmaline transducer *CD* are reversible. This is a direct experimental proof of the equality of the direct and converse piezoelectric moduli in tourmaline under hydrostatic pressure.

After Lippmann's theoretical proof of the equality of the direct and converse piezoelectric moduli, J. and P. Curie made an attempt to verify the equality for the modulus, d_{11} in quartz, but experimental difficulties enabled them to obtain only a very rough agreement between the two moduli. Since then, H. Osterberg and J. W. Cookson [4] have concluded that measurements of d_{11} by some observers who used the direct effect only have agreed on the whole with measurements of d_{11} by other observers who used the converse effect only.

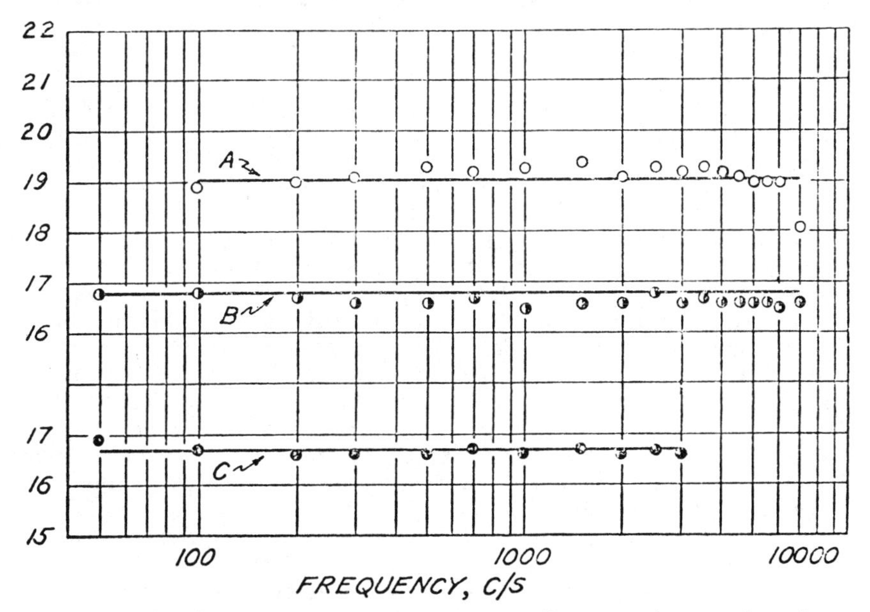

FIGURE 3.—*Graph showing reversibility of tourmaline transducer and condenser microphones.*

○ = 20 $\log_{10}(e_1/e'_2)(e'_1/e_2)$, and curve A = 20 $\log_{10}(C_2/C_1)$ at 1,000 c/s, for condenser microphone 2 and tourmaline transducer *CD*, in helium.

◑ = 20 $\log_{10}(e_1/e'_2)(e'_1/e_2)$, and curve B = 20 $\log_{10}(C_2/C_1)$ at 1,000 c/s, for condenser microphones 2 and 70, in helium.

● = 20 $\log_{10}(e_1/e'_2)(e'_1/e_2)$, and curve C = 20 $\log_{10}(C_2/C_1)$ at 1,000 c/s, for condenser microphones 2 and 102, in air.

Because of variations in the composition of tourmaline from different sources, it is not possible, in order to decide the equality of the two moduli for tourmaline, to compare the results of some observers who used the direct effect only with the results of other observers who used the converse effect only.

2. MICROPHONE CALIBRATIONS

The reversibility of condenser microphones 2 and 70 and of transducer CD having been established, the principle of reciprocity was applied to data obtained with these transducers to secure absolute pressure calibrations of microphones 2 and 70.

The results are shown in figures 4 and 5. The reciprocity calibrations are those obtained with condenser microphone 2 used as both source and microphone. The cavity volumes and gases which were used are indicated in the legends. The estimated probable error of measurement is 0.2 db. The volume effects of the microphone diaphragms and the cooling effect of the walls of the cavity have not been included in the responses, which were computed directly from eq 9 (the geometrical volume of the cavity being used for V). These omissions introduce systematic errors not much greater than the probable error stated above. The amounts of these are deduced in the following paragraph.

Measurements of the volume constant, κ, gave $\kappa_2=1.3 \pm 0.2$ cm^3, and $\kappa_{70}=0.3 \pm 0.2$ cm^3. These values are experimentally the same in air, hydrogen, and helium, and are experimentally independent of whether the sound source used in the measurements is a thermophone, a tourmaline crystal, or a condenser microphone. Measurements from which the above values of κ are obtained were made usually at cavity volumes of 15, 30, and 40 cm^3. The fact that the thermophone yields the same value for κ as does the other sound sources is surprising, since Ballantine [5] has shown that the diaphragm motion of a condenser microphone should react on the thermophone foil strips in such a way as to result in a net effect different for different gases and in any case much smaller than the experimental value found above for condenser microphone 2. The value of κ_2 found above is also somewhat greater than the value computed by L. J. Sivian [6] for microphones of the same type. Direct measurements of cos ϕ (where ϕ is the phase angle between the pressure applied to the microphone diaphragm and the voltage output of the microphone) showed that the values of κ_2 and κ_{70} are practically independent of frequency below 3,000 c/s. Because of the difficulty of measuring κ above about 3,000 c/s (although the results of the measurements of cos ϕ show that κ_2 and κ_{70} become zero at about 6,000 c/s and become negative at higher frequencies), it was decided to make the reciprocity measurements at cavity volumes of 40 cm^3 so as to keep the volume effects small. Omission of the volume effects means that the computed response of microphone 2 is low by 0.3 db and of microphone 70 is low by 0.1 db below 3,000 c/s. The cooling effect of the walls of the cavity must also be considered. It is usually incorrectly assumed that the temperature variation of the gas at the walls of the cavity is zero, on the grounds that the thermal conductivity of the metal walls is much greater than that of the gas. However, in a quasi-stationary temperature distribution (all temperatures sinusoidal functions of the time) the quantity of importance is the thermal diffusivity, which is the thermal conductivity divided by the product of density and specific heat. The thermal diffusivities of brass, air, hydrogen, and helium are, respectively, at room temperature and barometric pressure 0.34, 0.28, 1.67, and 1.78 cm^2 sec. From these values it can be deduced that a better boundary condition at the walls of the cavity is that the rms temperature at the walls is one-half the rms temperature in the gas at a distance from the walls. The inclusion of this boundary condition in the reciprocity calibrations raises the computed responses of microphones 2 and 70 by less than 0.1 db, for frequencies greater than 50 c/s and a cavity volume of 40 cm^3, and for any of the three gases mentioned above. The upshot of all the fore-

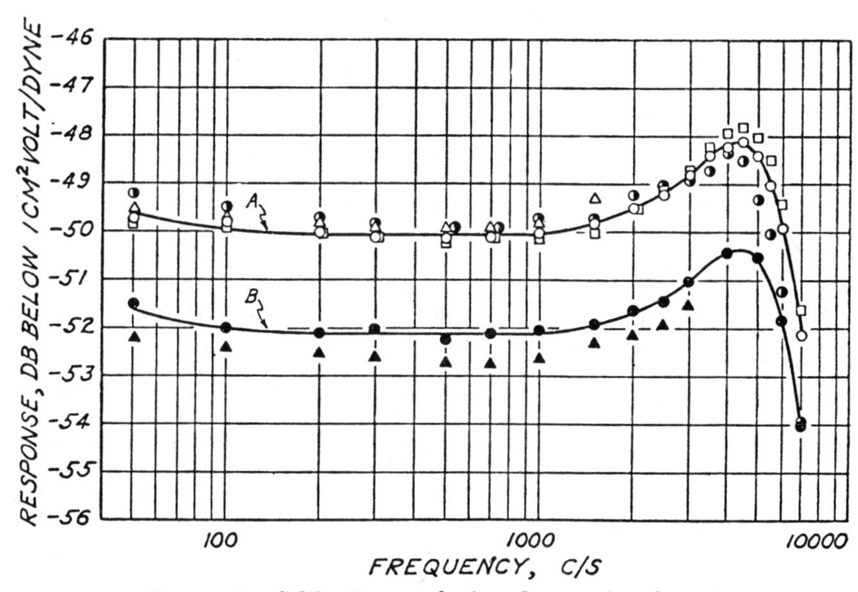

FIGURE 4.—*Calibration graph of condenser microphone 2.*

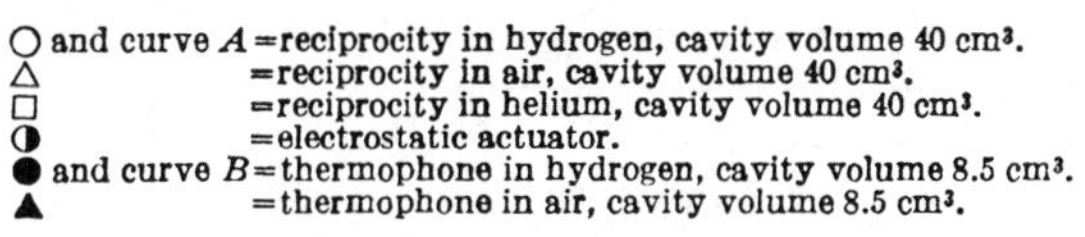

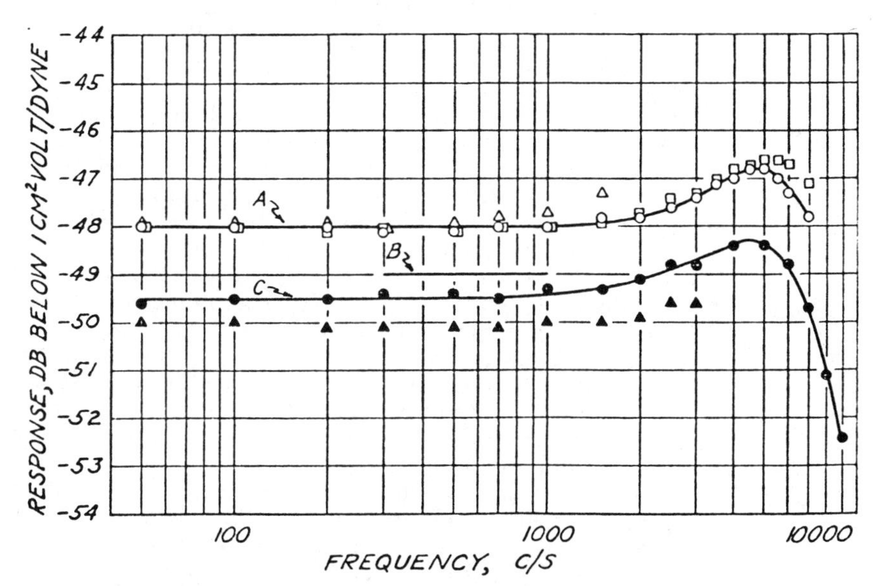

FIGURE 5.—*Calibration graph of condenser microphone 70.*

○ and curve *A* = reciprocity in hydrogen, cavity volume 40 cm³.
△ = reciprocity in air, cavity volume 40 cm³.
□ = reciprocity in helium, cavity volume 40 cm³.
Curve B = Rayleigh disk.
● and curve *C* = thermophone in hydrogen, cavity volume 10.8 cm³.
▲ = thermophone in air, cavity volume 10.8 cm³.

going is that the systematic errors introduced into the reciprocity calibrations by omission of the volume effects and cooling effects amount to a lowering of the response of microphone 2 by 0.3 db and of microphone 70 by 0.1 db.

The reciprocity calibrations of condenser microphones 2 and 70 obtained with the tourmaline transducer *CD* used as both sound source and microphone agree (within the estimated probable error) with the reciprocity calibrations obtained with the condenser microphone 2 used as both sound source and microphone.

Figure 4 shows also the results of electrostatic actuator calibrations of microphone 2. The actuator measurements were made with a "solid" grille having 25 holes each 0.7 mm in diameter and 3 mm long, and with a slotted grille having grille bars each 1.53 mm wide and 3 mm deep and grille slots each 0.88 mm wide. The pressure produced by the slotted grille was computed by means of nomographs made for such grilles by S. Ballantine [5]. The results with the slotted grille at frequencies higher than 1,000 c/s are plotted, in spite of the presence of numerous small "resonances" which appeared in the output of microphone 2 and which vitiate the results at such frequencies. Below 1,000 c/s the electrostatic actuator calibrations and the reciprocity calibrations in air, hydrogen, and helium of condenser microphone 2 all agree within the estimated probable error.

Figures 4 and 5 also show the results of gold foil thermophone calibrations of condenser microphones 2 and 70. The thermophone results were computed with the formula given by S. Ballantine [5], except for minor corrections which will be presently discussed. The thermophone cavity volumes are about the sizes used by several other observers. The estimated probable error of the individual thermophone determinations below 1,000 c/s is 0.2 db, but the estimated probable error of the difference between the air and hydrogen calibrations of a particular microphone is less than 0.2 db. The response obtained by the thermophone of a microphone of the same type as No. 2 had a value 0.3 db lower in helium than in air. It is plain that the thermophone calibrations in air, hydrogen, and helium disagree among themselves and disagree with the reciprocity calibrations in the regions of safe comparison below 1,000 c/s by amounts much greater than the probable errors of the differences. The inclusion of the volume effects will increase the responses of both microphones 2 and 70 below 3,000 c/s, so that the thermophone calibrations in air (which are more reliable than those in hydrogen and helium because of smaller systematic errors) will be below the reciprocity calibrations by 1.5 db for each microphone. This difference is still much greater than the estimated probable error. The disagreement between the thermophone calibrations in air, hydrogen, and helium is unchanged by inclusion of the volume effects.

The thermophone formula and calculations given by S. Ballantine [5] require correction because they are based on the assumption of no temperature variation in the gas at the walls of the cavity (this assumption was discussed above) and on the assumption that commercial gold foil has the same specific heat as pure gold. The specific heat of the gold foil, calculated from the results of a chemical assay, was 0.0328 cal/g° C, whereas that of pure gold is 0.0313 cal/g° C. The two corrections are in opposite directions and almost nullify one another, thus leaving calculations made from Ballantine's thermophone formula practically unchanged.

Pistonphone measurements were made by Guy Cook and the author on a condenser microphone of the same type as microphone 2. The measurements were made between 50 and 200 c/s in a cavity of 15 cm^3 at a sound pressure of about 5 dynes/cm^2. Air, hydrogen, and helium were used. The piston was driven by a loudspeaker coil, and its amplitude of motion was measured by means of an optical lever and autocollimator. The amplitude measurements were checked by a microscope having a calibrated graticule. The computed probable error of the pistonphone results was 0.3 db, and the results agreed with an electrostatic actuator calibration of the same microphone within the probable error.

Resonant-tube measurements were made by Guy Cook and the author on the same condenser microphone. The microphone was placed at one end of the resonant tube. The amplitude of motion of illuminated tobacco-smoke particles was observed (with a microscope having a calibrated graticule) at the central displacement loop in the gas in the resonant tube. Observations were made at 95 and 285 c/s at a sound pressure of about 75 dynes/cm^2. The estimated probable error of the resonant tube results was 0.5 db, and the results agreed with the electrostatic actuator calibration of the same microphone within the estimated probable error.

The disk curve B in figure 5 is placed 0.5 db above the thermophone calibration in hydrogen, and is the result of Rayleigh-disk and thermophone comparisons between 300 and 1,000 c/s on several other microphones of the same type as microphone 70. The Rayleigh-disk measurements were made in the free field by Guy Cook.

The conclusion is, that in the usually accepted thermophone formula there are systematic errors which are different for different gases, and which are much greater than the error of measurement. The trouble probably lies in the assumed modus operandi of the thermophone. It has been suggested that the gold foil possibly adsorbs and emits gas during each thermal cycle. Another suggestion is, that viscosity forces have an effect on the pressure produced by the thermophone. The computed responses in air, hydrogen, and helium are in the same order as the viscosity coefficients of these gases.

3. PIEZOELECTRIC MODULI

The results of acoustic measurements, based on the principle of reciprocity, of the adiabatic piezoelectric modulus $d_{33}+2d_{31}$ of California tourmaline are shown in figure 6. The tourmaline was crystal CD, which was used as both source and microphone.

The results of static testing-machine measurements (by use of the direct effect) of the modulus d_{33} and Voigt's [1] static measurements (also by use of the direct effect) on Brazilian tourmaline are also given in figure 6. The frequency scale of the abscissas does not apply to the horizontal lines representing the static tests; these lines were drawn merely to facilitate comparisons with the acoustic measurements. Static measurements yield isothermal moduli, whereas the acoustic measurements yield adiabatic moduli. However, D. A. Keys [7] has shown from thermodynamic considerations that the difference between the isothermal and adiabatic moduli is less than

0.3 percent. Below 500 c/s the acoustic measurements do not yield the adiabatic modulus, since the droop in the measured modulus $d_{33}+2d_{31}$ at low frequencies is due to the superposition of thermoelectric effects on the piezoelectric action. The droop above 3,000 c/s is caused by cavity resonances.

The testing-machine measurements of the modulus d_{33} were made at loads between 1,000 and 3,000 lb in a hydraulic testing machine. A preassigned load was applied to a crystal disk 3.8 cm in diameter and 0.15 cm thick, and the electric charge released from the crystal on sudden release of the testing-machine load was measured by a calibrated ballistic galvanometer. A check of the experimental setup

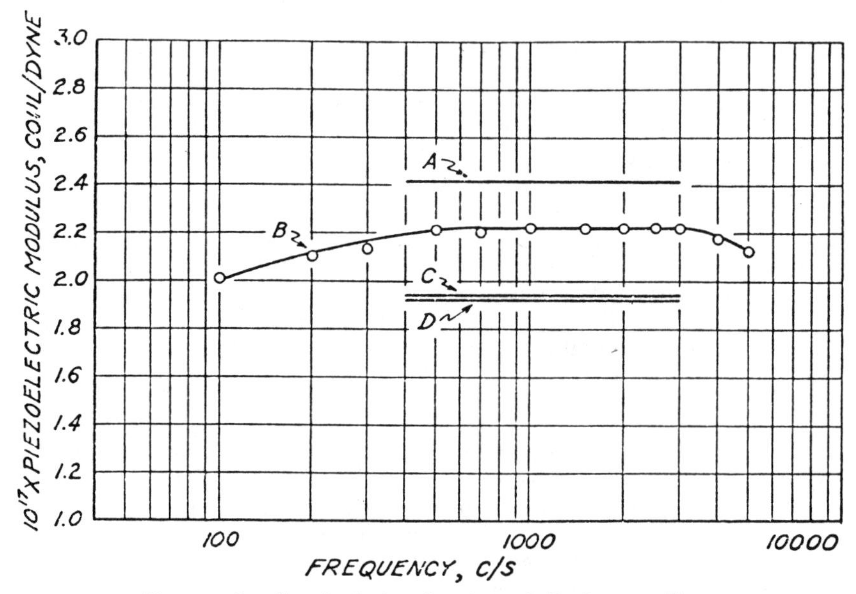

FIGURE 6.—*Graph of piezoelectric moduli of tourmaline.*

○ and curve B=acoustic measurements of $d_{33}+2d_{31}$ for California tourmaline.
Curve A=static measurements by Voigt of $d_{33}+2d_{31}$ for Brazilian tourmaline.
Curve C=static testing-machine measurements of d_{33} for California tourmaline.
Curve D=static measurements by Voigt of d_{33} for Brazilian tourmaline.

was made by measuring the piezoelectric modulus d_{11} of quartz, which is well known. The estimated probable error in the measured value of the modulus d_{33} for black California tourmaline is 2 percent.

The curious discovery was made that it is absolutely essential to load the crystal through rigid (e. g., steel) bearing blocks. Erroneous results were obtained when the tourmaline and quartz were loaded through Plexiglass (a methacrylate resin of low modulus of elasticity) bearing disks. It was established that these anomalies were not due to friction. Evidently the crystal is subjected to thermal actions and to large shearing stresses (in addition to the compressive stresses) when loaded through Plexiglass disks in contact with the crystal, and these thermal actions and shearing-stresses cause the anomalous piezoelectric effects.

The following measurements were made on California tourmaline, used as a sound source, in addition to the measurements of $d_{33}+2d_{31}$ given in figure 6. Three crystals (each 3.8 cm in diameter and 0.15

cm thick) cut from the same batch of tourmaline had identical piezoelectric moduli. When two of these disks were wrung together with vaseline so that the analogous ends of their crystal axes pointed in the same direction, the piezoelectric modulus was the same as that of a single disk, which shows that the acoustically measured modulus is independent of the crystal thickness. When the same two disks were wrung together so that the analogous ends of their axes pointed in opposite directions, the piezoelectric modulus was zero. Each disk suffered a volume change under the applied voltage, but the net volume change was zero, since the individual volume changes differed in phase by π. The piezoelectric modulus was independent of applied voltage in the range from 130 to 1,300 volts. The foregoing measurements were made in air, between 200 and 1,000 c/s, in a cavity of about 20-cm^3 volume. It is understood that the conclusions are valid only within the experimental uncertainty, which is about 2 percent of the modulus $d_{33}+2d_{31}$. In addition, a determination of the volume constant, κ, for a crystal yielded $\kappa=0$, within the experimental uncertainty of about 0.3 cm^3, which confirms the theoretical conclusion that a crystal used as a transducer should have no detectable volume effect.

A quartz-crystal disk, 3.6 cm in diameter and 0.46 cm thick, having its flat faces perpendicular to one of the three diagonal axes of the quartz, was used as a sound source in a cavity of about 20 cm^3. The measured piezoelectric modulus was about 1 percent of that of tourmaline at 500 c/s. At lower frequencies the disk had a small sound output, evidently due to electrocaloric effects superimposed on the piezoelectric. The conclusion is, that within the experimental uncertainty, the piezoelectric modulus of quartz under hydrostatic pressure is zero. This confirms experimentally the theoretical conclusion reached from considerations of crystal symmetry.

IV. INSTRUMENTS AND TECHNIQUE

The cavities were made of simple geometrical shapes (e. g., right circular cylinders), and the volumes were computed from measurements of the linear dimensions. The scale and micrometer calipers used in these measurements were compared with secondary standards kept in this Bureau. The estimated probable error in the measurement of volume is 0.4 percent.

The barometric pressure was measured by an aneroid barometer which from time to time was compared with a mercury barometer. The estimated probable error in the measurement of pressure by the aneroid barometer is 0.1 percent.

Microphone capacitances were measured by a strict substitution method. A variable air condenser, calibrated in the capacitance section of this Bureau, was used as a secondary standard. As a check on the technique, a particular condenser was measured in both the sound chamber and in the capacitance section. The estimated probable error in capacitance measurement is about 1 percent.

Voltage ratios were measured by means of an attenuation box to the nearest decibel, and by a copper oxide rectifier output meter to the nearest 0.1 db. The attenuation box was calibrated by comparison with a calibrated potentiometer, and had errors less than 0.1 db. The estimated probable error in the measurement of voltage ratio is

0.1 db. The output meter was used over a range not exceeding 0.5 db.

Nearly all the experimental measurements were made at room temperatures between 20° and 25° C, and at barometric pressures between 742 and 758 mm Hg. Within these limits there was no dependence of the responses of the condenser microphones on temperature or barometric pressure. Small variations in the responses (about 0.5 db in range) seemed to be entirely random in character.

Various tricks of technique were used to detect and eliminate spurious voltages in the microphone circuit. Such voltages are the commonest cause of trouble. For example, electrostatic coupling between the gold foil of the thermophone and the high-potential side of the tourmaline microphone causes a spurious voltage in the tourmaline-microphone voltage output. This will easily occur because the voltage applied to the thermophone is about 10^4 times greater than the voltage output of the tourmaline microphone. In another case, resistance coupling between the tourmaline-crystal sound-source circuit and the low-potential (ground) side of the condenser microphone causes a spurious voltage in the condenser-microphone voltage output. This might happen because the voltage applied to the crystal source is about 10^6 times greater than the voltage output of the condenser microphone. Reversal of the polarizing voltage on the condenser microphone (or, in the case of the thermophone, reversal of the polarizing current) will change the phase of the spurious voltage with respect to the voltage generated in the microphone by π, and the voltage output of the microphone will not be independent of polarity (as it should be), if the spurious voltage is appreciable. In crystal measurements, substitution of a dummy glass (or any nonpiezoelectric material) transducer for the crystal transducer will readily reveal spurious voltages.

V. SUMMARY

Application of a general principle of reciprocity to the action of a piezoelectric crystal or to a microphone yields absolute pressure calibrations of another microphone, and also yields absolute values of the adiabatic piezoelectric modulus under hydrostatic pressure. Absolute measurements of a volume, the barometric pressure and a capacitance are needed in addition to measurements of two voltage ratios; also, the ratio of specific heats of the gas used in the cavity is required.

Calibrations secured in this way are in good agreement with electrostatic actuator, pistonphone, and "smoke particle" calibrations of a condenser microphone. Thermophone calibrations in air, hydrogen, and helium disagreed among themselves and with the reciprocity calibrations of the same microphones by amounts much greater than the experimental uncertainty.

Absolute acoustic measurements of the adiabatic piezoelectric modulus of black California tourmaline under hydrostatic pressure gave $d_{33}+2d_{31}=(2.22\pm0.06)\times10^{-17}$ coul/dyne. It was experimentally proved that the direct and converse piezoelectric moduli $d_{33}+2d_{31}$ are equal, within the experimental uncertainty.

The following is a tentative set-up for securing absolute pressure calibrations of condenser microphones. A tourmaline-crystal source, calibrated by application of the principle of reciprocity, will be the primary standard. A condenser microphone will serve as a secondary-standard source, which will be calibrated from time to time by the primary tourmaline standard.

There are common crystalline household substances which are piezoelectric under hydrostatic pressure, and which might be used as sound sources. Such are sucrose (cane sugar) and tartaric acid (sour salt). The latter at room temperatures has a piezoelectric modulus about double that of tourmaline.

The theoretical conclusions and experimental results contained in this article were reported at the meeting of the Acoustical Society of America in Washington, D. C., in April 1940. Similar theoretical conclusions were reached by W. R. MacLean [8] and published in the July 1940 issue of the Journal of the Acoustical Society of America.

I thank Elias Klein, of the Naval Research Laboratory, for securing loans of tourmaline and equipment, and for his kind help and encouraging suggestions.

VI. REFERENCES

[1] W. Voigt, *Lehrbuch der Kristallphysik* (B. G. Teubner, Leipzig, 1910).
[2] Rayleigh, *The Theory of Sound* (Macmillan & Co., London, 1896).
[3] S. Ballantine, Proc. Inst. Radio Engrs. **17,** 929 (1929).
[4] H. Osterberg and J. W. Cookson, Rev. Sci. Instr. **6,** 347 (1935).
[5] S. Ballantine, J. Acous. Soc. Am. **3,** 319 (1932).
[6] L. J. Sivian, Bell System Tech. J. **10,** 96 (1931).
[7] D. A. Keys, Phil. Mag. [6] **46,** 999 (1923).
[8] W. R. MacLean, J. Acous. Soc. Am. **12,** 140 (1940).

WASHINGTON, August 14, 1940.

33

A NOTE ON VARIOUS RECIPROCITIES

H. B. Miller

This article was prepared expressly for this Benchmark volume

Fig. 1 Schottky Reciprocity

(a) Electromagnetic Transducer, above Resonance

$S_I = \frac{p_1^R}{I_1^L}$ — 0 dB/oct (freq.)

$M_E = \frac{E_2^L}{p_2^R}$ — −6 dB/oct (freq.)

$\frac{M_E}{S_I} \sim \frac{1}{f}$

(b) Piezoelectric Transducer, below Resonance

$S_I = \frac{p_1^R}{I_1^L}$ — +6 dB/oct (freq.)

$M_E = \frac{E_2^L}{p_2^R}$ — 0 dB/oct (freq.)

$\frac{M_E}{S_I} \sim \frac{1}{f}$

Fig. 2 Helmholtz – Rayleigh Reciprocity

$\frac{p_1^R}{\dot{V}_1^L}$

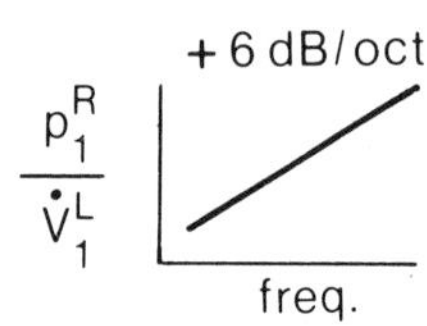

$\frac{p_2^L}{\dot{V}_2^R}$

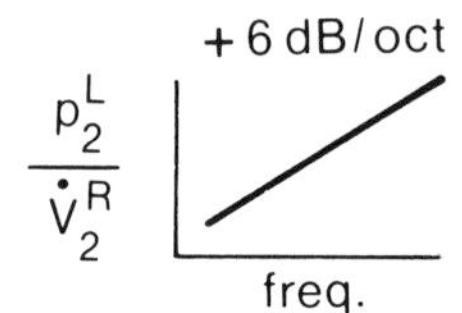

Note that $p = \frac{\rho f}{2d} \cdot \dot{V}$

Fig. 3 (a) El-Mag Loudspeaker — El-Mag Microphone (Faraday)

$\frac{\dot{V}_1^L}{I_1^L}$

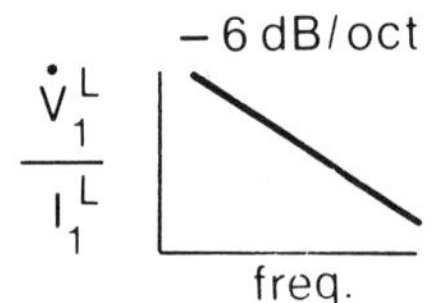

$\frac{E_2^L}{p_2^R}$

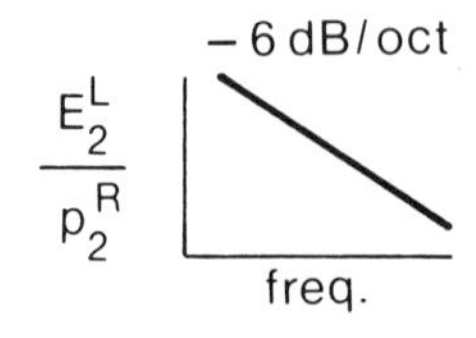

(b) Piezo-El Loudspeaker — Piezo-El Microphone (Lippmann) (Curies)

$\frac{\dot{V}_1^L}{I_1^L}$

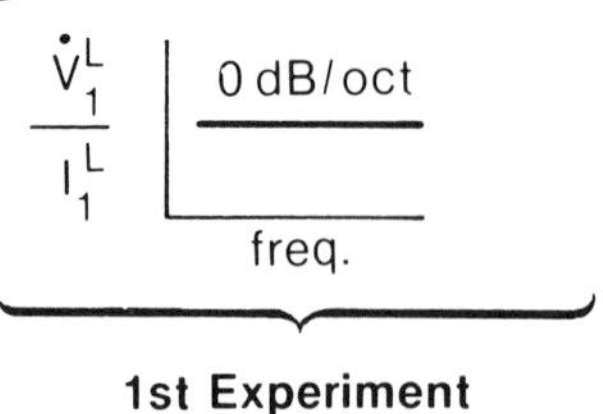

$\frac{E_2^L}{p_2^R}$

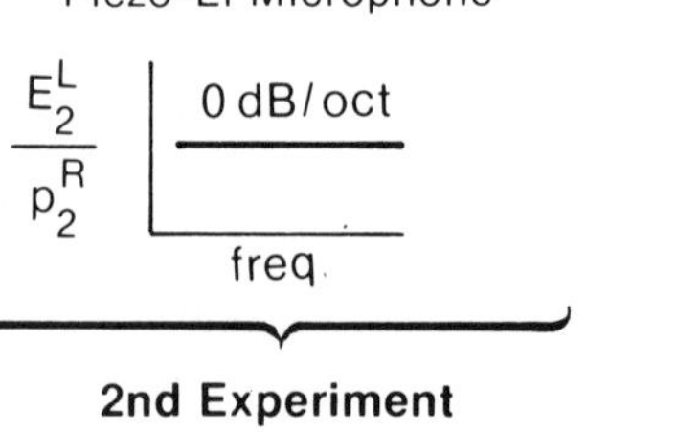

1st Experiment — **2nd Experiment**

A Note on Various Reciprocities

Figure 1a shows Schottky's two experiments. S_I is the loudspeaker sensitivity when current I_1 applied to a loudspeaker on the left side of a lawn generates pressure p_1 on the right side. M_E is the microphone sensitivity (of the reversible loudspeaker) when pressure p_2 on the right side of the lawn generates voltage E_2 at the microphone on the left side. Schottky realized that the ratio M_E/S_I must always be proportional to 1/frequency. He called this the "Law of Low Frequency Reception," since the microphone response increases as the frequency decreases, even though the loudspeaker response is flat versus frequency. Figure 1b shows that the ratio M_E/S_I is the same even when a nonflat loudspeaker is used.

Figure 2 shows Rayleigh's two experiments. A constant volume velocity radiating at the left side of a lawn generates a rising (6 dB/oct) pressure at the right side. The same volume velocity radiating at the right side of the lawn generates the same pressure response at the left side.

Figure 3a shows the two reciprocity experiments implied by Faraday's discovery for electromagnetic transducers that

$$F = BLI$$

or force equals flux density times length times current; and that

$$E = BLU$$

or voltage equals flux density times length times velocity.

Figure 3b shows the two reciprocity experiments implied by Lippmann's discovery for piezoelectric transducers that

$$x = d \cdot E$$

or: displacement equals d-constant times voltage. This is to be taken in conjunction with the Curie brothers' discovery that

$$q = d \cdot F$$

or: charge equals d-constant times force.

Returning to Figure 1: if we combine the function in Figure 2 (left) with the function shown in Figure 3a (left), we get the function shown in Figure 1a (left). Note that Figure 1a (right) comes directly from Figure 3a (right). And if we combine the function shown in Figure 2 (left) with the function shown in Figure 3b (left), we get Figure 1b (left). Note that Figure 1b (right) comes directly from Figure 3b (right).

MacLean realized that the equation $S_I = p_1/I_1$ is (from Fig. 2) merely $(\rho f/2d)\,\dot{V}/I_1$ and that $\dot{V}/I_1$ is (from Fig. 3) equal to E/p_2 or M_E. Hence, the ratio M_E/S_I would be $2d/\rho f$, thus turning Schottky's proportionality 1/f into an equality $2d/\rho f$; or $2d\lambda/\rho c$, which is the transfer admittance between p_1^R and $\dot{V}_1^L$.

One further comment is in order. In Figure 2, above, we see that free-field pressure $p_1 = (\rho f/2d)\dot{V}$. This can be written as $p_1 = (\rho\omega^2/4\pi d)V$. That is, pressure is proportional to volume acceleration $\omega^2 V$, where V is volume displacement. Now, in a closed cavity, the pressure $p_2 = K_{ac}V$, where K_{ac} is the cavity stiffness $\rho c^2/V_0$.

Hence $p_1/p_2 \sim \omega^2$, a 12 dB/oct law. This relation arose in the discussion of the thermophone paper of Arnold and Crandall (Paper 14), where p_1 followed a + 3 dB/oct law and p_2 followed a -9 dB/oct law. We can now see that the ratio *had* to follow a +12 dB/oct law.

ANNOTATED BIBLIOGRAPHY

Bobber, R. J., 1970, *Underwater Electroacoustic Measurements,* Naval research Laboratory, Washington, D.C., pp. 106, 107.
A good list of references on the reciprocity calibration method.

Curie, J., and P. Curie, 1881, *Compt. Rend.* 92:186, 352; 39:1137.
Experimental confirmation of Lippmann's predictions.

Faraday, M., 1839, *Experimental Researches in Electricity,* B. Quaritch, London.
This is a three-volume collection of Faraday's reprints. See vol. I, First Series (November 1831) through Second Series (January 1832). The Faraday pages show his struggle to find the converse of the law: blocked force = *Bl* times current. It was not current = (1/*Bl*) times blocked force. Rather, it turned out to be: voltage = *Bl* times velocity. Faraday expresses this voltage in terms of current and resistance.

Lippmann, G., 1881a, Principle of the Conservation of Electricity, *Compt. Rend.* 92:1049-1051, 1149-1152.

Lippmann, G., 1881b, Principle of the Conservation of Electricity, *J. Phys.* ser. 1, 10:381-394.

Lippmann, G., 1881c, Principle of the Conservation of Electricity, *Ann. Chim. Phys.* ser. 5, 24:145-178.
The Lippmann pages show his prediction of the converse piezoelectric effect, sometimes called the inverse effect, sometimes the reciprocal effect. He showed by thermodynamics that if the direct effect is expressed as: charge = d-constant times force, the inverse effect is not: force = (1/d-constant) times charge. Rather, it is: displacement = d-constant times voltage. (The Curie brothers confirmed Lippmann's predictions experimentally.)

Miller, H. B., 1977, Acoustical Measurements and Instrumentation, *Acoust. Soc. Am. J.* 61:274-282.

Part V

SOUND ABSORPTION COEFFICIENT MEASUREMENTS AND ACOUSTIC IMPEDANCE MEASUREMENTS

Editor's Comments on Papers 34 Through 41

In 1895 W. C. Sabine began his long-term study of reverberation, first reported on in 1900 in the famous paper (Sabine, 1922) wherein he set forth the relationship between the sound-absorbing power *A* of a large room and the reverberation time *T*. Thus

T sec = (0.05 sec/ft)(V/A). Here V is room volume in cubic feet and A is equal to αS, where α is the energy absorption coefficient of the exposed surface, and S is the area of this surface in square feet. The metric form of the equation is T sec = (0.161 sec/m) ($V/\alpha S$) where V is in cubic meters and S is in square meters. Equivalent relations are $A = 0.161$ (V/T) and $\alpha = 0.161$ (V/TS). Sabine had to use lots of absorbing material (that is, a large S) in achieving his desired A since his rooms had large V. Therefore it was, and still is, considered highly desirable to find a meaningful value of α from a direct measurement on a small sample of absorbing material.

One of the first people to successfully attack this problem was Professor H. O. Taylor of Cornell University, in 1913. His work is presented here as Paper 34. Taylor gave up the indirect, or reverberation, method and looked for a direct method of finding A and hence α. He found such a method by measuring the ratio of two intensities along a standing wave in a resonant tube. The method, in practice, is useful at only one frequency for a given tube. E. T. Paris, in Paper 35, carried the method considerably further, among other things using a tuned microphone so that the sound source itself did not have to be filtered clean. This is a good example of the application of reciprocity within a closed system that includes a microphone and a loudspeaker. Paris's method was used for years at the National Physical Laboratory, Teddington, England, as the standard British method for determing α.

E. C. Wente, meanwhile, was extending Taylor's method along a different path. In 1928 Wente and Bedell (Paper 36) repeated Taylor's method, using a resonant tube of fixed length but measuring at two points along the standing wave the absolute value of the individual pressures as well as their ratio and, in addition, confirming by calculation the value of every pressure used. They then proceeded to a tube of varying length, like a slide-whistle, which allowed measurements of the absorption coefficient to be made over a wide range of audio frequencies. They also showed how to extend the method to obtain the complex acoustic impedance of the porous material.

L. L. Beranek in 1947 (Paper 37) attacked anew the problem of measuring acoustic impedance via the standing-wave tube, a problem he had worked on previously (Beranek, 1940). He reviewed a number of previous methods for measuring acoustic impedance and presented his own modifications, greatly extending Wente's results in frequency range and in accuracy. We present a few pages from his 1947 paper. This is followed by the first two pages of W. K. Lippert's paper, Paper 38, which goes a little beyond Beranek's paper and book (Beranek, 1949). Basically, Lippert measures not one p_{max}/p_{min} but many of these. The resulting envelope curves provide an important correction factor in the measurement of the acoustic impedance of an absorbing panel.

P. B. Flanders in 1932 was looking for an impedance-measuring instrument that would be much more versatile than a standing-wave tube. His aim was to measure directly the complex acoustic input impedance of horns, tubes, and other "transmission lines." (G. W. Stewart (1926) had previously pioneered such a goal using a different approach.) In Paper 39 Flanders showed how the goal could be achieved via three pressure-microphone measurements taken at the junction of his radiating source-tube and, successively, (1) his unknown load (for example, a horn), (2) his first standard impedance, and (3) his second standard impedance. His beautiful experimental results, over a wide frequency range, compare closely with computed curves given in Olson (1947).

A. E. Kennelly, meanwhile, had been enthusiastically seeking to expand the applications of motional impedance, which is basically the transformed impedance of a mechanical or acoustic load as seen electrically by a telephone "receiver." Motional impedance was to be the ideal link between the acoustic test object and the electrical portion of the electromechanical driver. Kennelly's work pertinent to this volume is exemplified by two papers: the (1912) paper with G. W. Pierce and the 1921 paper with K. Kurokawa, presented here as Paper 41.

The famous 1912 paper treated motional impedance as an end in itself: how to analyze, electrically, the behavior of an electromechanical vibrator. As it happened, both Pierce and Kennelly were present in June 1912 at an AIEE meeting in Boston where a paper was presented by C. F. Meyer and J. B. Whitehead of Johns Hopkins University, entitled "The Vibrations of Telephone Diaphragms." During the discussion period Pierce got up and gave, in effect, a preview of the forthcoming Kennelly-Pierce paper. (Apparently Pierce came to the meeting armed with two lantern slides.) Pierce's contributions to this discussion period comprise Paper 40. A closing remark by Meyer is included. Clearly, an important discovery requires an observer who is not only lucky but also very alert.

Kennelly's 1921 paper with Kurokawa (Paper 41) treated motional impedance not as an end in itself but as a means—for measuring purely acoustical impedances. Kennelly tried to extend the motional impedance concept into a method and an instrument for the broad-band measurement of acoustic impedance. Now indeed, in the region of resonance the transducer is very "soft" and hence sensitive to changes in the acoustic load. Away from resonance, however, the transducer is very "stiff." The effect of the acoustic load on the stiff mechanical system is very small; the transducer is thus very insensitive, and the measurement accuracy is poor. This can be seen, in any book on transducers, from a look at a motional impedance or admittance "circle diagram" that covers a broad frequency range. The curve looks like a piece of string with one of more knots in it. Most of the knots are tight, but one of them may be loose enough to be called a loop. Only around a loop can the acoustic loading data be "reduced" with any accuracy. Attempts by later investigators to move a loop back and forth over the frequency band were not very successful (Fay and Hall, 1933).

REFERENCES

Beranek, L. L., 1940, Precision Measurement of Acoustic Impedance, *Acoust. Soc. Am. J.* **12**:3–13.

Beranek, L. L., 1947, Some Notes on the Measurement of Acoustic Impedance, *Acoustic. Soc. Am. J.* **19**:420–427.

Beranek, L. L., 1949, The Acoustic Transmission Line, *Acoustic Measurements,* Wiley, New York, pp. 317–336.

Fay, R. D., and W. M. Hall, 1933, The Determination of the Acoustical Output of a Telephone Receiver from Input Measurements, *Acoust. Soc. Am. J.* **5**:41–56.

Kennelly, A. E., and G. W. Pierce, 1912. The Impedance of Telephone Receivers as Affected by the Motion of Their Diaphragms, *Am. Acad. Arts Sci. Proc.* **48**:113–151. (Paper 34 in Acoustic Transducers, I. Groves, ed., Hutchinson Ross Publishing Company, Stroudsburg, Pa., 1981, pp. 338-352.)

Olson, H. F., 1947, Throat Impedance of Horns, *Elements of Acoustical Engineering,* 2nd ed., Van Nostrand, New York, pp. 100–112.

Sabine, W. C., 1922, Reverberation, *Collected Papers on Acoustics,* Harvard University Press, Cambridge, Mass. (Also Dover Publications, New York, 1964. The important first half of the paper appears as Paper 39 in *Acoustics: Historical and Philosophical Development,* R. B. Lindsay, ed., Dowden, Hutchinson & Ross, Stroudsburg, Pa., 1972, pp. 417–457.)

Stewart, G. W., 1926, Direct Absolute Measurement of Acoustical Impedance, *Phys. Rev.* **28**:1038–1047.

ANNOTATED BIBLIOGRAPHY

Crandall, I. B., 1926, Sound Transmission in Tubes, *Theory of Vibrating Systems and Sound*, Van Nostrand, New York, pp. 95–107.
Fine discussion of transmission-line problems. Refers frequently to the analogous behavior of electric lines. Reminiscent of Rayleigh's treatment.

Fleming, J. A., 1927, *Propagation of Electric Currents in Telephone and Telegraph Conductors*, 3rd ed., Constable & Son, London.
A famous basic reference for transmission-line behavior; and for good reason.

Hunt, F. V., 1954, Motional Impedance, *Electroacoustics*, Harvard University Press, Cambridge, Mass., pp. 94–102, 117–133, 155–167, and 202–208.
This is probably the finest discussion of motional impedance that has been written.

Kennelly, A. E., 1923, Acoustic Impedance, *Electrical Vibration Instruments*, Macmillan, New York, pp. 167–190.
Good expansion of Kennelly's various motional impedance papers.

Meyer, E., and E. Neumann, 1972, Transmission Line Theory, *Physical and Applied Acoustics*, Academic Press, New York, pp. 29–44.
The fine translation by J. M. Taylor makes it possible for English-speaking readers to catch the flavor of Meyer's originality in his treatment of acoustical problems.

Morse, P. M., 1948, The Flexible String, *Vibration and Sound*, 2nd ed., McGraw-Hill, New York, pp. 133–147.
A classic in its treatment of acoustic transmission-line problems. The first sharp departure from Rayleigh's treatment. (See also, Propagation of Sound in Tubes, pp. 233–261 in the same volume.)

Richardson, E. G., 1940, Acoustic Impedance, *Sound*, 3rd ed., Arnold and Co., London, pp. 227–250.
Good discussion of acoustic transmission lines.

Stewart, G. W., and R. B. Lindsay, 1930, Distributed Acoustic Impedance, *Acoustics*, Van Nostrand, New York, pp. 132–158, 190–196.
Good background material for the transmission-line papers by Taylor and Wente et al. Reminiscent of Rayleigh's treatment.

34

Reprinted from pages 270–274 and 277–287 of *Phys. Rev.* **2**:270–287 (1913)

A DIRECT METHOD OF FINDING THE VALUE OF MATERIALS AS SOUND ABSORBERS.

By Hawley O. Taylor.

THE sound-absorbing quality of materials has been recognized for some time as having an important bearing upon architectural acoustics. In 1873, T. Roger Smith, M.R.I.B.A., a leading British architect, wrote:[1]

"Where there is too much resonance in a room, carpets or curtains may be advantageously employed to lessen it; in fact, to absorb the injurious excess of sound, and where there is an echo, a curtain judiciously hung will have the effect of deadening or stopping the sound before reaching the echoing surface. This expedient is well known, and often successfully employed; but carpets, as a remedy for excessive resonance, have perhaps not been so frequently made use of."

Since 1895, extensive investigations have been carried on at Harvard University by Professor Wallace C. Sabine[2] to determine some quantitative relations between the size, reverberation and sound-absorbing power of a room, and he placed his results in the well-known form:

$$a = \frac{.171\,V}{t},$$

where a is the absorbing power of a room of volume V, and t is the time of decay of the residual sound of an organ pipe whose intensity in terms of "minimum audibility" had been previously determined. As determined by Sabine, a room is most satisfactory for speaking purposes when t has a value of about 1.5 seconds, thus the proper amount of absorbing power, a, is limited. (The above equation does not apply to echoes and defects due to the shape of a room.) Since the sound-absorbing power of an auditorium is thus limited to a narrow range of values, the consideration of absorption is of great importance in auditorium design and construction.

The amount of absorption is equal to the sum of the absorbing powers of the different materials composing the bounding surfaces and the

[1] T. Roger Smith, "Acoustics of Public Buildings," p. 39.

[2] Prof. W. C. Sabine, Am. Arch., Vol. LXVIII., April 7, 21, May 5, 12, 26, June 9, 16, 1900.

furniture of the room; the absorbing power of the material is equal to the product of its area and the amount of sound absorbed by unit area. Sabine rated the amount of sound absorbed by unit area of open window as 1, then the amount absorbed by a material which reflects a part of it is always less than 1. This fraction is the coefficient of absorption of sound for a given material.

Besides the power of materials to absorb sound reflected from their surfaces, it is desirable also to know the power of materials to absorb sound transmitted through them. The application of this to architecture extends beyond the acoustics of a single room and has to do with the isolation of one room from sounds produced in another. The interior finish of rooms is of no consequence here, but the filling in the partitions and in the space under the floors is the important thing. The fraction of sound absorbed by unit thickness of a given material may be called the coefficient of absorption of the material for transmitted sound, or the insulation coefficient of the material for sound.

To find the coefficient of absorption of a material for sound reflected from it the method usually employed is to bring a known amount of the material in question into a room and make observations for the time of decay of the residual sound, applying the formula above stated. This may be called the reverberation method. The purpose of this investigation is to devise a direct method of finding the coefficient of absorption of materials for sound. Such a method is formulated and following is the theoretical basis:

When a train of progressive waves, P (Fig. 1), of amplitude, p, moves in the positive direction, say, then all points of the medium in the path of the waves move through a distance $2p$, as shown by the shaded portion of Fig. 1.

If the progressive waves strike an object, S (Fig. 2), and a portion of

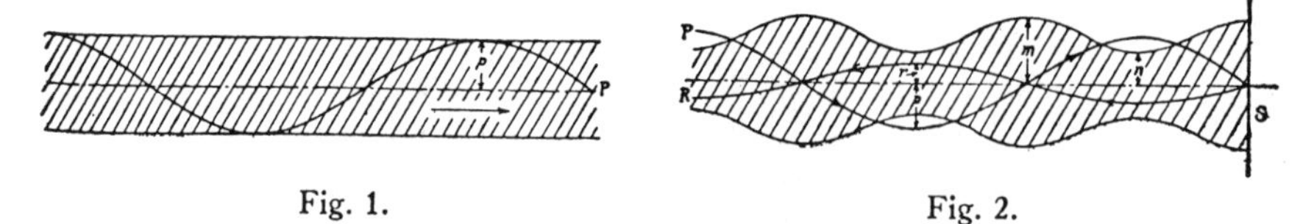

Fig. 1. Fig. 2.

each is reflected, these reflected waves, R, of amplitude, r, will travel in the negative direction, and the two trains of waves will meet and combine. Some points of the medium will then move through a greater and others through a less distance than before, as shown by the shading in Fig. 2. The maximum amplitude (where crest meets crest and trough meets trough) will be $p + r$, and the minimum amplitude (where the crests of P meet the troughs of R, and vice versa) will be $p - r$.

A special case of this phenomenon is when the progressive and reflected waves have the same amplitude, p. The resultant maximum amplitude is then $2p$, and the minimum amplitude is zero. This is the ordinary stationary or standing wave. In order to have no motion at the nodes the two waves must be equal in amplitude, which is only the case when there is no absorption at the reflecting surface.

The distance through which the medium moves varies from point to point, as represented by the amplitude of the envelope in Fig. 2. In the case of sound waves, this movement is the longitudinal vibration of the air particles, the intensity of the sound being greatest where the amplitude of vibration is greatest. The coefficient of absorption of sound is a function of sound intensity and therefore of the squares of the amplitudes of the progressive and reflected waves and of the resultant stationary wave (the envelope, Fig. 2). The intensity of sound in the progressive wave is $(kp)^2$, and the intensity of sound in the reflected wave is $(kr)^2$ (where k is a proportionality constant). This gives a coefficient of reflection,

$$\rho = \frac{r^2}{p^2}.$$

If the maximum intensity of sound (corresponding to the maximum amplitude of the envelope, Fig. 2) is called m, and the minimum intensity (corresponding to the minimum amplitude of the same envelope) is called n, then

$$m = (p + r)^2,$$

$$n = (p - r)^2,$$

from which

$$p = \frac{m^{\frac{1}{2}} + n^{\frac{1}{2}}}{2},$$

$$r = \frac{m^{\frac{1}{2}} - n^{\frac{1}{2}}}{2}.$$

Substituting these expressions for p and r in that for the coefficient of reflection of sound, we have

$$\rho = \frac{(m^{\frac{1}{2}} - n^{\frac{1}{2}})^2}{(m^{\frac{1}{2}} + n^{\frac{1}{2}})^2}.$$

The coefficient of absorption of sound, α, is the fraction of sound not reflected, or

$$\alpha = 1 - \rho = 1 - \frac{(m^{\frac{1}{2}} - n^{\frac{1}{2}})^2}{(m^{\frac{1}{2}} + n^{\frac{1}{2}})^2} = \frac{4}{m^{\frac{1}{2}}/n^{\frac{1}{2}} + n^{\frac{1}{2}}/m^{\frac{1}{2}} + 2}.$$

To find the coefficient of absorption, α, of a material, means must be provided (I.) for isolating a train of sound waves having a single period or pitch and a constant amplitude, (II.) for causing this train of waves to be reflected upon the surface whose absorption is desired, and (III.) for measuring the maximum and minimum intensities, m and n, of the resultant stationary wave.

Method of Procedure.

I. A tone of uniform intensity was produced by supplying an organ pipe with air from a pneumatic tank. This tank consisted of two sheet-metal cylinders closed at one end, the smaller of which was about 9 inches in diameter and 18 inches high and was placed, open end downward, into the larger which was half full of water. Air was drawn off by means of a faucet on top of the smaller tank and conducted to the organ pipe through ¼-inch tubing. The pressure of the air was regulated by adjusting a stop against which a rod soldered to the valve of the faucet would strike when the faucet was open the desired amount. The inner tank was weighed down by about 25 pounds of iron, thus making the pressure uniform for almost the entire air capacity.

To obtain a simple tone sound was passed through what might be called a tone screen which absorbed all the overtones leaving nothing but the fundamental. This tone screen is based upon Quincke's modification of Herschel's interference tubes described by Rayleigh.[1]

If *CD*, Fig. 3, is tuned to the pitch of sound waves traveling from *A*

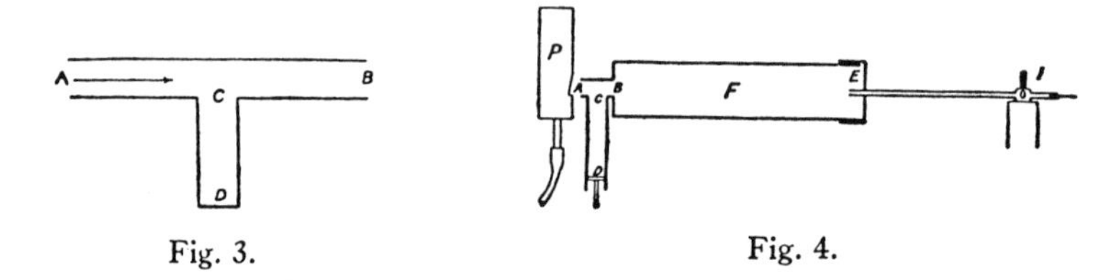

Fig. 3. Fig. 4.

to *B*, and the opening at *C* is as large as the section of the tube *AB*, no sound will reach *B*—it is completely absorbed by the tube *CD*.

The truth of this was tested by taking a wooden flue, *F* (Fig. 4), of square cross-section 9 centimeters on a side and 115 centimeters long, placing a cap over end, *E*, through which projected a ½-inch glass tube from a sound-measuring instrument, *I*, and placing the interference tube or tone screen at end *B*. The tube *AB* was an inch hole bored in a block of wood 6 inches thick, and tube *CD* was a one-inch glass tube, the end *D* being a cork piston which allowed tuning of the tube. When the tube *CD* was removed, sound from the stopped organ pipe, *P*, passed through tube *AB* into the flue, *F*, and the measuring instrument indicated

[1] Lord Rayleigh, Theory of Sound, Vol. II., p. 210.

a large deflection. When the tube *CD* was in place and tuned to the organ pipe, the deflection of the measuring instrument was zero. (The tube *CD* was tuned to the fundamental of the organ pipe and hence also to its retinue of overtones, thus all of the sound of the organ pipe was absorbed.) When the tube was placed slightly out of tune with the organ pipe, a small deflection was obtained, and the deflection became larger as the dissonance was increased.

To isolate the fundamental tone of the organ pipe in the flue, *F*, four tubes similar to *CD*, tuned one to each of the first four overtones of the organ pipe, were placed around the passage *AB*. Since any overtones above the fourth for the organ pipe used were very weak if present at all, this constituted a very efficient tone screen. The tone in the flue was found to be simple by finding the sound intensity for each centimeter throughout the length of the flue and plotting a curve of intensity against distance. The curve was that of a simple tone.

A tone of uniform intensity having been obtained from an organ pipe, and its fundamental isolated in the flue, *F*, we must now ascertain whether or not the amplitude of the wave in the flue is constant throughout the length of the flue, as it is in Figs. 1 and 2. To this end, the first requisite is a flue of uniform cross-section to prevent the wave-front from spreading out and thus diminishing in amplitude. But if the flue lining absorbs the wave, the effect will be the same as a widening flue—the wave-front will spread out and both the progressive and reflected waves will diminish in amplitude as they advance.

Any detrimental presence of this lining absorption may be found by observing the maximum intensities all along the flue; they should be practically equal, otherwise the lining of the flue is absorbing sound.

Following are some readings made with a wooden flue and with the flue lined with felt (Fig. 5).

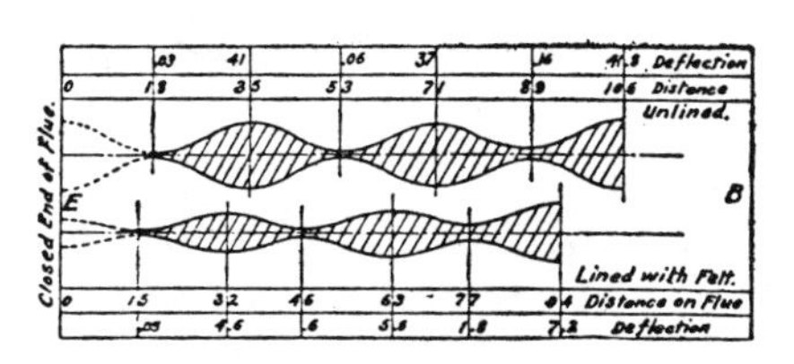

Fig. 5.

Lining absorption occurs in both cases, but a marked diminution from *B* toward *E* is seen in the case of the felt. A smooth, painted wood lining shows practically uniform maxima.

In connection with flue lining, the table of observations indicates that the wave-length shortens as the lining absorption increases. This means a decrease in the velocity of sound with increase of absorption.

[*Editor's Note:* Material has been omitted at this point.]

Any other "open end pipe" effect of the flue must be stopped. If the flue is a metallic cylinder, 4 or 5 inches in diameter, all crevices at the end, *E* (Fig. 9), may be done away with by means of the construction shown—a shrunk-on collar, *s*, over the finished surface of which the metallic cap, *c*, slides on and off. The absorbing material, *S*, is placed at the end of the flue, backed by any material, *G*, such as wood, that it is desired to use.

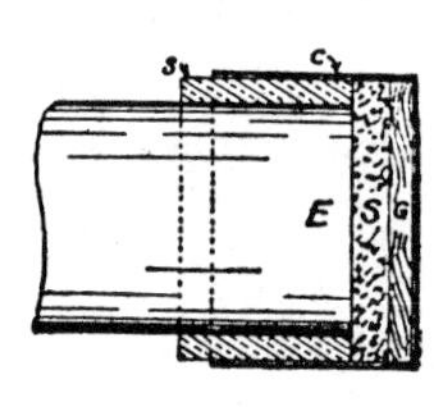

Fig. 9.

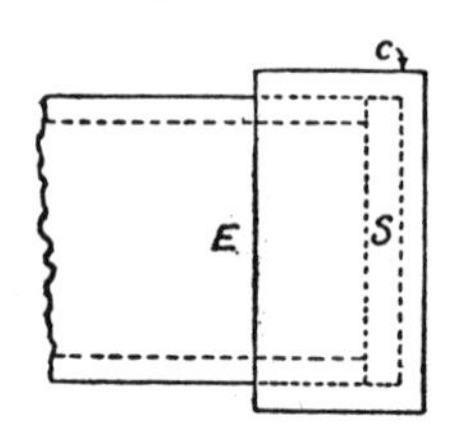

Fig. 10.

The flue used in the experimental work was made of wood of the dimensions already given. Its end, *E* (Fig. 10), was provided with a wooden cap, *c*, which served to hold the material, *S*, snugly against the flue and to close the end as nearly air-tight as possible.

III. In the search for means for measuring the intensity of sound tests were made of everything of any promise, and telephone receivers and transmitters,[1] strong and weak field galvanometers, molybdenite and silicon rectifiers, barretters[2] and microradiometers[3] all figured. The Rayleigh disc[4] was finally adopted as the most reliable and sensitive sound measuring instrument.

[1] Prof. G. W. Pierce, Proc. Am. Acad., Vol. XLIII., 13, p. 377, 1908.

[2] Prof. A. E. Kennelly, Trans. Internat. Elect. Cong. St. Louis, 1904, Vol. III., pp. 415–437, 1905; Bela Gati, Engr., 105, p. 24, Jan. 3, 1908.

[3] Prof. F. R. Watson, PHYS. REV., 28, p. 385, 1909.

[4] Lord Rayleigh, Phil. Mag., Vol. XIV., p. 186, 1882.

The disc used was made of mica, 1 cm. in radius, suspended on a quartz fiber about 10 cm. long, a small mirror being fixed to the top of it like the mirror of a D'Arsonval galvanometer, but when the disc was turned through an angle of 45°, the mirror faced the front. It hung in a little wooden box with a glass front and two one-inch glass tubes led into opposite sides of the box so that, virtually, the disc was suspended near the center of a one-inch glass tube. This tube was stopped at both ends with corks, and was of such a length as to be in resonance with the sound employed. A one-half-inch glass tube about three feet long (the length adjusted to resonance with the pitch employed) pierced one of the corks and served to deliver the sound energy to the disc. A sound at the end of the long tube causes the air in the tube to vibrate, and an antinode, or region of greatest velocity of air particles, is formed where the disc hangs. As is well known, the disc tends to set itself at right angles to the direction of motion of the air stream, and turns through an angle which is a measure of the intensity of the sound at the end of the long tube.

The equation of the torque acting upon the suspended disc, as worked out by W. König,[1] is

$$M = 4/3\rho r^3 v^2 \sin 2\theta,$$

where M is the moment of the couple due to the stream of density ρ, flowing with a velocity v, and r is the radius of the disc whose normal makes an angle θ with the direction of the undisturbed stream. In the case of the sound vibrations the velocity v would be alternating and the air particles would move with simple harmonic motion, thus the energy due to the vibrations would be proportional to the mean square velocity. W. Zernov has shown[2] that König's equation holds for sound vibrations in free air, the v^2 being proportional to the intensity of the sound and equal to the mean square velocity of the air particles.

For use in measuring sound intensity, the Rayleigh disc is suspended in a resonance tube. It has been shown by Stewart and Stiles[3] that the intensity of sound in a resonance tube is proportional to the intensity of the exciting sound in the free air at the mouth of the tube. The following simple experiment also brings out this fact: A disc was hung near the mouth of a tube and a source of sound (an organ pipe) placed near the end of the tube (Fig. 11). A stop at d made the tube a resonator for the sound and deflections were made for several different distances of the organ pipe from the tube. The stop was then moved to d' thus

[1] W. König, Wied. Ann., XLIII., p. 51, 1891.

[2] W. Zernov, Ann. der Phys., XXVI., p. 79, 1908.

[3] Stewart and Stiles, PHYS. REV., April, 1913, p. 309.

throwing the tube out of resonance. The sound vibrations which affected the disc were then not due to resonance but were of the character of reverberation in a room and the disc responded as in free air. Deflections were again taken for the same distances of the organ pipe from the end of the tube. Following is the result — showing the same law to hold in both cases:

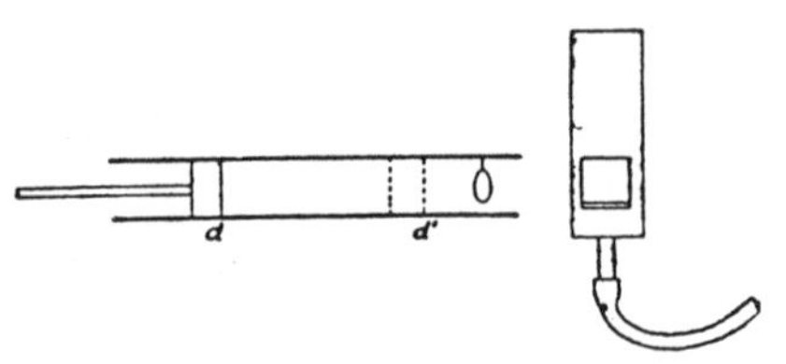

Fig. 11.

Case of Reverberation:

Organ Pipe Distance.	Deflection.
1 cm.	20.1
2 cm.	10.2
3 cm.	6.0

Case of Resonance:

Organ Pipe Distance.	Deflection.
1 cm.	36.0
2 cm.	18.4
3 cm.	10.8

The similarity in the behavior of the disc in the two cases is clearly shown by the similarity in the shape of the curves (Fig. 12).

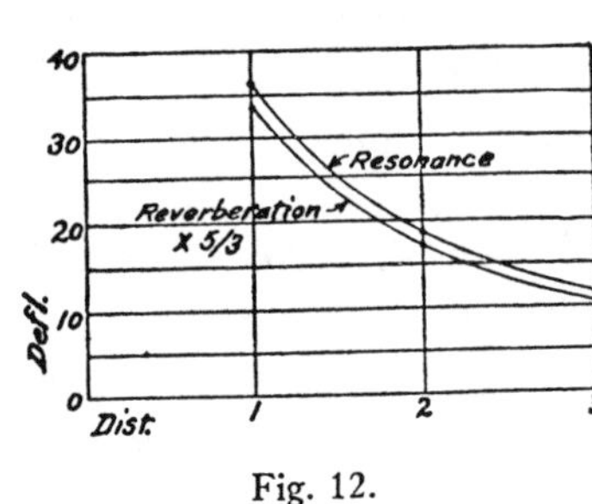

Fig. 12.

Thus the intensity of sound in a resonance tube is proportional to the intensity of the sound in free air at the mouth of the tube, and the deflections of a disc in a resonance tube are a measure of the intensity of the sound at the mouth of the tube.

With König's equation as a basis, the calibration of the disc for sound intensity may be worked out as follows:

When responding to sound, the disc turns through an angle which we will represent by ϕ. Since the apparatus was so constructed that $\theta = 45°$ when there is no sound, then $\phi = 45° - \theta$, and $2\phi = 90° - 2\theta$. When the return torque of the suspending fiber is balanced by the couple acting on the disc due to sound vibrations, then

$$L\phi = M = kv^2 \sin 2\theta,$$

where L is the moment of torsion of the suspending fiber, and $k = 4/3\rho r^3$. (ρ, the density of the air, is a constant at an antinode of sound vibration.)

$$\cos 2\phi = \cos (90° - 2\theta) = \sin 2\theta,$$

therefore

$$L\phi = kv^2 \cos 2\phi = qi \cos 2\phi,$$

where i (the intensity of the sound) is proportional to v^2, q being equal to k times the proportionality factor.

Then

$$i = \frac{L\phi}{q \cos 2\phi} = \frac{b2\phi}{\cos 2\phi},$$

where

$$b = \frac{L}{2q}.$$

Since the deflection, δ, on the scale is equal to $s \cdot \tan 2\phi$, where s is the distance of the mirror from the scale, then $2\phi = \tan^{-1} \delta/s$, and

$$i = \frac{b \cdot \tan^{-1} \delta/s}{\cos (\tan^{-1} \delta/s)} = \frac{b \cdot \tan^{-1} \delta/s}{s/(\delta^2 + s^2)^{\frac{1}{2}}}$$

$$= \frac{b(\delta^2 + s^2)^{\frac{1}{2}} \tan^{-1} \delta/s}{s} = h(\delta^2 + s^2)^{\frac{1}{2}} \tan^{-1} \delta/s,$$

where $h = b/s$.

For small deflections, δ^2 may be neglected in comparison with s^2, and the angle and its tangent may be considered equal, then the intensity of sound is proportional to the deflection of the disc, or

$$i = k\delta.$$

The error involved in using this approximate formula may be brought out by the following example:

Let the distance from mirror to scale, s, be 100 cm. and two deflections be $\delta_1 = 40$ cm. and $\delta_2 = 0.87$ cm.

Using the exact formula for computing relative sound intensity:

$$\frac{i_1}{i_2} = \frac{h \cdot 107.71 \cdot 21.8\pi/180}{h \cdot 100.004 \cdot \pi/360} = 46.96.$$

Using the approximate formula involves merely a comparison of the deflections:

$$\frac{i_1}{i_2} = \frac{40}{.87} = 45.84.$$

Thus the percentage error is $\frac{46.96 - 45.84}{46.96} \times 100$ or 2.2 per cent.—the approximate formula giving the smaller value for the louder sound.

The error involved in using the approximate formula for the computation of the coefficient of absorption, α, may be shown by using these results in the formula for α already derived:

For the correct ratio of sound intensities, $m = 46.96$, and $n = 1$; then

$$\alpha_0 = \frac{4}{m^{\frac{1}{2}}/n^{\frac{1}{2}} + n^{\frac{1}{2}}/m^{\frac{1}{2}} + 2} = \frac{4}{6.8527 + 0.1459 + 2} = \frac{4}{8.9986} = 0.4445.$$

For the approximate ratio of sound intensities, $m = 45.84$, and $n = 1$; then

$$\alpha_1 = \frac{4}{6.7702 + 0.1447 + 2} = \frac{4}{8.9179} = 0.4485.$$

The percentage error for the coefficient of absorption is

$$\frac{0.4485 - 0.4445}{0.4445} \times 100 = 0.9 \text{ per cent.}$$

As the ratio of sound intensities approaches unity, the percentage error decreases, therefore, for the computation of the coefficient of absorption, the approximate formula for relative sound intensity is close enough for material of high absorbing power but should not be used when the absorbing power is less than 0.20.

In agreement with the above calibration of the disc deduced from theoretical considerations, the result of the following experiment may be cited.

The disc was suspended in a resonance tube an inch in diameter and ¾-wave-length (about 30 cm.) long, closed at both ends except for four small glass tubes of one-eighth-inch bore lying across one end, each tube pierced with a little hole which opened into the resonance tube (Fig. 13). The tubes were mounted in a wooden cap which fitted nicely over the end of the resonance tube, and the whole arrangement constituted an adaptation of Quincke's interference tube. A sound passing through one or all of the four little tubes would be swallowed up by the resonance tube, if its pitch is that to which the latter is tuned. One end of each of these four tubes was fitted into the wooden cap closing end, *E*, of the wooden flue already described which served as a horn to catch the sound from an organ pipe placed at the end, *B*, and four little wooden rods were made of such a size as to slip into the glass tubes and stop all sound from coming into them.

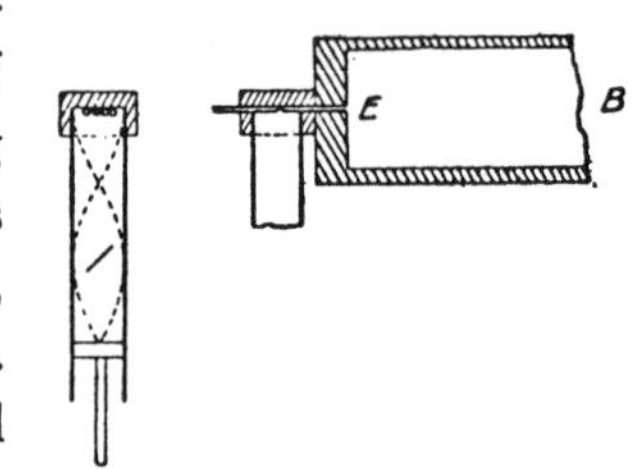

Fig. 13.

With one rod removed, the organ pipe was sounded and a deflection of the disc was noted. The resonance tube in which the disc hung had absorbed all of the sound which was passing through one little tube, and the disc felt the impulse of the sound vibration. The same was done with two tubes open, then three and then all four. The following is a sample of the deflections received:

Number of Tubes.	Deflection.
1	.12
2	.46
3	.98
4	1.96

The deflections are seen to run about in the following proportion: 1 : 4 : 9 : 16. Variations may be due to slight differences of bore in the tubes. The sound in the first tube produced a certain amplitude in the resonance tube; the sound in two tubes gave twice as great an impulse or produced twice the amplitude in the resonance tube, etc. All of the sound delivered to the resonance tube was very weak and could produce no great crowding of air particles, and so the amplitudes were able easily to be superimposed one upon another without loss of motion, and since the intensity of sound varies as the square of the amplitude, it would vary, also, in this case, as the square of the number of tubes delivering sound to the resonance tube. The resulting deflections very nicely confirm this, and since the deflections were all small they would come within the approximation made above which gave the final calibration formula, $i = k\delta$, deduced from König's law of the Rayleigh disc. Thus, for the Rayleigh disc, it is theoretically and experimentally established that, for small deflections, the intensity of sound varies as the deflection; for large deflections, the intensity may be computed from the more extended formula,

$$i = h(\delta^2 + s^2)^{\frac{1}{2}} \tan^{-1} \delta/s.$$

The calibration of the Rayleigh disc provides the necessary means for measuring the ratio of the maximum and minimum sound intensities in the sound flue (Fig. 4) and makes possible the calculation of the coefficient of absorption of sound by the use of the formula:

$$\alpha = \frac{4}{m^{\frac{1}{2}}/n^{\frac{1}{2}} + n^{\frac{1}{2}}/m^{\frac{1}{2}} + 2}.$$

When the Rayleigh disc is used to measure sound intensity the range of pitch is limited by the fact that the disc must lie in an antinode, and density variation of the air must be absent. For very short waves the disc would swing out of the region of constant density and thus complicate the calculation of sound intensity. For very high pitches, therefore, very small discs must be used. For a disc one millimeter in diameter there would be no great error when used with a wave two centimeters long, which would correspond to a frequency of about 16,500 vibrations per second, or a pitch of about C_9—six octaves above middle C (C_3—256 vibs./sec.). An octave and a half above this would take us to the upper limit of audibility, showing that the apparatus is applicable to any probable range of pitch.

The different parts of the apparatus used for making observations for sound absorption have already been described. The parts were assembled as shown in Fig. 14. The flue, *F*, moved along a graduated

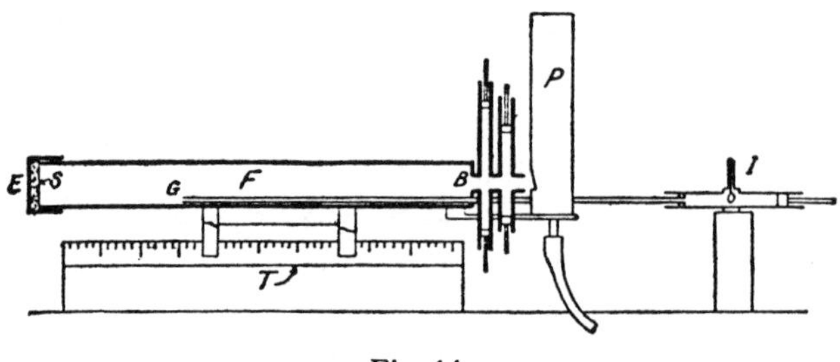

Fig. 14.

track, *T*. The one-half-inch glass tube, *IG*, from the suspended disc passed through the tone screen at *B* and projected into the flue. The intensity of sound at the end, *G*, of this tube causes the air in the tube to vibrate and this in turn produces a deflection of the disc.

In taking observations, the flue, *F*, with the organ pipe, *P*, attached to it was moved along the track, *T*, in steps one centimeter long. Observations for intensity were made at each step, and at the region of maximum and minimum intensities for shorter steps, and thus the maximum and minimum intensities, m and n of the formula, were found. Applying these values in the formula, the coefficient of absorption of sound for the material, *S*, closing the end, *E*, of the flue, *F*, was calculated as already shown by an example.

Following is the coefficient of absorption of a few materials computed by using the approximate formula:

Material.	m	n	α
Smyrna Rug	6.55	.04	.26
Brussels Carpet	7.30	.03	.23
Hair felt (1-inch thick)	24.70	.78	.51
Ceilinite (½-inch thick)	57.50	.22	.25
Asbestos roll fire felt (¾-inch thick)	54.50	.30	.26
Compressed cork (1½-inch thick)	27.00	.25	.32

The last four materials were kindly furnished by the Johns-Manville Co. of New York. The values of the absorbing power given here for the first three materials check fairly well with values obtained by the reverberation method.

The absorbing power of material is increased by increasing the space between it and the wall behind it. This may be explained by considering sound absorption as a retardation of the motion of the vibrating air particles; hence, by placing the absorbing material out a little from the wall (where the air is at rest) it becomes more effective in retarding motion and thus more efficient as a sound absorber. Following are the coefficients for compressed cork taken for several different distances between it and the board behind it:

Compressed Cork.

Distance from Board.	m	n	α
2 mm.	25.00	.25	.34
5 mm.	25.50	.34	.38
10 mm.	18.30	.46	.47
20 mm.	13.70	.74	.61

The relation is also shown by the curve (Fig. 15):

An important application of the flue to auditorium acoustics is the following: Sabine has found[1] that the reverberation of a room varies for different parts of the musical scale. To correct the acoustics of a room, therefore, absorbing material must be so applied as to make the reverberation for each tone normal. Since the absorbing power of a material varies with its distance from the wall, and this distance varies for the maximum absorption of each tone, it is possible to so graduate the distance of the material from the wall as to make the reverberation of the room for each tone of the entire musical scale normal. The flue makes the determination of the absorbing power of materials for variation of tone and distance from wall a simple matter.

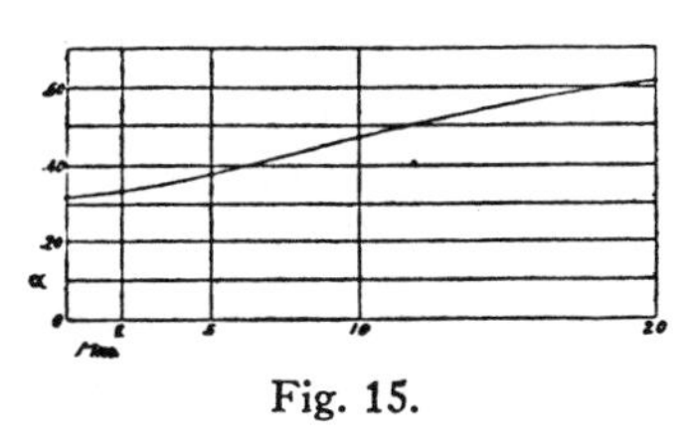

Fig. 15.

The flue method of finding the value of a material as a sound absorber eliminates the effect of the interference system of a room on the absorbing power of the material. As shown by Sabine[1] a material of high absorbing power placed in a room where, due to interference, the intensity of sound, or sound pressure, is very small would have little effect on the sound in the room, but if placed in a region of large relative sound pressure it would diminish the sound perceptibly.

If the absorbing power of this material were obtained by the reverberation method in this room in the first position mentioned, it would be found to be relatively small, and in the second position, relatively large. If one of these is the correct absorbing power, the other is not, and the probability is that neither is correct.

[1] Prof. W. C. Sabine, Arch. Quar. of Harvard Univ., March, 1912, p. 20.

Conversely, if the correct absorption of a material is found by means of the flue, and this material is used to reduce the reverberation of an auditorium, its value will depend somewhat upon its position in the room. Thus the true power of absorption of a material will be useful in connection with auditorium acoustics only when used with due precaution—and then it is of much value. It is useful also as an index to give the material a rating with relation to other sound absorbing materials.

The power of materials to absorb sound transmitted through them was found as follows:

The wooden flue already described (Fig. 4) was stopped at end, *E*, by a board one-half-inch thick, through the center of which passed a one-half-inch glass tube from the Rayleigh disc, *I* (Fig. 16). The tone screen was removed from end *B* of the flue, and a closely fitting tin box slipped into the end. The two opposite ends of this box, through which the sound must pass to enter the flue, were made of cheese cloth and their distance apart was adjustable.

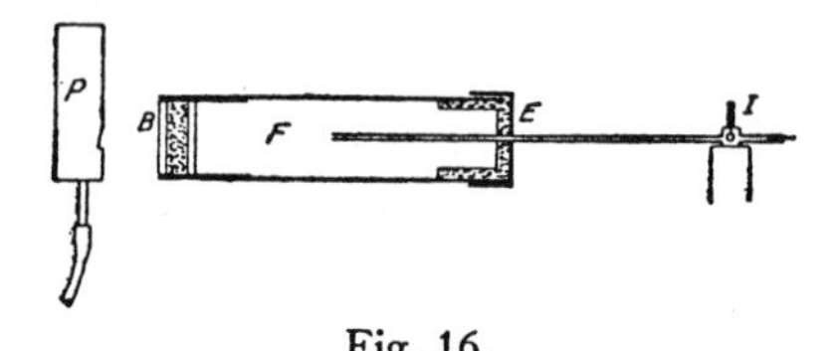

Fig. 16.

With this box in the end of the flue, *F*, and an organ pipe, *P*, placed immediately behind it, the first reading was made giving the intensity of the sound in the flue when end, *B*, was not closed with the material whose absorption was wanted. Then the cheese cloth ends of the tin box were adjusted for several different distances apart and for each one the box was filled with absorbing material and slipped into the end of the flue for a reading. Following are the readings made for water glass crystals loosely shaken down but not packed, the organ pipe sounding the note *si* (240 vibs./sec.):

Thickness. cm.	Deflection. δ.	Fraction of Sound	
		Transmitted.	Absorbed.
0	44.20	1.000	0.000
2	24.90	.564	.436
4	15.50	.352	.648
6	11.50	.260	.740
10	5.50	.124	.876
16	1.10	.025	.975
20	.53	.012	.988

The intensity of sound varies as the deflection, $i = k\delta$. If I represents the sound intensity transmitted through zero thickness of the material, and i the intensity transmitted through thickness, t, then i/I is the fraction of sound transmitted through thickness, t, and $1 - i/I$ is the fraction absorbed by thickness, t. The following equation gives the relation, when t is small:

$$i/I = (1 - \beta)^t,$$

where β is the fraction of sound absorbed by unit thickness (1 cm.) of the material. This fraction is the insulation coefficient for sound for the material.

The insulation coefficient for three materials follows:

Material.	β
Water Glass Crystals	.211
Hair Felt	.226
Compressed Cork (1-inch thick)	.500

The curve between observed thickness and sound transmitted shows the experimental relation (Fig. 17):

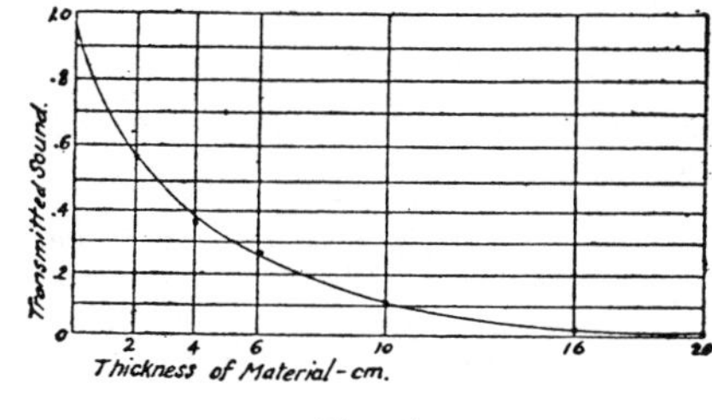

Fig. 17.

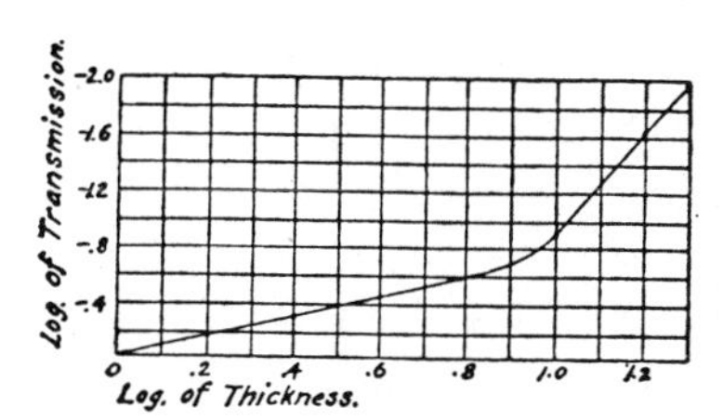

Fig. 18.

The curve between the logarithm of the fraction of sound transmitted and the logarithm of thickness shows that the logarithmic relation holds only when t is small, and that a distinct change occurs in the law of absorption in a certain region (Fig. 18):

The absorbing lining at the end, E (Fig. 16), of the flue makes internal reflection in the flue impossible. When sound strikes the outer surface of the material at the end, B, it is, in general, partially reflected, thus the amplitude of the wave which actually enters the material is not that of the incident wave. The coefficients given above express the relation between the incident and transmitted sound. Architecturally, this coefficient would probably be more useful than the coefficient based upon the sound which actually enters the material, since it is the intensity present in a room and incident upon a wall that one hears, and the wall is constructed to isolate this sound from adjoining rooms. Thus this is properly the insulation coefficient of a material for sound.

The intensity of sound which actually enters a material may be obtained by finding the coefficient of reflection, ρ, of the material when it is in place at the end, B, of the flue, F, and taking the product of this and the first deflection, δ_0, in the above table; δ_0 is proportional to the incident sound, therefore $\rho\delta_0$ is proportional to the reflected sound, and $\delta_0 - \rho\delta_0$ is proportional to the intensity of the sound entering the material. Simplifying, we have $\delta_0(1 - \rho) = \delta_0\alpha$, since $1 - \rho = \alpha$; therefore the intensity of the sound actually entering the material is proportional to the product of the deflection when no material covers the end, B, of the flue and the coefficient of absorption of the material when it is in place at the end of the flue with a column of air behind it. The fraction of sound transmitted when $\delta_0\alpha$ is used as a basis may properly be called the transmission coefficient of the material for sound, and it may be computed by using $\delta_0\alpha$ in place of δ_0 in the computations for $(1 - \beta)$.

Reflection from the outer surface of the material at the end, B, of the flue may be prevented by a covering of acoustically black material—loose felt. The sound transmitted by this covering may be found by making observations with it alone, and $\delta_0\alpha$ thus obtained experimentally.

The methods for finding the absorption of sound as outlined above may be applied in a number of ways; following is a partial list of suggestions:

(*a*) Determining the relation between the absorbing power of a material and its distance from the wall behind it.

(*b*) Determining the relation between the absorbing power of a material and pitch of sound.

(*c*) Determining the relation between the absorbing power of a material for transmitted sound and the thickness of the material.

(*d*) Determining the relation between the absorbing power of a material for transmitted sound and pitch of sound.

(*e*) Determining the relation between the absorbing power of a material and its composition or texture for a range of different pitches of sound.

(*f*) Determining the relation between the sound absorbing quality and elasticity of a material.

(*g*) Determining the absorbing power of materials by means of the effect of absorption upon the pitch of an organ pipe.

The Rayleigh disc may be used:

(*h*) For the detection of the presence of overtones in a sound by suspending it in a tube tuned to the pitch of the overtone.

(*i*) To bring psychological observations of ear accommodation up to a greater degree of completeness and accuracy.

PHYSICAL LABORATORY,
CORNELL UNIVERSITY,
May 26, 1913.

EDITOR'S NOTES

Page 271: Figure 2 shows two separate phenomena: the progressive wave system of *displacement* waves *p* and *r* and the standing-wave system of *pressure* waves, which is shifted by $\lambda/4$ from a standing-wave system of displacement waves (not shown). If the two displacement waves were exactly equal, the standing-wave pressure minimum at *n* would be exactly zero. With an absorbing panel at one end of the tube, however, the minimum is greater than zero.

Taylor's method does not require the measurement of absolute displacements or pressures or whatever. Therefore, Taylor only had to measure the ratio of some maximum to a corresponding minimum. Since he used a Rayleigh disc (see p. 278), he was actually measuring particle velocity-squared maximum to particle velocity-squared minimum. This indeed gave him the ratio of two intensities (see also p. 272). Paris's paper (Paper 35), portions of which are reprinted here, claims to be measuring pressure. But Paris used a hot-wire microphone, another particle velocity-squared device whose readings correlate with pressure only in the far field of a free-field wave system. However, Paris took the square root of his readings and then used only ratios, like Taylor, and so achieved correct results.

Page 273: In Figure 3, *CD* is a "wave trap" comprised of a quarter-wave branch tube. The trap does not "absorb" the sound, it diverts the sound.

Page 287: Suggestion (*b*) had to wait for Wente (Paper 36, Fig. 3).

35

Reprinted from pages 269 and 272–273 of *Phys. Soc. London Proc.* **39**:269–295 (1927)

ON THE STATIONARY-WAVE METHOD OF MEASURING SOUND-ABSORPTION AT NORMAL INCIDENCE.

By E. T. Paris, *D.Sc.*, *F.Inst.P.*

Received January 12, 1927.

Abstract.

A description is given of apparatus employed for measuring coefficients of sound-absorption by the stationary-wave method. The apparatus differs from that used by earlier workers in the use of (1) a small tuned hot-wire microphone for determining relative pressure-amplitudes in the sound-waves ; (2) the employment of a steady valve-driven source of sound with arrangements for maintaining the strength at a constant value ; (3) the screening of source and experimental pipe from disturbances due to the movements of the observer. By the employment of a certain procedure the relation between the response of the microphone and the amplitude of the pressure-variation in the sound-wave is eliminated. Some examples of the employment of the apparatus for determining the coefficients of absorption at normal incidence of acoustic plasters and hair-felt are given.

[*Editor's Note:* Material has been omitted at this point. On the following pages, Equation 2.7 shows the symmetry between adding two progressive waves to achieve a standing wave and adding two standing waves to achieve a progressive wave. Equation 2.91 provides additional insight by presenting an alternative formulation of Equation 2.7. Both points of view are useful when a "node" is only a quasi-node.]

The total potential within the pipe is

$$\Phi=\varphi+\varphi' \quad \ldots \ldots \ldots \quad (2.3)$$

If the reflecting surface at $x=0$ were perfectly rigid, the condition to be satisfied by Φ at $x=0$ would be

$$\left(\frac{d\Phi}{dx}\right)_{x=0}=0 \quad \ldots \ldots \ldots \quad (2.4)$$

expressing the condition that there is no motion in the plane $x=0$, so that no vibrational energy can be transmitted into the reflecting substance. Under these circumstances, the amplitudes of the incident and reflected waves are equal ($A=B$) and $\varepsilon=0$. The result is a stationary wave

$$\Phi=2A \cos kx \cos kat \quad \ldots \ldots \ldots \quad (2.5)$$

within the pipe.

If part of the energy of the incident sound is lost at reflexion we have

$$\Phi=A \cos k(at-x)+B \cos k(at+x+\varepsilon) \quad \ldots \ldots \quad (2.6)$$

which, by a simple transformation ($t'=t+\frac{1}{2}\frac{\varepsilon}{a}$, $x'=x+\frac{1}{2}\varepsilon$), may be written

$$\begin{aligned}\Phi&=A \cos k(at'-x')+B \cos k(at'+x')\\ &=(A+B) \cos kx' \cos kat'\\ &\quad +(A-B) \sin kx' \sin kat' \quad \ldots \ldots \ldots \quad (2.7)\end{aligned}$$

so that the motion within the pipe can be regarded as being due to two superimposed stationary waves, one of amplitude $(A+B)$ and the other of amplitude $(A-B)$, the nodes and loops of the first being one quarter of a wavelength distant from the nodes and loops of the second. The result is that there is within the pipe a series of positions of maximum and minimum pressure-variation proportional respectively to $(A+B)$ and $(A-B)$, spaced at distances of one quarter of a wavelength. It is the ratio of the amplitudes at these maxima and minima which is found in the experiment.

To calculate the coefficient of absorption we note that the energy-flux in the incident wave is proportional to A^2, and the energy-flux in the reflected wave is proportional to B^2. The fraction of the incident sound-energy lost at reflexion is thus proportional to $(A^2-B^2)/A^2$, and this is, by definition, the coefficient of absorption (α). Let the observed ratio of the maximum amplitude to the minimum amplitude be a/b. Then to find α in terms of ratio a/b we have, since

$$\frac{a}{b}=\frac{A+B}{A-B}, \quad \ldots \ldots \ldots \quad (2.8)$$

$$\alpha=\frac{A^2-B^2}{A^2}=\frac{4ab}{(a+b)^2} \quad \ldots \ldots \ldots \quad (2.81)$$

Or,

$$\alpha=\frac{4}{2+a/b+b/a} \quad \ldots \ldots \ldots \quad (2.82)$$

This last expression is that given by Hawley Taylor, and is the most convenient form for computing α from the observed ratio a/b.

[*Editor's Note:* In line 9 above, ϵ is a phase-shift factor with dimensions of length.]

Alternatively, let—

$$A=(1+m)B, \qquad (2.9)$$

where m is positive, since $A>B$. Then—

$$\Phi=B\cos k(at'-x')+B\cos k(at'+x')+mB\cos k(at'+x')$$
$$=2B\cos kx'\cos kat'+mB\cos k(at'+x'). \qquad (2.91)$$

Alternatively, therefore, the motion within the pipe may be regarded as being that due to a stationary wave of amplitude $2B$, on which is superimposed a progressive wave of amplitude mB. The loops and nodes of the stationary wave are at the positions of maximum and minimum pressure-variation which occur when an absorbent substance is placed at the end of the pipe.

Also, according to (2.91), the positions of the maxima and minima differ from the position of the nodes and loops which would occur if a perfect reflector were used to close the pipe, by a distance $\frac{1}{2}\varepsilon$. No special attempt was made to measure this shift in the experiment here described, but it may be of interest to note that it was of the order of 1 cm. for a substance with an absorption-coefficient of about 0·3 at 512 variations per second. The shift was in the direction x positive, and thus indicated a virtual node of the stationary vibration in (2.91) about 1 cm. within the surface of the absorbing material. It is possible that the measurement of this shift might be of service in investigating the mechanism by which absorption takes place in some materials. In the case of resonant reflexion it might happen that the virtual node would be in front of the reflecting surface.

[*Editor's Note:* Material has been omitted at this point.]

36

Reprinted with permission from pages 1–8 and 10 of *Bell Syst. Tech. J.* **7**: 1–10 (1928)

The Measurement of Acoustic Impedance and the Absorption Coefficient of Porous Materials

By E. C. WENTE and E. H. BEDELL

SYNOPSIS: Various ways of determining the acoustic impedance and the absorption coefficient of porous materials from measurements on the standing waves in tubes are discussed. In all cases the material under investigation is placed at one end of the tube and the sound is introduced at the other end. Values of the coefficient of absorption of a number of commonly used damping materials as obtained by one of the methods are given. Several types of built-up structures are shown to have a greater absorption coefficient for low frequency sound waves than is conveniently obtainable by a single layer of material.

THE most commonly used method of determining the sound absorption coefficient of a material is that devised by the late Professor W. C. Sabine. In this method the reverberation time of a room is measured before and after the introduction of a definite amount of the material. This method has the great merit that the values so determined usually apply to the materials precisely as they are ordinarily used in rooms for damping purposes. However, it is tedious and requires a very quiet room and large samples of the materials. A simpler scheme has been devised by H. O. Taylor,[1] in which the absorbing material is placed at one end of a tube. The coefficient of absorption is determined from a measurement of the ratio of maximum to minimum pressures of the standing waves within the tube when sound is introduced at the open end. Thus only a small sample of the material is required and with suitable apparatus the measurements can be made with great facility. In this paper several modifications of Taylor's tube method are discussed; in addition, it is shown that by a similar method it is possible to determine not only the absorption coefficient but also the acoustic impedance, a quantity which is playing an important part in present day applied acoustics.

General Theory

Consider a tube of length l, which is filled with a medium having a propagation constant $P = \alpha + i\beta$ and a characteristic acoustic im-

[1] *Phys. Rev.*, II, 1913, p. 270.

pedance [2] equal to Z_0 per unit area. At one end, O, let the velocity be uniform over the whole cross-section and equal to $\dot{\xi}_1 e^{i\omega t}$. At a distance l from O let the tube be terminated by the material which is to be investigated, and the acoustic impedance of which may be represented by $Z_2 = R_2 + iX_2$ per unit area. Under these conditions, the pressure, p, at any point in the tube at a distance x from O, is by analogy with the electrical transmission line

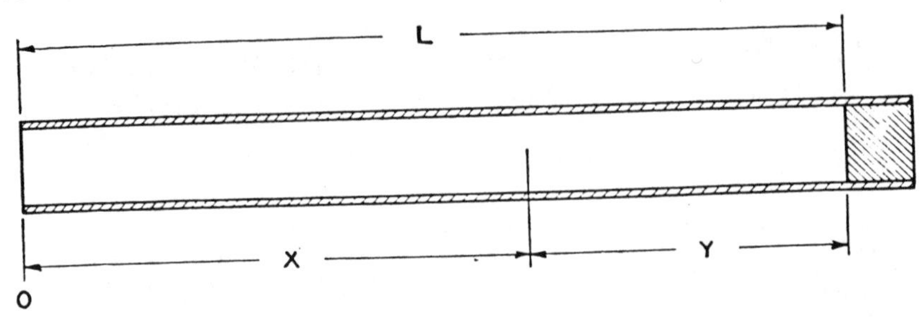

Fig. 1

$$p = \dot{\xi}_1 e^{i\omega t} \left[\frac{Z_2 \cosh Pl + Z_0 \sinh Pl}{Z_0 \cosh Pl + Z_2 \sinh Pl} \cosh Px - \sinh Px \right] Z_0.^3 \quad (1)$$

If there is no attenuation along the tube, we get, on dropping the time factor,

$$p = R\dot{\xi}_1 \left[\frac{Z_2 \cos \beta l + iR \sin \beta l}{R \cos \beta l + iZ_2 \sin \beta l} \cos \beta x - i \sin \beta x \right], \quad (2)$$

where $R = c\rho$, the product of the velocity of propagation along the tube and the density of the medium, and

$$\beta = \frac{2\pi f}{c}.$$

Equation (2) indicates numerous possible ways of determining Z_2, e.g., from the values of $\dot{\xi}_1$ and of p at any point in the tube; from the pressures for two values of either x or l, if $\dot{\xi}_1$ is constant; from the pressures at any point in the tube for the unknown and for a known value of Z_2; from the magnitude of p as a function of either x or l. However, we shall confine our discussion to three methods, which appear to be most practicable.

[2] The term acoustic impedance as here used may be defined as the ratio of pressure to volume velocity; the characteristic impedance is this impedance if the tube were of infinite length.

[3] J. A. Fleming, "Propagation of Electric Currents in Telephone and Telegraph Conductors," page 98; 3d Ed.

(a) *Pressure Measured at Two Points in the Tube*

It has already been pointed out that the impedance Z_2 can be determined if the relative phase and magnitude of the pressures at any two points in the tube are known. However, from the standpoint of convenience and precision it appears best to measure the pressures at the reflecting surface and at a point a quarter of a wave-length away. We then have at the reflecting surface $x = l$ and

$$p_2 = R\dot{\xi}_1 \left[\frac{R_2 + iX_2}{R \cos \beta l + iZ_2 \sin \beta l} \right],$$

and for the point $x = l - \frac{\lambda}{2} = l - \frac{\pi}{2\beta}$,

$$p_1 = R\dot{\xi}_1 \left[\frac{iR}{R \cos \beta l + iZ_2 \sin \beta l} \right],$$

so that

$$\frac{p_2}{p_1} = \frac{X_2 - iR_2}{R} \equiv Ae^{i\varphi}.$$

Hence

$$\left. \begin{aligned} R_2 &= -AR \sin \varphi, \\ X_2 &= AR \cos \varphi, \\ |Z_2| &= AR. \end{aligned} \right\} \tag{3}$$

If the coefficient of reflection is expressed as [4]

$$Ce^{i\psi} = \frac{Z_2 - R}{Z_2 + R}, \tag{4}$$

we get

$$C = \left[\frac{1 + 2A \sin \varphi + A^2}{1 - 2A \sin \varphi + A^2} \right]^{1/2},$$

where

$$\varphi = \tan^{-1} \frac{2A \cos \varphi}{A^2 + 1}. \tag{5}$$

The absorption coefficient, which is generally defined as the ratio of absorbed to incident power, is equal to $1 - |C|^2$.

(b) *Tube of Constant Length; the Absolute Value of the Pressure Measured at Points along the Tube*

The method discussed under this section is that adopted by H. O. Taylor for measuring the absorption coefficient of porous materials.

[4] I. B. Crandall, "Theory of Vibrating Systems and Sound," page 168.

For the absolute value of the pressure at any point in the tube we get from equation (2)

$$|p| = \left[\frac{R_2^2 + X_2^2 + R^2 + (R_2^2 + X_2^2 - R^2)\cos 2\beta y + 2X_2R\sin 2\beta y}{R_2^2 + X_2^2 + R^2 - (R_2^2 + X_2^2 - R^2)\cos 2\beta l - 2X_2R\sin 2\beta l}\right]^{1/2} R\xi_1, \quad (6)$$

where $y = l - x$.

$|p|$ has maximum or minimum values when

$$\tan 2\beta y = \frac{2X_2R}{X_2^2 + R_2^2 - R^2}; \quad (7)$$

for the maximum value $2\beta y$ lies in the first and for the minimum, in the third quadrant. We therefore get

$$\frac{|p|_{\max}}{|p|_{\min}} = \left[\frac{X_2^2+R_2^2+R^2+\sqrt{(X_2^2+R_2^2-R^2)^2+4X_2^2R^2}}{X_2^2+R_2^2+R^2-\sqrt{(X_2^2+R_2^2-R^2)^2+4X_2^2R^2}}\right]^{1/2} \equiv A. \quad (8)$$

Let y_1 be the value of y for which the pressure is a maximum; we then have from (7) and (8) and (4)

$$R_2 = \frac{2AR}{(A^2+1) - (A^2-1)\cos 2\beta y_1}, \quad (9)$$

$$X_2 = \frac{R(A^2-1)\sin 2\beta y_1}{(A^2+1) - (A^2-1)\cos 2\beta y}, \quad (10)$$

$$C_1 = \frac{A-1}{A+1}, \quad (11)$$

$$\Psi = 2\beta y_1.$$

The relation (11) can be derived more simply on the classical theory, as it was done by H. O. Taylor. A derivation of (11) is given by Eckhardt and Chrisler,[5] which differs from that of H. O. Taylor. From their derivation it would appear that for (11) to be valid the length of the tube should be adjusted for resonance and that the change in phase at the reflecting surface should be small. The derivation here given shows that (11) is general; it implies only that the waves be plane and that there be no dissipation of power along the tube.

[5] Scientific Paper of the Bureau of Standards, No. 526, page 56.

(c) Tube of Variable Length. Pressure Measured at the Source

The absolute value of the pressure at the driving end of the tube according to (2) is

$$|p_1| = \left[\frac{R_2^2 + X_2^2 + R^2 + (R_2^2 + X_2^2 - R^2)\cos 2\beta l + 2X_2R\sin 2\beta l}{R_2^2 + X_2^2 + R^2 + (R_2^2 + X_2^2 - R^2)\cos 2\beta l - 2X_2R\sin 2\beta l}\right]^{1/2} R\dot{\xi}_1$$

and $|p_1|$ is a maximum or a minimum when

$$\tan 2\beta l = \frac{2X_2R}{R_2^2 + X_2^2 - R^2}.$$

For the maximum value $2\beta l$ lies in the first and for the minimum, in the third quadrant. We therefore have

$$\frac{|p_1|_{\max}}{|p_1|_{\min}} = \frac{X_2^2 + R_2^2 + R^2 + \sqrt{(X_2^2 + R_2^2 - R^2)^2 + 4X_2^2R^2}}{X_2^2 + R_2^2 + R^2 - \sqrt{(X_2^2 + R_2^2 - R^2)^2 + 4X_2^2R^2}} \equiv A.$$

By analogy from the equations derived in section (*b*) above, we see that

$$R_2 = \frac{2\sqrt{A}R}{(A+1) - (A-1)\cos 2\beta l_1},$$

$$X_2 = \frac{R(A-1)\sin 2\beta l_1}{(A+1) - (A-1)\cos 2\beta l_1},$$

$$C = \frac{\sqrt{A} - 1}{\sqrt{A} + 1},$$

$$\Psi = 2\beta l_1,$$

where l_1 is the length of the tube when p_1 has a maximum value.

Discussion of the Precision of the Methods

Of the three methods of measuring impedance discussed above, the first is undoubtedly the simplest and most convenient, if an a.c. potentiometer is available. Theoretically, in this case the impedance may be determined with a high degree of precision. However, the method presupposes that the points where the pressures are measured are exactly a quarter of a wave-length apart; a more detailed analysis shows that, if A is small, variations in this distance will have a large effect on both the ratio of the pressures and their phase difference. It therefore is necessary to keep the temperature of the tube accurately constant or else to determine the distance corresponding to a quarter of a wave-length before each measurement. A precise determination of the point a quarter of a wave-length from the reflecting surface may be made by placing a smooth metal block at the reflecting end and finding then the position in the tube at which the pressure is a minimum.

In the other two methods it is relatively less important that the temperature be maintained constant, for the ratio of pressures is affected very little by any temperature variations. In the third method, where the length of the tube is varied, the expressions for R_2 and X_2 are the same as in (*b*), except that in place of the ratio of pressures they involve the square root of this ratio. For small values of pressure ratios the precision is therefore somewhat greater. However, for high values of reflection the ratio becomes very large and great care is required in the experimental set up to prevent errors creeping into the measurements through extraneous vibrations and stray electromotive forces in the measuring circuit. The main advantage of the method in which the pressure at the source only is measured is that a short length of exploring tube is required. If measurements down to a frequency of 60 cycles are made, the tube length must be at least 8 feet. An exploring tube reaching the whole length would ordinarily introduce too much attenuation if it were of sufficiently small bore to prevent resonance effects at the lower frequencies.

Experimental Procedure

In the case of the experimental results here reported the measurements were all made by the method outlined in section (*c*), i.e., the pressures were measured at the source while the length of the tube was varied. The experimental set up is shown in Fig. 2. A piece of Shelby

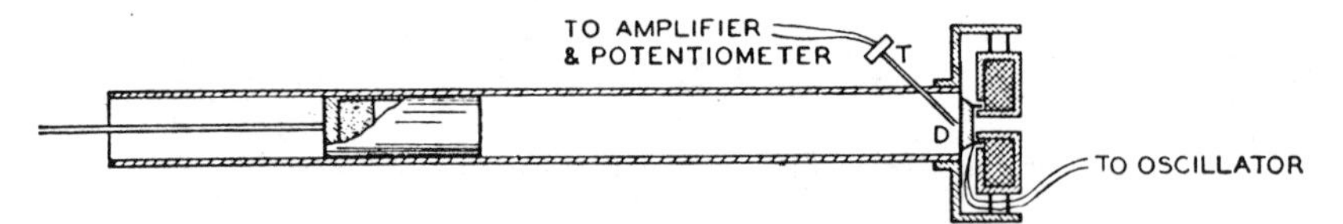

Fig. 2—Diagram of apparatus

steel tubing, 9 feet long, of 3″ internal diameter, and with 1/4″ wall, was fitted with a piston carrying the absorbing material. This piston was made up of a brass tube one foot long with a wall 1/64″ thick, the far end of which was closed with a one-inch brass block. To insure the propagation of plane waves and a constant velocity at the source, the diaphragm at *D* had a diameter of 2⅞″, and a mass of about 100 grams. This was driven with a coil 2″ in diameter situated in a radial magnetic field. The annular gap between the edge of the diaphragm and the interior of the tube was closed by a flexible piece of leather. To prevent vibrations of the magnet from getting to the tube, the magnet was held in position by flexible supports. The exploring tube *t* was about 5″ long with a 1/16″ bore which led to the transmitter, *T*. The voltages generated by the transmitter were measured with an amplifier and an a.c. potentiometer. The potentiometer was used because with it small voltages can be measured and errors due to harmonics are avoided. The proper functioning of the apparatus was determined by measuring the coefficient of reflection with no absorbing material in the piston. Theoretically the reflection should then be practically 100 per cent. The pressure ratios that were actually observed were of the order of 12,000 which corresponds to a reflection coefficient of 98 per cent. Evidently some extraneous pressures or voltages were still present. However, no attempt was made to reduce these further as the materials tested had a reflection coefficient considerably less than this value.

Experimental Results

A brief study was made of the absorption of hair felt, as there is an appreciable variation in the data given by various investigators on the absorption frequency characteristic of felts of presumably the same

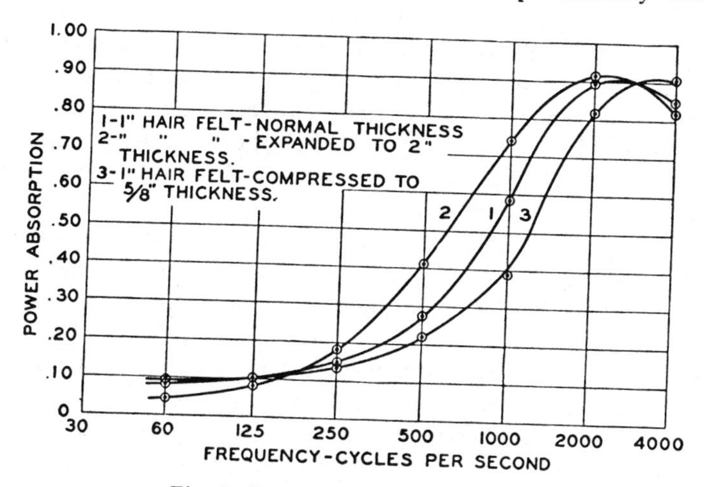

Fig. 3—Power absorbed by hair felt

type. After measurements on several samples it was evident that concordant results could not be expected as the absorption varied considerably with the packing of the felt. This point is illustrated by the curves shown in Fig. 3. These curves were all obtained on the same piece of hair felt but with different degrees of packing. It is thus evident that a felt which has become loosened by handling may have an absorption frequency characteristic quite unlike that of a new piece.

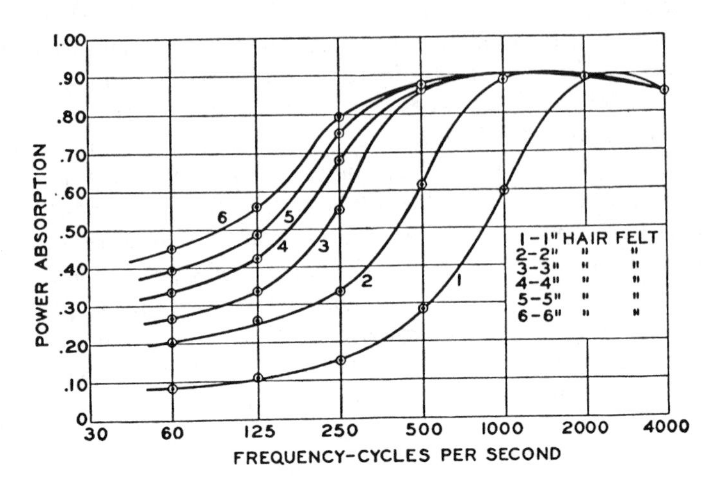

Fig. 4—Power absorbed by hair felt

In Fig. 4 are given the absorption coefficients for various thicknesses of hair felt. These values are in general agreement with those obtained by the reverberation method according to published results. Exact agreement is not to be expected, for the values here given apply only to sound waves having a perpendicular incidence on materials solidly backed by a hard surface. When the materials are applied in a room, the support is often more flexible and the absorption is partly due to inelastic bending. However, the agreement between the sets of values is sufficiently good to show that the results obtained by the simpler tube method may be used to get a good approximation to the values of absorption of the materials when applied in rooms for damping purposes.

Measurements have been made on a large number of porous materials. Although most of these materials are very good absorbers at the higher frequencies, none of them were found to be very efficient in the lower frequency region. Uniform absorption over most of the frequency range was found only in materials which are relatively inefficient absorbers. High absorption at the lower frequencies was obtained only when the thickness of the material was greatly increased. This fact is typically illustrated by the curves of absorption for hair felt given in Fig. 4.

When a sound wave of low frequency is reflected from a wall covered

[*Editor's Note:* Page 9, which consists entirely of Table I, absorption coefficients of various sandwich structures, has been omitted.]

with a porous material, the velocity of the air particles near the reflecting surface is small and hence there can be but little absorption. We may look at the phenomenon of reflection in still another way. In order to have a small coefficient of reflection the mechanical impedance of the wall per unit area should, as nearly as possible, be equal to the acoustic impedance of the air per unit area. The reason for the high reflection at low frequencies by a rigid wall covered with a porous material lies in its high stiffness reactance. At a given frequency this reactance can be compensated by loading the air near the reflecting surface. This may be accomplished in various ways. One of these ways is to place at a short distance from the wall a second wall which is porous or perforated. This arrangement has the effect of covering the wall with a multiplicity of resonators, which may be given any desired resonance frequency by properly proportioning the size, length and number of perforations and the spacing of the walls. The surface of the walls forming the air space should be absorbing or else the space should be provided with absorbing material.

To get a wider absorption band two or more perforated walls with proper spacing may be used, as this arrangement is equivalent to an aggregate of multiple resonators. The values of absorption coefficients of a number of structures of this type are given in the accompanying table. The measurements refer to sound which is incident from right to left as the structures are given in the table. The building board referred to in the table is a commercial type of insulating-board one inch thick with 400 1/4 inch by 3/4 inch holes per square foot. The felt in all cases is one-inch hair felt. These values show that relatively high absorption may be obtained at low as well as at high frequencies without an excessive amount of absorbing material. The use of combinations of absorbing materials, such as are given in the table, offers the advantage that more uniform damping at all frequencies can be obtained, and the degree of damping can be readily controlled by covering the proper area of surface. These two factors have become increasingly important in studio and auditorium design, with improved technique in recording and reproducing speech and music.

EDITOR'S NOTES

Page 2: Footnote 2 is confusing. Today's definition of the characteristic acoustic impedance of a medium is not $p/\dot{V}$ (volume velocity), which equals $\rho c/S$ or $Z_{acoustic}$ but $p/\dot{\xi}$ (linear velocity), which equals ρc. Hence, when Wente says Z_0 per unit area, which requires division by area S giving $\rho c/S^2$, he means Z'_{ac} for unit area, which implies multiplication by area S. Then $p/\dot{V}$ times S will equal the desired ρc or traditional Z_0.

Page 2: An understanding of Equations 1 and 2 will be helped by J. A. Fleming's Equation 67, *Propagation of Electric Currents in Telephone and Telegraph Conductors*, 3rd. ed., Constable & Son, London, 1927; also by P. M. Morse's *Vibration and Sound*, 2nd ed., McGraw-Hill, New York, 1948, pp. 133–143. Morse is more difficult than Fleming, but almost essential for the later papers in this portion of the volume.

Page 3: Note that in Equation 3, A is $|p_2|/|p_1|$.

Page 5: In method (c), since $\beta\ell = \omega\ell/c$, we can vary either ℓ or ω and get the same result. Thus when Wente varies ℓ, he gets a [tan $\beta\ell$] function, or a [-cot $\beta\ell$] function, and so on. Flanders, in paper 39, keeps ℓ fixed but varies ω and gets the same curve.

Page 6: Figure 2 shows that Wente is clearly measuring the pressure component of the standing-wave system, unlike Taylor or Paris.

Page 8, Middle of the page: Note that the absorption coefficient for oblique incidence has still not been solved.

Page 10, line 8: Wente's comments on "high stiffness reactance" also apply to the insensitivity of Kennelly and Kurokawa's instrument when used away from resonance (see Paper 41).

37

Reprinted from pages 420 and 423 of *Acoust. Soc. Am. J.* **19**:420–427 (1947)

Some Notes on the Measurement of Acoustic Impedance

LEO L. BERANEK*

I. GENERAL

IN an earlier paper an apparatus was described for the measurement of the normal acoustic impedance of small diameter samples of acoustical materials.[1] It was pointed out that the curve-width (either variable-length or variable-frequency) method of measurement is limited to the determination of impedances which do not approach closely the impedance of free air. Also it was pointed out that the ambient temperature was the limiting factor affecting the precision of the method at the higher frequencies and that it seemed to be difficult to obtain accurate results without great care at frequencies near 100 c.p.s. Other observers in the field have suggested that the standing-wave method[2] of measurement is more straightforward and that there is a possibility that systematic errors exist when using the curve-width method which have not been taken into account in the mathematical analysis. Hence, a comparison of the two types of measurement would seem to be in order.

As part of a research program which will be described in a later paper, it was decided to design a measuring apparatus with which either the curve-width or standing-wave type of measurement could be performed without disturbing the sample under test, and one which avoided the previous sources of inaccuracy mentioned in the first paragraph. A further object of the investigation was to determine the effect of sample size on the shape of the impedance curve, because it has been recognized that the binding effects of the measuring tube on the edges of the sample might be serious.

II. DESIGN OF THE NEW MEASURING TUBE

The essential features of the new tube are shown in Fig. 1. The sample is held in a thin-walled tube fastened to and backed by a heavy piston which can be moved in the tube by the precision screw shown on the right. The tube itself is a three-inch I.D. piece of Shelby precision steel tubing having a total length of 111 inches. The source of sound is mounted on the opposite end of the tube from the sample and is a modified form of the W.E. 555 loudspeaking receiver unit. The entire tube is surrounded by a circulating water jacket which can be warmed by a thermostatically controlled heating element. In practice it is found that for control at a temperature a few degrees above that of the room, the necessary heat is adequately supplied by the friction of movement of the liquid through the system.

Two different microphones are provided for the measurements. One is fixed in position and is located at the edge of the source-end of the tube. The other is mounted on a small carrier such that it can be moved along the longitudinal axis of the tube and its position read from a scale shown at the left side of Fig. 1. Details of the mechanism for advancing the movable microphone can be seen in Fig. 2c.

* Now at the Massachusetts Institute of Technology.
[1] L. L. Beranek, "Precision Measurement of Acoustic Impedance," J. Acous. Soc. Am. **12**, 3 (1940).
[2] W. M. Hall, "An Acoustic Transmission Line for Impedance Measurement," J. Acous. Soc. Am. **11**, 140 (1939).

[*Editor's Note:* Material has been omitted at this point.]

III. GRAPHICAL AIDS TO CALCULATION OF IMPEDANCE

One of the chief obstacles to the use of impedance as a method of specifying the properties of acoustical materials has been the difficulty of calculating the impedance from the quantities measured. Fortunately, developments in the microwave field during the war have introduced charts which can be applied to our needs and which greatly simplify the calculations. Using the curve-width method the acoustic impedance, $Z=R+jX=\rho c\gamma \exp(j\phi)$, is calculated from the equation:[1]

$$\gamma \exp[j(\phi+k_m/\omega_m)]=\coth[(k-k_{na})\times(l/c)-j(\omega/c)\{(1-k^2/2\omega^2)l-l_0\}]=\coth[C-jD] \quad (1)$$

where l and k are the resonant length and damping constant with the sample present; l_0 and k_{na} are the same with the sample replaced by a rigid wall; $k=\pi f(l''-l')/l$; l'' and l' are the lengths of the tube at the half-intensity points; $k_m=k-k_{na}$; $\omega_m=(\omega^2-k^2)^{\frac{1}{2}}$, and ω is the frequency at which the measurement is made. Usually $k^2 \ll \omega^2$ so approximately,

$$\gamma \exp(j\phi)=\coth[k_m l/c-j(l-l_0)\omega/c]. \quad (2)$$

Using the standing-wave method, the acoustic impedance is calculated from the equation,[3]

$$\gamma \exp(j\phi)=\coth[\coth^{-1}(P_{max}/P_{min})-j(L-L_0)\omega/c] \quad (3)$$

where (P_{max}/P_{min}) is the ratio of the pressure at the first maximum in front of the sample to the pressure at the first minimum corrected for tube dissipation; L_0 is the distance of the first minimum from a rigid termination; and L is the distance of the first minimum from the surface of the sample under test. To correct for dissipation in the tube, Sabine gives a curve which relates the true value of P_{max}/P_{min} (in decibels) to the values of (P_{max}/P_{min}) determined by obtaining the values of P_{min} at two successive nodal points. From Eqs. (2) and (3) it is obvious that,

$$\coth^{-1}(P_{max}/P_{min})=(k_m l/c)=C, \quad (4)$$

and

$$(l-l_0)\omega/c=(L-L_0)\omega/c=D. \quad (5)$$

[3] H. J. Sabine, "Notes on Acoustic Impedance Measurement," J. Acous. Soc. Am. **14**, 143 (1942).

In order to convert (4) and (5) into $R+jX$, use is made of a "Transmission Line Calculator"[4] which is reproduced in Fig. 5a. This calculator is operated by setting the value of $20 \log_{10}(P_{max}/P_{min})$ on a rotatable transparent arm (Fig. 5b) which pivots about the center of the calculator and at the same time aligning the hair line of the transparent arm with one of the numbers around the edge labeled "wave-lengths toward the load." The impedance is read from the intersection of the crosslines on the rotatable arm. The "wave-lengths toward the load," W, is determined by:

$$W=0.25+(l-l_0)/\lambda=0.25+D/2\pi. \quad (6)$$

A similar chart prepared by the author giving γ and ϕ is shown in Fig. 6. The rotatable member drawn in Fig. 6b is not the same as that supplied with the Emeloid calculator, but has been drawn so that the hyperbolic cotangent of $(k_m l/c)$ need not be determined..

For high values of impedance (i.e., $P_{max}/P_{min} > 10$ at the same time that $D<0.1$) the following formulas may be used for the calculation of impedance:

$$R/\rho c=\frac{(P_{min}/P_{max})(1+D^2)}{(P_{min}/P_{max})^2+D^2}, \quad (7)$$

$$X/\rho c=-\frac{D[1-(P_{min}/P_{max})^2]}{(P_{min}/P_{max})^2+D^2}. \quad (8)$$

[*Editor's Note:* Material has been omitted at this point.

Equation 1 refers to the "curve-width method," analogous to finding the Q of a simple resonant circuit in the case of lumped electrical circuits.

The work described after Equation 2 was carried further by Beranek and is discussed in his book, *Acoustic Measurements* (Wiley, New York, 1949, Chapter 7). A necessary background can be obtained from Morse's *Vibration and Sound* (2nd ed., McGraw-Hill, New York, 1948, pp. 240–247.]

[4] P. H. Smith, "An Improved Transmission Line Calculator," Electronics **17**, 130, January (1944). The calculator described by Smith is manufactured by the Emeloid Company, Arlington, New Jersey. Graph paper imprinted as shown in Fig. 5a is also available.

38

Reprinted from pages 153–154 of *Acustica* 3:153–160 (1953)

THE PRACTICAL REPRESENTATION OF STANDING WAVES IN AN ACOUSTIC IMPEDANCE TUBE

by W. K. R. LIPPERT

Division of Building Research, C.S.I.R.O., Melbourne, Australia

Summary

It is shown that the envelope curves of the standing waves along the whole of an acoustic impedance tube with attenuation can be represented with the aid of diagrams. Suitable diagrams for easy application to impedance measurements are given and their usefulness is discussed in some specific cases.

The position of the pressure minima are measured with different orifices of the probe tube and the resultant shifts of position are discussed.

Special test measurements with a rigid terminal are proposed for specifying the tube performance.

Sommaire

On montre que les enveloppes des ondes stationnaires dans un tube absorbant à impédance acoustique peuvent être représentées à l'aide de diagrammes. On donne des diagrammes utilisables dans les mesures d'impédance acoustique, et on démontre l'utilité de la méthode proposée par quelques exemples expérimentaux.

Les positions des minima des ondes stationnaires ont été mesurées avec différents orifices du petit tube sonde; on discute les variations correspondantes de position.

On propose de faire quelques expériences propres à caractériser la qualité d'un tube acoustique.

1. Introduction

The transmission line method is the most widely used one for measuring the absorption coefficient or the specific acoustic impedance of a material. A smooth rigid-walled tube has a source of sound at one end and the sample under test at the other and the acoustic impedance can, as is well known, be found by measurements of the standing waves along the tube.

The method requires the measurement of the pressure in the tube without disturbing the sound field, and several approaches to the solution of this problem are known [1]···[6]. That of TAYLOR [1], improved by SCOTT [5], uses a long smallbore probe tube to measure the sound pressure in the main tube. The probe tube is moved along the main tube and is terminated by a microphone outside the main tube.

This method is frequently used because of its simplicity and accuracy. SCOTT has shown that it is not permissible to assume that the tube attenuation is negligible when precise measurements are required and he derived the relevant theory. The practical application of this theory is lengthy, particularly when the standing wave has been explored by recording sound pressure by a high speed level recorder. BERANEK [7] discussed SCOTT's theory and showed how to derive the specific acoustic impedance from measured data from part of the standing wave close to the sample. However, in the course of the present work it was found preferable to use the record along the whole of the tube, and suitable diagrams as presented below were found particularly useful. With these diagrams a sensitive method is available for detecting irregularities in the operation of the equipment, extrapolation of minima to zero distance is easier and the whole provides an improvement in the specification of tube performance.

2. The envelope function

Acoustic impedance tube theory for a tube with attenuation is given in text books [7] and it is necessary only to quote the results here. The ratio of the Mth pressure maximum to the Nth pressure minimum is

$$\left(\frac{|P_{max}|}{|P_{min}|}\right)_{M,N} = \tag{1}$$

$$\sqrt{\frac{2\cosh^2(\alpha D_M+\psi_1)-\frac{\alpha^2}{2k^2}\sinh^2 2(\alpha D_M+\psi_1)}{2\sinh^2(\alpha d_N+\psi_1)+\frac{\alpha^2}{2k^2}\sinh^2 2(\alpha d_N+\psi_1)}}$$

and the distance d_N of the Nth pressure minimum (counted from the sample) is

$$\frac{d_N}{\lambda}=\frac{2N-1}{4}-\frac{\psi_2}{2\pi}-\frac{\alpha}{2k^2\lambda}\sinh 2(\alpha x+\psi_1), \tag{2}$$

wherein α is the attenuation constant for the sound wave due to energy losses at the side walls of the tube and in the gas,

$f=\omega/2\pi$ is the frequency,
c is the wave velocity,
λ is the wavelength,
$k=\omega/c=2\pi/\lambda$ is the wave number,
x is the distance from the sample surface,
d_N is the distance of the Nth minimum from the sample surface,
D_M is the distance of the Mth maximum from the sample surface,
P_i is the pressure of the incident wave,
P_r is the pressure of the reflected wave and $\psi=\psi_1+\mathrm{j}\psi_2$ is given by

$$P_r/P_i=\mathrm{e}^{-2\psi}=\mathrm{e}^{-2(\psi_1+\mathrm{j}\psi_2)}. \tag{3}$$

The values of ψ_1 and ψ_2 are calculated by substitution of experimental values of $(P_{max}/P_{min})_{M,N}$ and d_N/λ in equations (1) and (2) and the acoustic impedance ratio is then

$$\frac{Z}{\varrho c}=\frac{R}{\varrho c}+\mathrm{j}\frac{X}{\varrho c}=\frac{1}{1-\mathrm{j}\frac{\alpha}{k}}\coth(\psi_1+\mathrm{j}\psi_2)\approx \tag{4}$$

$$\approx\coth(\psi_1+\mathrm{j}\psi_2),$$

wherein Z is the acoustic impedance,
R is the acoustic resistance,
X is the acoustic reactance,
and ϱ is the density of the medium.

The ratio of the maximum to the minimum pressure in the tube is commonly measured by moving the probe tube at constant speed and recording the sound pressure with a high speed level recorder. A typical record is given in Fig. 1. The curves

$$P=A\sqrt{2\cosh^2(\alpha x+\psi_1)-\frac{\alpha^2}{2k^2}\sinh^2 2(\alpha x+\psi_1)} \tag{5}$$

and

$$P=A\sqrt{2\sinh^2(\alpha x+\psi_1)+\frac{\alpha^2}{2k^2}\sinh^2 2(\alpha x+\psi_1)}, \tag{6}$$

where A is a constant depending upon the sound pressure level, will enclose the whole of the recorded trace and hence are envelope curves.

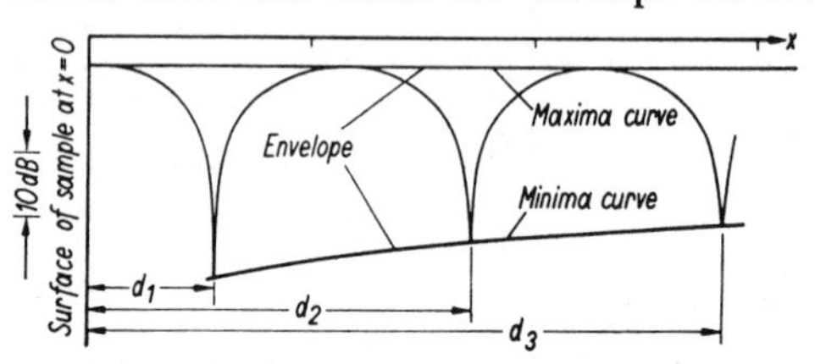

Fig. 1. A typical record of the sound pressure with a high-speed level recorder.

These curves have been superimposed on Fig. 1. The curve which encloses the maxima (cf. eq. (5)) is practically a straight line since αx is usually small. The envelope curves (cf. eq. (5) and (6)) will be separated by an ordinate (in dB) which may be called the envelope value $L(x)$ such that

$$L(x)= \tag{7}$$

$$20\log\sqrt{\frac{2\cosh^2(\alpha x+\psi_1)-\frac{\alpha^2}{2k^2}\sinh^2 2(\alpha x+\psi_1)}{2\sinh^2(\alpha x+\psi_1)+\frac{\alpha^2}{2k^2}\sinh^2 2(\alpha x+\psi_1)}}$$

which can be readily transformed to

$$L(x)= \tag{8}$$

$$20\log\left[\coth(\alpha x+\psi_1)\sqrt{\frac{1-\frac{\alpha^2}{k^2}\sinh^2(\alpha x+\psi_1)}{1+\frac{\alpha^2}{k^2}\cosh^2(\alpha x+\psi_1)}}\right].$$

It will be observed that the shape of the envelope curve for a given tube and frequency depends on the tube attenuation and the impedance of the sample. The value of ψ_2 (the phase change caused by the sample) does not affect the envelope; a change in ψ_2 affects both the value and position of the maxima and minima in such a way that the envelope curves are unaltered.

Before discussing the envelope curves further it is desirable to distinguish two cases: case A including highly reflecting to normally absorbent samples, say $\psi_1\leqq 0.6$, and case B with highly absorbent samples, say $\psi_1>0.6$.

[*Editor's Note:* In the original, material follows this excerpt.]

39

Reprinted with permission from *Bell Syst. Tech. J.* **11**:402–410 (1932)

A Method of Measuring Acoustic Impedance *

By P. B. FLANDERS

An apparatus is described whereby acoustic impedances may be measured in terms of a known acoustic impedance and the complex ratios of two electrical potentiometer readings to a third. As a known impedance, there is chosen the reactance of a closed tube of uniform bore which is an eighth wave-length long. The electrical readings are obtained by balancing the amplified output of a condenser transmitter against the electrical input of the source of sound. The condenser transmitter picks up the acoustic pressure at the junction of the sound-source and the attached impedance. A balance is made for each of three successively attached impedances: (1) a closed tube an eighth wave-length long, (2) a rigid closure of the sound-source, and (3) the impedance to be measured. The unknown acoustic impedance Z is then calculated in terms of the known acoustic impedance Z_0 by means of the equation $Z = Z_0 \frac{z_1 - z_2}{z_3 - z_2}$, where z_1, z_2 and z_3 are, respectively, the three electrical impedance settings of the potentiometer. As indicated by this equation, the constants of the electrical circuit are involved only as ratios, so that the response characteristics of the source of the sound, condenser transmitter and amplifiers (provided they are invariable) do not affect the measurement.

Illustrations are given of impedance measurements on a closed tube of uniform bore, a conical horn, an exponential horn, an " infinite " tube, and a hole in an " infinite " wall.

THE progress in acoustics during the past few years has caused acoustic impedance measurement to have the same relative importance that impedance measurement in electrical work has had for many years. The concept of acoustic impedance is derived from the analogy [1] that exists between electrical and acoustic devices, as shown by the analogous differential equations describing their action. Acoustic impedance is usually defined as the complex ratio of pressure to volume velocity (or flux) but it is sometimes more convenient to deal with ratios of pressure to linear velocity or force to linear velocity. The magnitudes of these are interrelated, of course, by powers of the area involved.

The earliest efforts to measure acoustic impedance seem to have been made by Kennelly and Kurokawa.[2] In their method, electrical measurements were made of the motional impedance of a telephone receiver, with and without an attached acoustic impedance. Except for frequencies near resonance, the method was inaccurate because the acoustic impedance was associated with a relatively large mechanical impedance.

* Presented before Acous. Soc. Amer., New York City, May 3, 1932.

[1] This analogy was first pointed out by A. G. Webster in *Nat. Acad. of Science*, **5**, 275 (1919).

[2] *Proc. Am. Ac. Arts and Sc.*, **56**, 1 (1921).

Later, a direct method was described by Stewart[3] who measured the change in acoustic transmission through a long uniform tube when the unknown impedance was inserted as a branch.

The apparatus to be described in this paper measures acoustic impedance directly in terms of a known acoustic impedance and three balance readings of an electrical potentiometer. The only assumptions involved in the method are that the elements of the apparatus be invariable during a measurement, and that the value of the comparison acoustic impedance be known accurately.

Apparatus

Fig. 1 shows the general arrangement of the apparatus. An oscillator feeds electrical energy into a loud speaker where a portion is converted into acoustic energy which travels along the tube and into an

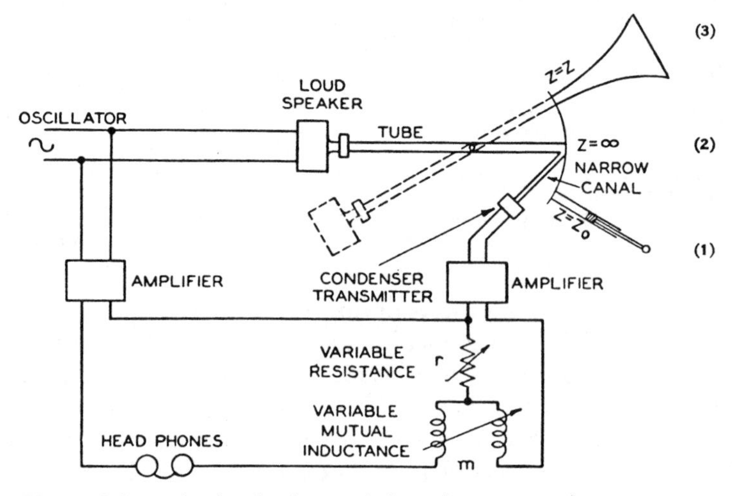

Fig. 1—Schematic circuit of acoustic impedance measuring apparatus.

attached impedance. A canal about 0.06 inches in diameter picks up the sound pressure at the junction of the tube and the attached impedance, and passes it along to a small condenser transmitter. A corresponding voltage, generated by the transmitter, is amplified and the current output of the amplifier passed through a variable resistance in series with the primary of a variable mutual inductance.

The same oscillator also feeds energy through a second amplifier at the output of which the voltage is balanced (by the null method) against the voltage drop across the variable resistance and the secondary of the mutual inductance.

[3] *Phys. Rev.*, **28**, 1038 (1926).

At the end of the tube from the loud speaker are three different impedances. One is the reactance offered by a closed tube of uniform bore. The closure is formed by a well-fitting plunger whose position in the tube may be adjusted. The second is the infinite impedance offered by a rigid wall closing the end of the tube from the loud speaker. The third is the impedance to be measured. All three impedance elements are fixed in position. The loud speaker, tube, condenser transmitter and associated amplifiers are, however, mounted together on a carriage which can be rotated so as to bring any one of these impedances into alignment with the tube. For brevity, reference to these three positions will hereinafter be to positions 1, 2, and 3.

For any one frequency a balance is obtained for each of the three positions. These three electrical readings and the reactance value of the closed tube are sufficient to determine the impedance being measured.

A photograph of the apparatus is shown in Fig. 2.

Fig. 2—View of acoustic impedance equipment, showing loud speaker, tubes, condenser transmitter amplifiers and small horn in position for measurement.

Theory

Thevenin's theorem [4] states that, in an invariable electrical network, the current in any branch is equal to the current that would flow in a simple series circuit composed of an electromotive force and two impedances. The electromotive force is the voltage that would obtain at the branch terminals on open circuit. The impedances are the impedance at the terminals looking back into the source of power, and the impedance of the branch.

[4] K. S. Johnson's "Transmission Circuits for Telephone Communication," Ch VIII.

Since the differential equations of acoustics are analogous to those of electrical lines and networks, the theorem may be applied to the action of this apparatus with a considerable saving in labor.[5]

By Thevenin's theorem then, the tube, loud speaker, oscillator, etc. may be replaced by one pressure and one impedance. The pressure is the "open-circuit" pressure at the end of the tube, or in other words the pressure that would be exerted on a rigid wall if placed there. The impedance is the complex ratio of pressure to velocity at the end of the tube which would exist if acoustic energy were sent into it toward the loud speaker, the oscillator being shut off. In electrical terms, this would be called the impedance looking into the source. The velocity or acoustic current that flows into an impedance attached to the end is then the current that would flow in an analogous circuit composed of this vibromotive force or pressure and the two impedances in series. This impedance diagram is given in Fig. 3.

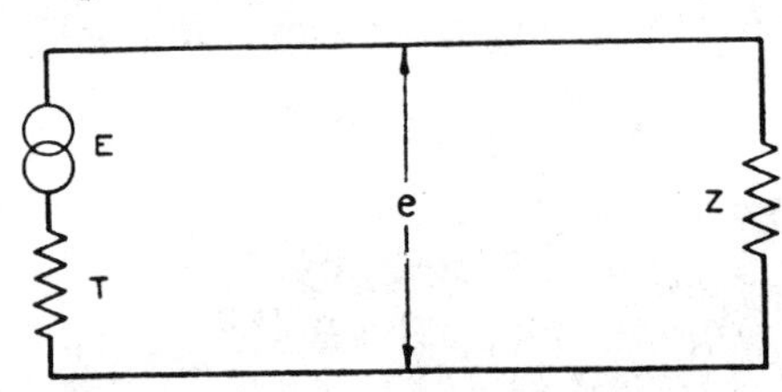

Fig. 3—Impedance diagram for Thevenin's theorem.

E is the open-circuit voltage or pressure, T the impedance looking into the source of sound at the junction, and Z the attached impedance. The pressure e at the junction of T and Z is, of course, the velocity-current $\frac{E}{T+Z}$ in the loop, multiplied by Z. The three equations for three values of attached impedance are

$$Z = Z_0, \qquad e_1 = \frac{EZ_0}{T+Z_0},$$

$$Z = \infty, \qquad e_2 = E,$$

$$Z = Z, \qquad e_3 = \frac{EZ}{T+Z}.$$

The two unknown quantities, E and T, can be eliminated giving one equation

$$\frac{Z}{Z_0} = \frac{\dfrac{e_2}{e_1} - 1}{\dfrac{e_2}{e_3} - 1},$$

[5] A more direct proof of Thevenin's theorem as applied to acoustics is given by W. P. Mason in *B. S. T. J.*, **6**, 291 (1927).

whereby Z may be calculated in terms of Z_0 and two ratios of pressures at the junction.

Referring now to Fig. 1 it will be seen that the current through the resistance and mutual primary is proportional to the pressure at the junction. The drop in voltage across the secondary and the resistance is equal in magnitude and opposite in phase to a voltage proportional to E, when no current passes through the head-phones. If k signifies the circuit constant and if z be the impedance value of the resistance and the mutual inductance, then $ez = kE$; and the above equation becomes

$$\frac{Z}{Z_0} = \frac{\dfrac{z_1}{z_2} - 1}{\dfrac{z_3}{z_2} - 1} = \frac{z_1 - z_2}{z_3 - z_2} = \frac{(r_1 - r_2) + j\omega(m_1 - m_2)}{(r_3 - r_2) + j\omega(m_3 - m_2)},$$

where r is the resistance component of z and m is the mutual inductance.

The reactance of a closed tube of uniform bore whose length is one-eighth the wave length of sound for the measuring frequency is chosen as the known impedance. If dissipation in the tube be neglected, the impedance is readily calculated [6] to be a pure negative reactance of 41 mechanical ohms per square centimeter [7] at a temperature of 20° C. This value is chosen because it is of the same order of magnitude as most acoustic impedances. By mechanical ohms per square centimeter is meant the complex ratio of pressure to the linear velocity of the air. The justification for assuming negligible dissipation will be apparent when measurements made on a closed tube, several wave-lengths long, are described.

In making measurements, the three impedance values necessary for balance are read for the three impedance conditions in the 2-1-3 or 2-3-1 order. Afterwards, as a check to ensure that the circuit constant has not changed during the measurement, condition 2 is measured again. This series of four measurements is repeated for each frequency.

Application

Fig. 4 shows the results of measuring the reactance of a closed tube. The tube was 2.4 inches long and 0.7 inch in diameter. The comparison impedance was the calculated reactance of this same tube in the one-eighth wave-length condition, assuming no dissipation. The impedance was also calculated,[8] taking into account viscosity and

[6] See I. B. Crandall's "Theory of Vibrating Systems and Sound," p. 104.
[7] See definitions 8007 and 8011 in "Standardization Report of I. R. E." in "Year Book of I. R. E.," 1931.
[8] See Rayleigh, "Theory of Sound," Vol. II, pp. 318 and 325.

losses through heat conduction, for frequencies near the half wave-length anti-resonance, where dissipative effects are most pronounced. It will be seen from Fig. 5 that there is a close agreement between the theoretical curve and the measured points. It seems reasonable, therefore, to assume that the value chosen for the comparison impedance is quite accurate.

Fig. 6 shows the impedance of a conical horn and Fig. 7 that of an exponential horn. In both cases, the mouth of the horn projected through a window into open air, so as to minimize reflection effects

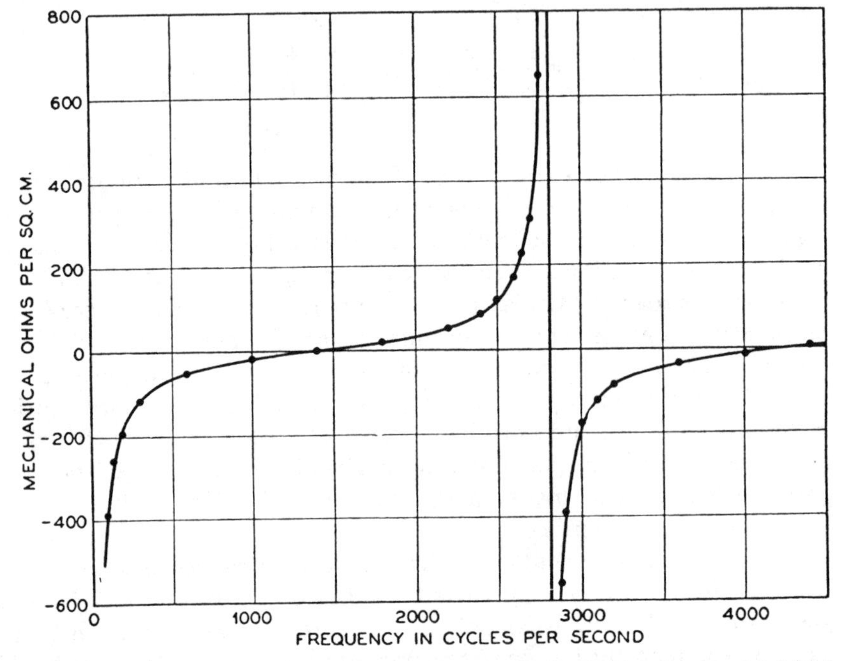

Fig. 4—Acoustic reactance of closed cylindrical tube, 2.4 inches long and 0.7 inch in diameter.

from external objects. Reflection effects from the mouth, where there is a change in impedance, are present, however, and these appear as oscillations of the impedance about a mean which is the characteristic impedance of the horn. By characteristic impedance is meant the impedance that would obtain looking into the throat of the horn were it infinite in length.

Fig. 8 is the impedance of an "infinite" tube. The tube was actually 112 feet long and coiled into a helix. At low frequencies, where the dissipative losses are small, reflection effects from the open end are

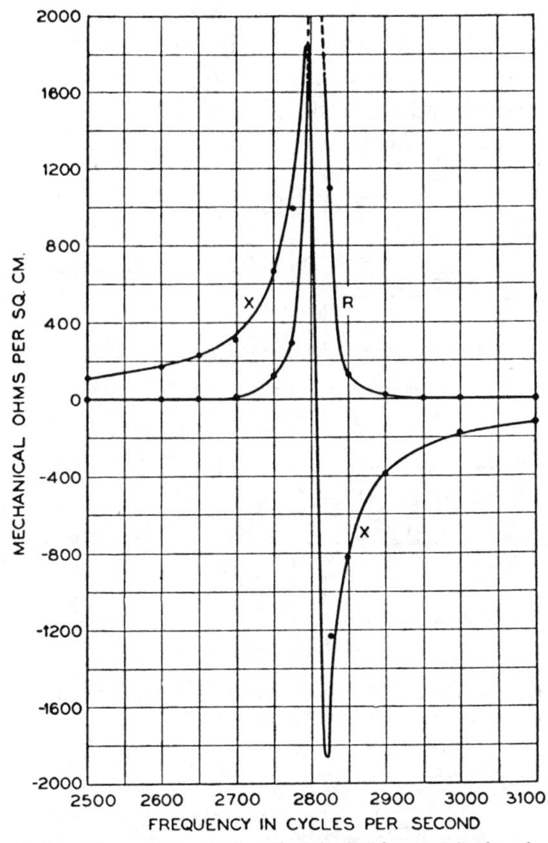

Fig. 5—Acoustic impedance of closed cylindrical tube, 2.4 inches long and 0.7 inch in diameter, showing agreement between measured and calculated values.

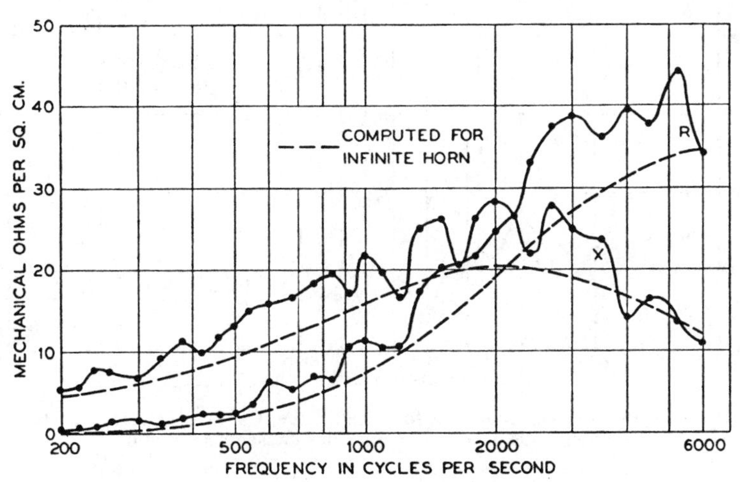

Fig. 6—Acoustic impedance of 38 inch conical horn, having end diameters of 0.7 inch and 28 inches.

observed as oscillations of the impedance at about 5-cycle intervals. An examination of the measurements in this oscillatory region (Fig. 9) will make evident the precision of the apparatus.

Fig. 10 shows the radiation impedance of a hole, 0.7 inch in diameter and surrounded by a flange which approximates an infinite wall for

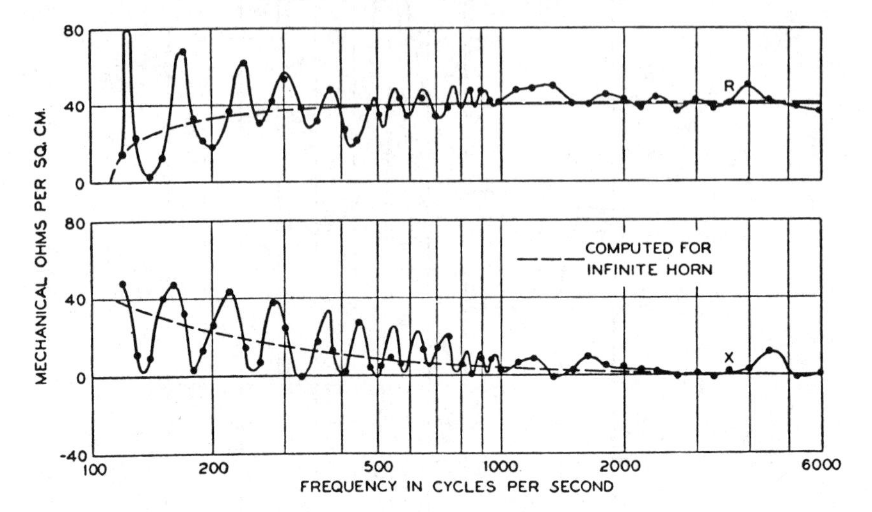

Fig. 7—Acoustic impedance of 6 foot exponential horn, having end diameters of 0.7 inch and 30 inches.

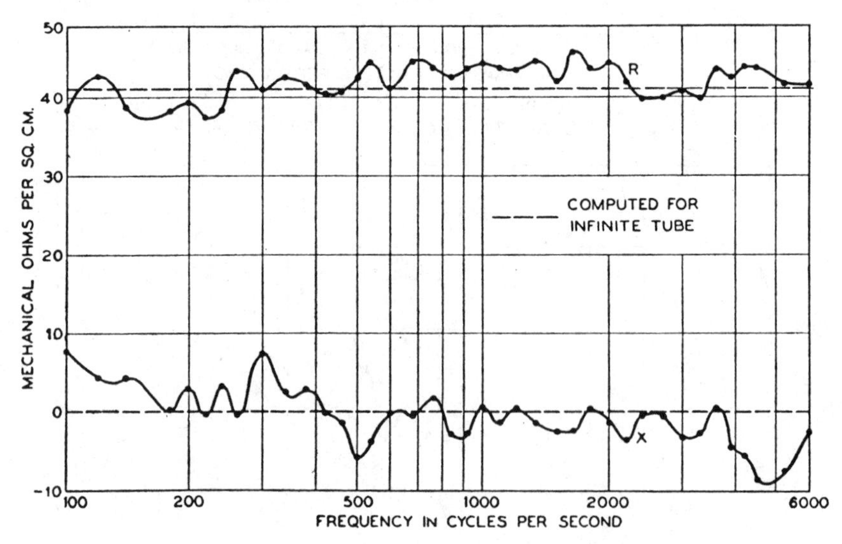

Fig. 8—Acoustic impedance of 112 foot open tube, 0.7 inch in diameter, coiled into helix. Measuring frequencies chosen at random.

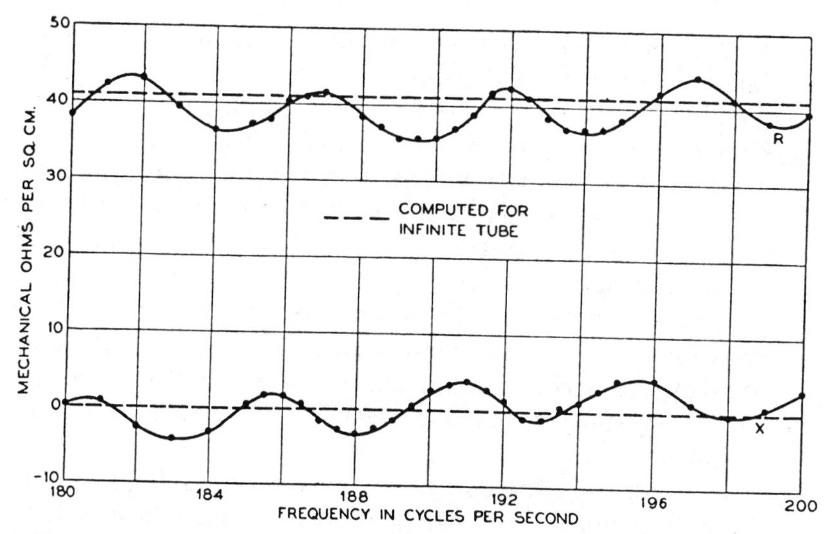

Fig. 9—Acoustic impedance of 112 foot open tube, 0.7 inch in diameter, coiled into helix. Measuring frequencies chosen at half and one cycle intervals to show oscillatory character of impedance.

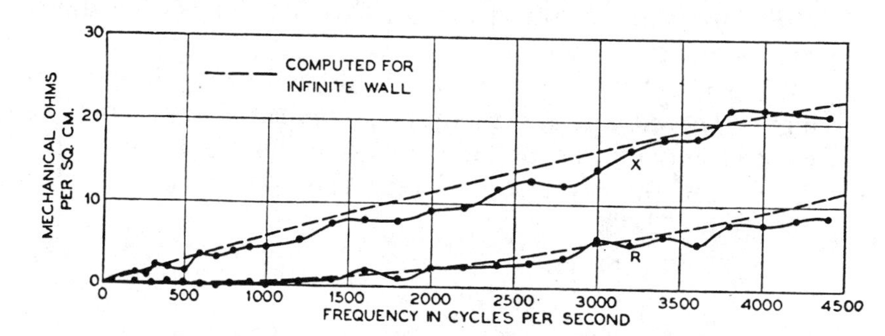

Fig. 10—Acoustic radiation impedance of hole, 0.7 inch in diameter, in flange having diameter of 6 inches.

the frequencies of interest. The dotted lines are the resistance and reactance as calculated by the equations of Rayleigh.[9]

[9] "Theory of Sound," Vol. II, p. 164.

EDITOR'S NOTES

Page 403: Figure 1 shows the radiating source: a loudspeaker coupled to a long open tube. This source is then loaded consecutively by three different acoustic loads: a horn, an infinite impedance, and a $\lambda/8$ closed tube. But why not use the loudspeaker alone, without the long tube, as the source? Recall that Thevenin's theorem requires that the acoustic load shall look to the left, toward the loudspeaker, to find the Thevenin generator impedance, T. The bare loudspeaker would present an impedance T in the form of a mechanical (or acoustical) impedance $R + jX$ that rises or falls steeply with small frequency changes Δf. By adding a long tube in front of the loudspeaker diaphragm, Flanders created a buffering or decoupling network, which tended to smooth out the wild variations of R and of jX—that is, the Thevenin impedance T, seen by the acoustic load, became the smoother impedance $R' + jX'$. (Ideally it would be nice if T were a simple constant resistance.)

Page 407: Figure 4 is the curve of $[-\rho c \cot \beta x]$ (see comments on Wente's paper). The $\lambda/4$ situation occurs at about 1,400 cps, and hence the $\lambda/8$ situation occurs at about 700 cps. Then βx is $\pi/4$ and $-\cot \beta x$ is -1. So $[-\rho c \cot \beta x]$ should be -41 ohms, using the value 41 for ρc. This, indeed, is what the experimental graph shows.

Page 409: The curve for R in Figure 7, top, is striking. It agrees remarkably well with the theoretical curves for finite horns calculated and plotted by H. F. Olson in his book *Elements of Acoustical Engineering*, 2nd ed., Van Nostrand, New York, 1947, pp. 102–109.

40

Reprinted from pages 2343–2347, 2348, 2349 and 2351 of *Am. Inst. Electr. Eng. Proc.* **31**:2320–2351 (1912)

DISCUSSION ON "THE VIBRATIONS OF TELEPHONE DIAPHRAGMS" (MEYER AND WHITEHEAD)

G. W. Pierce, F. Wenner, and C. F. Meyer

[*Editor's Note:* In the original, material precedes this excerpt. Only the discussion of the above authors is included here.]

George W. Pierce: I have been very much interested in this paper, particularly as Professor Kennelly and I have recently been making some experiments[1] which include as a part of the work the determination of the period of vibration of the telephone diaphragm. First, as to temperature. One of our

1. Published in full in Proc. Am. Acad. (Boston), Vol. 48, No. 6, 1912.

laboratory mechanics, Mr. Greaves, who was trying to make a telephone of particular pitch, found when he breathed on the diaphragm he changed the period perhaps fifty per cent. The period changes enormously if the diaphragm is rigidly clamped into a metallic frame, but if the diaphragm is loosely clamped, as in the case of the ordinary telephone receiver, with washers or gaskets, the temperature does not have so great an effect.

In regard to the occurrence of the dimple in two or three of Meyer's and Whitehead's oscillograms, also with some oscillograms similar to these, I found that the dimple occurred when the current was large, independent of the pitch, and it seems to me to be the phenomenon that Professor Shepardson has mentioned, of the preponderance of impressed force over the directing force of the permanent magnet. To avoid that effect I found that, if you use a permanent magnet for the field and attach a coil to the diaphragm, so as to link with the magnetic flux of the field magnet, and send alternating current through the coil, you can vary the pull on the diaphragm to any extent, yet get no dimple at all, because there was no change in the B_0 magnetic field by the effect of the impressed current.

In regard to the period of the diaphragm, Professor Kennelly and I have been measuring resistances and inductance of the telephone receiver at different frequencies, and the result is very interesting and indicates the marked effect of periodicity of the diaphragms as Messrs. Meyer and Whitehead have found. If you measure the resistance and the inductance with the telephone diaphragms damped you get one value, and then, if you take your finger off the diaphragm, and allow it to vibrate, you get a different value of inductance and resistance.

With the diaphragm vibrating, the inductance and resistance may differ by 50 per cent from the inductance and resistance with the diaphragm damped. Let us call the excess of the inductance or resistance when the diaphragm is free over the inductance or resistance when the diaphragm is damped *the motional inductance or resistance* of the receiver. The motional inductance multiplied by the angular velocity we shall call the *motional reactance.* If now we plot motional resistance and motional reactance against the angular velocity of impressed e.m.f. we get —for a particular receiver—the curves marked "reactance" and "resistance" in Fig. 1. At the pitch of 5700 radians per sec. the vibration of the diaphragm increases the resistance by a large amount (22 ohms). At a slightly different pitch—5900—the vibration of the diaphragm decreases the resistance by (45 ohms). The reactance may follow a curve somewhat like the resistance curve but in general with the change of pitch, the vibration of the diaphragm causes the reactance to decrease to a minimum and then to go up again, as shown by the curve marked "reactance."

Now, if we measured the inductance, calculated the reactance and measured the resistance, we had the impedance, also we knew the current—and, multiplying the square of the current by the resistance, you get the power, and taking the difference between the power when the telephone is free and vibrating, and the power when the telephone is damped, you get the curve marked "power" of Fig. 1. The change-of-power curve, the amount of power supplied to the telephone when free in excess of the amount when damped is very large when you approach the resonant period of the diaphragm, and attains its maximum in the neighborhood of natural period of the diaphragm, which in the case of Fig. 1 was 5820 radians per second. This change of power amounted to as much as 68 per cent, when the telephone was making a pretty good noise. That is, if you put your finger on

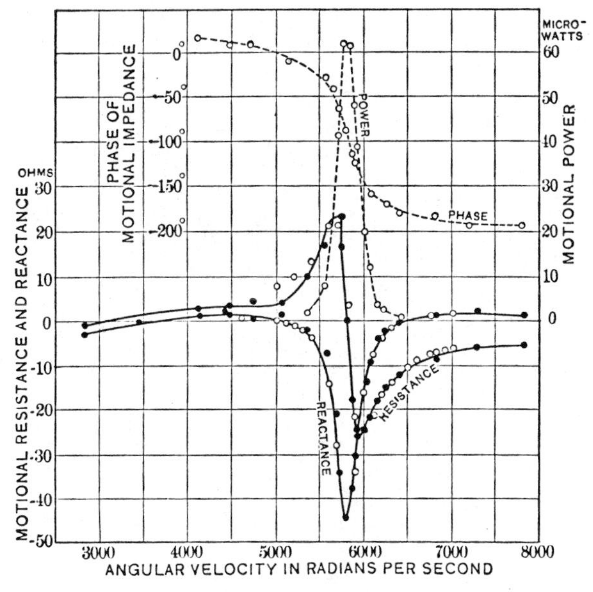

FIG. 1

the diaphragm and measured the power input at constant voltage, and take your finger off the diaphragm, and then measure the power input, you will find that the power input has increased by 68 per cent.

The telephone was free in a room, and the room was full of sound, when we were using current of frequency near the resonance point, and the sound interfered by reflection producing stationary waves in the room. If an assistant walked through the room, so as to go through the different points of maxima and minima of sound, the effect is a reaction on the telephone, so that its inductance and resistance changes with the position of the assistant in the room, and if you had a bridge balance for inductance and resistance and allowed a man to walk through the room, the bridge would be variously thrown in and out of balance. The reflected sound, coming back and striking the diaphragm, determine in part the work by the diaphragm, so that a shift of the stationary wave system in the room, affected inductance, resistance and power. With a given e.m.f. the resistance of the diaphragm would usually increase with the amount of work done by the diaphragm, and that would depend upon the stationary wave system.

If you plot the motional reactance against the motional resistance—meaning by the motional values the excess of reactance or resistance when the diaphragm is free over the corresponding value when it is damped—if you plot one of these quantities against the other, R being plotted horizontally and L being vertical, you get a circular locus, as in Fig. 2. The resistance and inductance change in relation to each other. The position of the center of the circle is determined by the mechanical and electrical constant of the diaphragm, and if you plot angular velocities of impressed e.m.f. around the circle, you begin at the origin with angular velocity zero, and as the angular velocity increases to infinity the vector "motional" impedance goes once around the circle in a negative direction. The frequency of e.m.f. which gives this point of the impedance circle diametrically opposite to origin is the natural frequency of the diaphragm; and the periods of the diaphragm differ in different instruments and also the sharpness of these resonance curves differ in different instruments. If you take a curve like that, which Messrs. Meyer and Whitehead have in their paper—or the resonance curve of the excess power put into the telephone when the diaphragm is free, you find that with an ordinary Bell receiver, the change of frequency to throw the diaphragms well out of resonance may be as much as 100 radians per second, but if you take a diaphragm rigidly clamped around the periphery by a heavy metallic clamp a change of period of 10 radians per second in five thousand would throw it completely out of resonance and reduce the amplitude of vibration so as to give almost complete silence. If you breathe on the rigidly-clamped diaphragm when it is actuated by a resonant current, the heat of the breath may so change the natural mechanical period of the diaphragm as to produce complete silence.

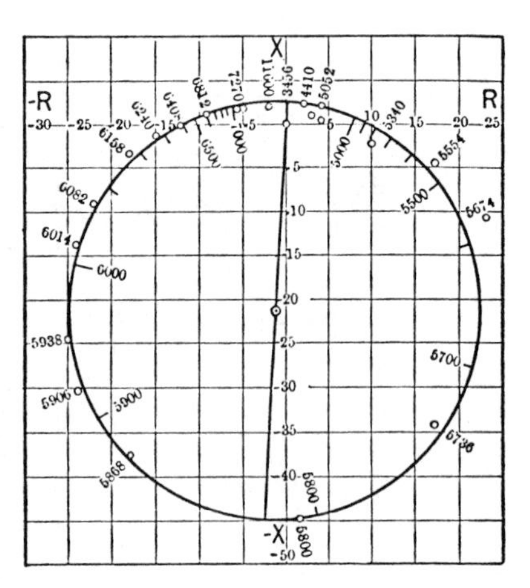

FIG. 2

Professor Kennelly and I find that the curves that I have shown agree closely with theoretical values obtained by considering the reaction of the moving membrane on the magnetic circuit. The theoretical treatment resembles the theoretical discussion of refraction and absorption in the neighborhood of the absorptive band in optics, except that in the telephone diaphragm problem account had to be taken of the fact that the magnetization of the iron lags behind the magnetizing forces consequently the shift in phase in the telephone problem differs from that in the optical problem. We determined experimentally the shift of phase, and found that it was equal to twice the angle of lag of the magnetization behind the magnetizing force as was demanded by the theory.

[*Editor's Note:* Material has been omitted at this point.]

Frank Wenner: There is one point which I should like to call attention to, and that is that the motion of the diaphragm produces an e.m.f. in the circuit just as in any other dynamo electric machine the relative motion of a part of the magnetic field and the winding results in an induced or generated e.m.f. This e.m.f. is a back e.m.f. and at the resonating frequency of the diaphragm may amount to as much as 90 per cent of the impressed e.m.f.

This is a matter to which I wish merely to call your attention now, as I expect to take it up in the discussion of the vibration galvanometer, the telephone receiver being one form of vibration galvanometer.

George W. Pierce: This whole effect of the shifting of reactance and inductance is the reaction of the diaphragm on the coils and that is taken account of in our theoretical treatment.

I have no doubt the treatment is the same as for the vibration galvanometer, except for the lag in the iron, which will shift our change of resistance curve to where our change of induction curve ought to be if there were no magnetic lag; that is, the double angle of lag amounts to enough to almost interchange these two curves.

As to this amount of power, perhaps my illustrations were not clearly explained. I said that this 68 per cent increase of power occurred when we got at the resonant point of the diaphragm. That does not mean that we have the measure there of the energy of the moving diaphragm; it merely means that with a given e.m.f. we draw more power under one condition than under another condition. The excess of power drawn amounted to a considerable quantity, and no doubt a good deal of it went into sound and energy radiation from the diaphragm, a part of which is not sound. The experiments were made with 0.3 of a volt in the telephone circuit, *i. e.*, under conditions which were somewhere near normal. The strength of the current in the telephone varies with the frequency, but it is in the neighborhood of one milliampere.

This reaction of the absorption of the sound upon the sounding body is a matter that Professor Sabine, at the physical laboratory, of Harvard, has emphasized and pointed out to me by means of an experiment of that kind. He had a tuning fork electrically driven in a room and measured the energy of sound all over the room, everywhere, and integrated it so that he got the total amount of energy in the room. He then put felt, which is an absorber of sound, in the room, with the intention of reducing the sound and measuring it again. He measured it again, and he had more energy than before. Although the felt absorbed a lot of energy, there was more energy than before. The explanation was that, putting the felt in the room shifted the wave zones in the room. He set his tuning fork at a constant amplitude, but the power drawn electrically happened to be increased by the shift of waves due to the introduction of the felt.

On the question of the absorption of energy by an observer, Professor Sabine has measured the absorption of energy by a person—that is, he got the energy in the room, and measured it, and then got the energy absorbed per person. It is interesting that he found that a woman absorbs more sound than a man, the difference being due to the difference in the character and amount of clothing of the woman and the man.

[*Editor's Note:* Material has been omitted at this point.]

Charles F. Meyer:

The observation of Mr. Pierce upon the variation of power consumed by the telephone, accompanying a variation of the system of standing waves in the room, is very interesting. He did not state the proportionate magnitude of the change in power. In the present investigation no effect of the standing waves upon the amplitude was observed. The photographs were all obtained with the experimenter in one position, which was necessary for manipulating the apparatus, but no variation in amplitude due to walking around the room was visually observed. It is thought that a variation of ten per cent could not have escaped detection.

[*Editor's Note:* In the original, material follows this excerpt.]

EDITOR'S NOTES

Page 2346: In Figure 2, "once around the circle in a negative direction" means "in a clockwise direction." Note that this is an offset circle whose center is not at the origin of the coordinate system. Note also that the "axis" of the circle, joining the origin with the point diametrically opposite, has been tilted clockwise around the origin through an angle $2\beta \simeq 90°$. This causes much of the circle to lie in the $-R$ domain. This result is even more striking in Figure 1. Pierce explains (p. 2347) that β is "the angle of lag of the magnetization behind the magnetizing force." This is due to hysteresis and to eddy currents. If these effects were corrected for, the depression angle 2β in Figure 2 could be made 0°, and most of the circle would swing up into the $+R$ domain. Simultaneously, the curves in Figure 1 would change radically, would almost interchange, in fact (see Pierce's remark at top of p. 2349). The new reactance curve would look almost like the old resistance curve, and the new resistance curve would look like a mirror image of the old reactance curve.

Page 2348: A remark by a Mr. Wenner is included because it is interesting and pertinent. "Back emf" is nowadays treated merely as a component of motional impedance, but back emf is sometimes a useful concept on its own.

Page 2351: Meyer gave the closing rebuttals. The reason he missed the boat was that his receiver was probably facing downward, whereas Pierce's was facing upward. (See F. V. Hunt, *Electroacoustics*, Wiley, New York, 1954, p. 96, on this point.)

41

Reprinted from pages 8-9, 14-15, and 23-24 of *Am. Acad. Arts Sci. Proc.* **56**:1-42 (1921)

ACOUSTIC IMPEDANCE AND ITS MEASUREMENT

A. E. Kennelly and K. Kurokawa

[*Editor's Note:* In the original, material precedes this excerpt.]

Again,

$$Z' = \frac{A\dot{x}}{I} = \frac{A^2}{\mathbf{z}'} \qquad \begin{array}{l}\text{electric absohms } \angle \\ \text{or C.G.C. magnetic} \\ \text{units of resistance } \angle\end{array} \qquad (14)$$

This means that in the motional impedance circle of the receiver, at any impressed frequency, the vector motional impedance Z', as obtained from electrical measurements with a Rayleigh bridge, is equal to the square of the complex force factor A, divided by the total mechanic impedance $\mathbf{z}'$ at that frequency. At the frequency of apparent resonance, this becomes

$$Z'_0 = \frac{A^2}{r''} \qquad \text{electric absohms } \angle \qquad (15)$$

since $\mathbf{z}'$ degrades at resonance into the total mechanical resistance r''. The slope of the motional impedance Z' will then be the same as the slope of A^2, or $-2\beta°$. Z'_0 is thus the diameter of the motional impedance circle, in the ordinary simple case, where the telephone is tested for its motional impedance.

From (14) it follows that, at any impressed frequency, the total mechanic impedance $\mathbf{z}'$ is

$$\mathbf{z}' = \frac{A^2}{Z'} = A^2 Y' \qquad \begin{array}{l}\text{dynes per kine } \angle \\ \text{or mechanic absohms } \angle\end{array} \qquad (16)$$

Consequently, if we divide the complex constant A^2 by the measured motional impedance Z', we obtain the size and slope of the total mechanic impedance of the diaphragm at this frequency.

The total mechanic impedance $\mathbf{z}'$ is made up of several mechanic impedances; namely,

(1) The mechanic impedance of the diaphragm

$$\mathbf{z}_d = r_d + j\left(m_d\omega - \frac{s_d}{\omega}\right) \qquad \text{mechanic absohms } \angle \qquad (17)$$

(2) The virtual mechanic impedance of the diaphragm[6] due to its motion in the permanent magnetic field

$$\mathbf{z}_v = j\,\frac{pA}{\omega} = r' + \frac{js'}{\omega} = \frac{p\,|A|}{\omega}\sin\beta + j\,\frac{p\,|A|}{\omega}\cos\beta$$

$$\text{mechanic absohms } \angle \qquad (18)$$

The real part of $\mathbf{z}_v$, although varying inversely with ω, is ordinarily taken as constant to a first approximation, the working range in ω being usually small.

(3) The acoustic impedance $\mathbf{z}_1$ of the air in the chamber behind the diaphragm, and which varies, in some manner similar to (17), as the impressed frequency is varied.

(4) The acoustic impedance $\mathbf{z}_2$ of the air in front of the diaphragm, and which constitutes the load to be varied.

Consequently,

$$\mathbf{z}' = \mathbf{z}_d + \mathbf{z}_v + \mathbf{z}_1 + \mathbf{z}_2 \qquad \text{mechanic absohms } \angle \qquad (19)$$

Keeping the impressed frequency constant, $\mathbf{z}_d$, $\mathbf{z}_v$, and $\mathbf{z}_1$ will remain constant; but $\mathbf{z}_2$ can be made to vary by changing the acoustic load. The changes in the electric impedance, measured on changing the acoustic load, enable the latter to be computed. Moreover, by observing the changes in Z' when the frequency is varied, if allowance can be made for the changes thereby produced in $\mathbf{z}_d$, $\mathbf{z}_v$ and $\mathbf{z}_1$, the variations in the acoustic impedance $\mathbf{z}_2$ may also be computed.

[*Editor's Note:* Material has been omitted at this point.]

[6] A. E. Kennelly and H. Nukiyama. "Electromagnetic Theory of the Telephone Receiver, with Special Reference to Motional Impedance." *Proc. A. I. E. E.*, March, 1919. *Bulletin No. 17, Research Div., El. Eng'g. Dept., Mass. Inst. Tech.*

The next step is to connect the receiver to its acoustic load. In the case of Figure 2, the load is the cylindrical column of air in the tube TT. The resistance and inductance of the telephone are now meas-

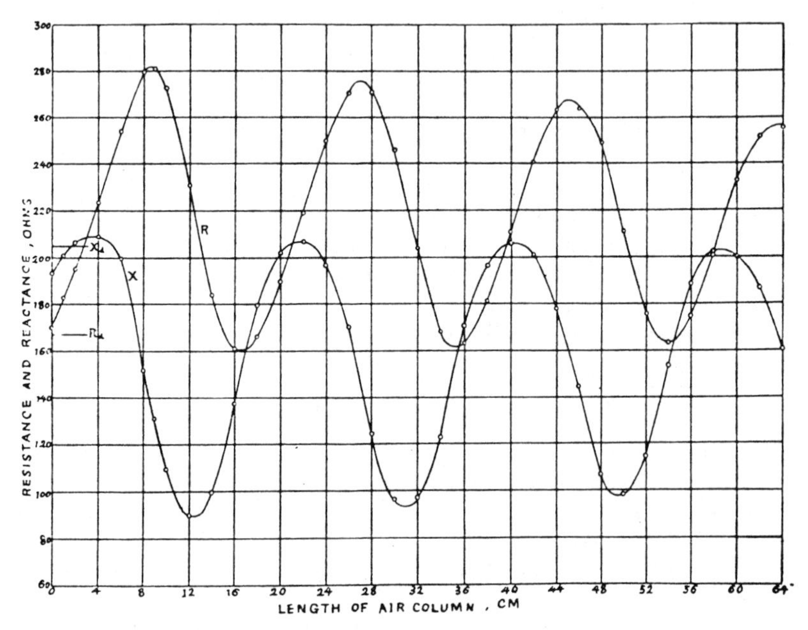

FIGURE 7. Curves of Apparent Resistance and Reactance of Receiver when attached to Air Column of successively varied Length.

ured at successive positions of the piston P, holding the frequency f, and alternating-current strength, constant. The values of receiver resistance and inductance so measured, are found to fluctuate periodically as the length L of the air column in the acoustic load changes. Figure 7 gives a pair of wavy curves in one test, at $f = 921 \sim$ and $I = 1.575$ milliamperes rms. in the receiver.

It is important that the frequency and the temperature in such a test should be held as nearly constant as possible. The frequency should be near to the resonant frequency f_0. If the frequency is remote from the resonant frequency, the mechanic reactance of the diaphragm is likely to be so large that the addition of the acoustic load to be measured will not affect the vector motional impedance materially; whereas at resonance, a small change in added acoustic load will be likely to affect the motional impedance considerably.

The variations in R and X indicated in Figure 7, represent corresponding variations in the electric impedance $Z = R + jX$. These are due to variations in the acoustic impedance $\mathbf{z}_2$, which is the subject of investigation. The curves of R and X in Figure 7, represent total apparent resistance and reactance, in the receiver, at varying air-column lengths. The values R_d and X_d indicated on the left-hand side of the figure, are respectively the damped resistance and reactance of the receiver at this frequency. With reference to the impedance $Z_d = R_d + jX_d$, as origin, the values of motional impedance Z', corresponding to the successive air-columns, are plotted vectorially in Figure 8. The numbers marked on this spiral are air-column lengths in cm. It will be seen that at L = 0 cm., or with the piston within 1.5 mm. of the diaphragm, the motional impedance was $12.5 \nwarrow 75^\circ.4$ ohms. This value was nearly repeated with L = 19 cm. At L = 10 cm., however, the motional impedance had increased to $142.8 \nwarrow 42^\circ.1$ ohms. The curve is a slowly contracting spiral, with a pitch of 18.7 cm., which is half a wave length at 921 ∾; since the velocity of sound at 20°C being 34,430 cm. per sec., the wave length $\lambda = 34{,}430/921 = 37.38$ cm.

In order to utilize the motional-impedance diagram of Figure 8 for determining the acoustic impedance of the load in front of the receiver diaphragm, we may refer to equations (16) and (19). Equation (16) shows that we must find Y', the vector reciprocal of the motional impedance, and multiply by the square of the vector constant A of the receiver, in order to derive the total mechanic impedance $\mathbf{z}'$ on the diaphragm. This means that we must invert the diagram of Figure 8, or find the locus of the reciprocal spiral.

[*Editor's Note:* Material has been omitted at this point.]

The exact
shape of this graph O D E F is not regarded as of great importance; because, if the instrument was placed against the ear with less or more pressure, or if it was centered differently on the ear; or if some other ear was selected, the shape of the graph would be likely to be altered appreciably. To a first approximation, however, the graph O D E F may be regarded as an approach to the dotted circular motional impedance O G E H, which may be described as the motional circle of

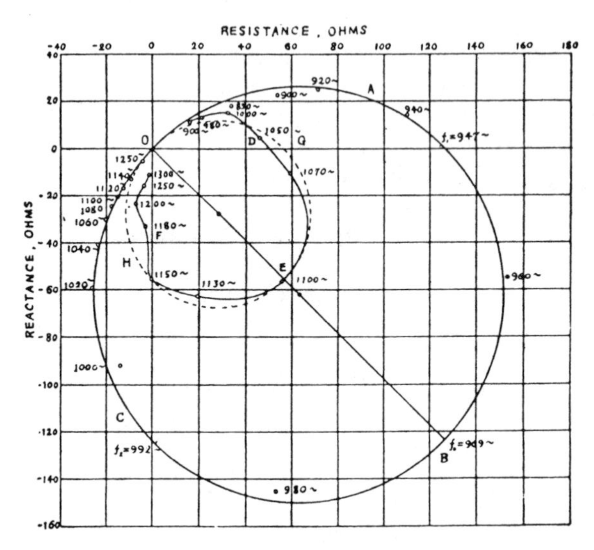

FIGURE 12. Motional Impedance Graphs of Receiver in Room and pressed against Listener's Ear.

reference, with the instrument applied to this ear. It will be observed that the diameter of this circle of approximation is 78.5⦣ 44° ohms. The diameter O E has, therefore, substantially the same slope as the free-air motional impedance diameter O B; but has been reduced in length somewhat more than 50 per cent. The frequency of apparent resonance has also been changed from 969 ∾ to 1100 ∾, by applying the receiver cap to the ear.

Figure 13 gives the corresponding mechanic impedance graphs for this receiver and ear. The straight line *a b c*, parallel to the reactance axis, corresponds to the motional impedance circle O A B C of Figure 12, taken in the free air of a room. This means that the total mechanic impedance of the receiver free to the room, was a constant mechanic resistance of 149 dynes per kine, plus or minus a mechanic reactance, depending on the impressed frequency. At 969 ∾, the frequency of apparent resonance, this reactance disappears, leaving the resistance of 149 dynes per kine as the residual mechanic impedance.

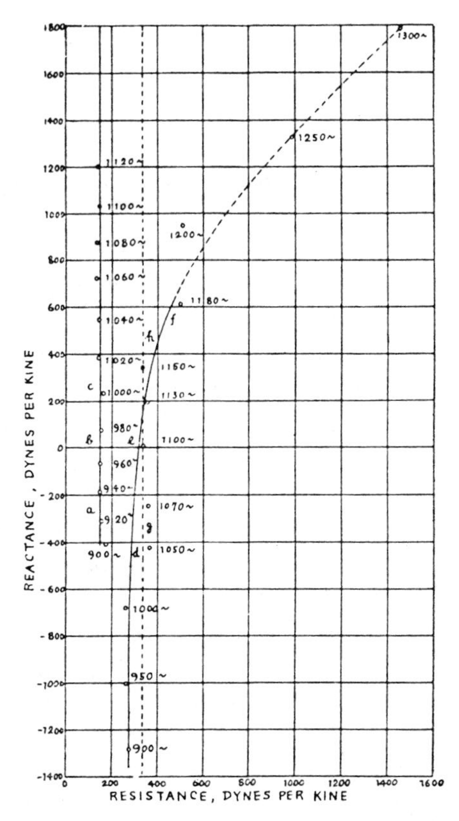

FIGURE 13. Mechanic Impedance of Receiver in Room and when pressed against Listener's Ear, by inversion from Fig. 12, and multiplying by A^2.

It should be observed that any motional-impedance circle of a telephone receiver, passing through the origin of coördinates, corresponds to such a straight-line graph of total mechanic impedance parallel to the reactance axis, no matter what the depression angle of the motional-impedance diameter may be. The application of the factor A^2, in formula (16), cancels out the diametral depression angle.

[*Editor's Note:* In the original, material follows this excerpt.]

EDITOR'S NOTES

Page 8: Equation 14 (where A is $B\ell$) is based on the two relations: $F = B\ell I$ and $E = B\ell\dot{x}$. Then motional $Z' = E/I = B\ell\dot{x}/I = (B\ell\dot{x}) \cdot (B\ell/F)$ or $(B\ell)^2/z'_{mechanical}$. Then $z'_{mech} = (B\ell)^2/Z'_{mot}$.

Page 9: The paragraph following Equation 19 calls attention to the key to the success or failure of this method: whether or not motional Z' and mechanical z_d, z_v, and z_1 can be accurately determined.

Pages 14 and 15: In Figure 7, if we vary L and consider only the region from $L = 0$ to $L = 17$ cm, we will have covered one cycle of the input impedance $Z_{in} \simeq \rho c S \tanh(\alpha + j\beta)L$, where β, or ω/c, is related to the reactance X and α is related to the maximum value of the resistance R. It can be seen that one cycle of this function also describes the input impedance components of a lumped parallel-resonant circuit. That is, the curve for Z_{in} of a line is the same whether we vary ω or L.

If now we replot this one cycle in the complex plane as X versus R, we will get a motional impedance "circle diagram," which would be seen in Figure 8 (not shown). Note that Figure 7 plots only electrical (that is, motional) resistance and reactance. We must still find $z'_{mech} = (B\ell)^2/Z'_{mot}$. This is discussed in the last paragraph of page 15.

Pages 23 and 24: The first line on page 23, referring to Figure 12, should begin: "The exact shape of this graph ODEF . . . ." The inversion of the Z'_{mot} circle into the z'_{mech} straight line is best seen in Figures 12 and 13, where now the air-column length L is held constant and the frequency ω is varied. It is frequently demonstrated that any straight line perpendicular to the abscissa will conformally transform into a circle, by inversion. Hence it is not surprising that a second inversion will transform the circle back into a straight line.

In his closing paragraph (pp. 26, 27), omitted here, Kennelly suggested that this method could be "applied to architectural acoustics; as, for example, in the measurement of the sound-reflection coefficient of draperies. . . . Relatively feeble reflected sound waves, falling on the diaphragm, would thus appreciably affect the motional impedance, and enable the corresponding acoustic variation to be measured." In practice, this application has never proved useful except in a very narrow frequency band, the region around resonance. Elsewhere the diaphragm is too stiff to respond to variations in the acoustic loading impedance.

Part VI

PHASE DISTORTION AND TRANSIENT DISTORTION IN ELECTROACOUSTIC SYSTEMS

Editor's Comments on Papers 42 Through 47

The system function or transfer function $H(\omega)$ is a complex function that can be resolved into an "amplitude response" $|H(\omega)|$ or log $|H(\omega)|$ and a "phase response" exp $- j\theta(\omega)$. In this book, amplitude response and amplitude distortion are not stressed, both for lack of space and because of the widespread knowledge of their behavior, much discussed since the 1920s. Instead, the less widely discussed phase response and phase distortion is treated along with the resulting transient distortion. Transient distortion, resulting from both amplitude and phase distortion in a system, has been displayed as a simple oscillogram of amplitude versus time (Paper 43; McLachlan, 1960) and also as a three-dimensional presentation showing amplitude versus both time and frequency (Shorter, 1946; Paper 46).

Phase distortion in acoustics was not much worried about before the 1930s, due in part to both Helmholtz's writings and Ohm's acoustic law. Then in the July 1930 issue of the *Bell System*

Technical Journal, three rather remarkable papers appeared, discussing the subject of phase distortion in telephone apparatus (Papers 42 and 43; Steinberg, 1930). The commercial impetus had been twofold: (1) the requirements of long-distance wire transmission of radio programs for the large radio networks and (2) the emergence of wire-borne television. J. C. Steinberg (1930) showed that phase distortion had a great influence on the intelligibility of speech; but the influence on music was still not widely worried about. (Indeed, music is often "intelligible" when only the rhythm survives.) Concern about music was to come somewhat later (McLachlan, 1960; Shorter, 1946; Paper 45; Yenzer, 1949).

Another commercial impetus was simultaneously being supplied by the requirements of wireless television. The tools developed for the attack on visible distortion (Wheeler, 1939; Goldman, 1948; DiToro, 1948) were subsequently taken up for the attack on audible distortion (Heyser, 1967; 1969; 1971; Preis, 1976; 1977; Papers 45 and 46).

Paper 42 by H. Nyquist and S. Brand of the Bell Telephone Laboratories treats the measurement of phase distortion in telephone apparatus via a tone burst (gated sine wave) that moves through a system in discrete frequency steps. Because of space constraints, only the method using modulation is given here. The original article is recommended to the reader for an interesting method of determining envelope delay from impedance measurements. Paper 43, also on telephone distortion, consists of a two-page excerpt from the paper by C. E. Lane, also of Bell Labs. This excerpt could almost be a part of the Nyquist paper.

Paper 44 by F. M. Wiener, on phase distortion in electroacoustic systems, was written when Wiener was working at Harvard with F. V. Hunt. Hunt, of course, had an enthusiastic interest in the whole chain of components involved in sound recording, including microphones as well as phonograph pickups and loudspeakers. Space constraints limit us to a two-page excerpt—enough, it is hoped, to show Wiener's originality and careful workmanship as an experimenter.

Paper 45 by M. S. Corrington of the Radio Corporation of America treats the transient testing of loudspeakers via tone bursts. His theoretical treatment uses the complex frequency plane and the Laplace transform.

Paper 46 by J. M. Berman and L. R. Fincham of KEF Electronics in England treats the transient testing of loudspeakers via a repeated impulse in conjunction with signal averaging. The theoretical treatment uses the real frequency plane and the Fourier transform. A radical innovation by these authors is the storage of the impulse response or its Fourier transform in a computer memory system and the later interrogation of this numerically simulated "loudspeaker" regarding its behavior toward signals overlooked previously.

Paper 47 by R. C. Heyser (California Institute of Technology), who analyzed loudspeaker theory as a labor of love, treats the accurate measurement of the phase response and the amplitude response of a loudspeaker and the significance of this information as applied, e.g., to transient distortion. This pioneer paper, using analog techniques, provided the impetus for Berman and Fincham to progress even further using digital techniques. Heyser's theoretical treatment uses the real frequency plane and the Fourier transform. One of his teaching contributions is the demonstration of the usefulness of the Hilbert transform in assessing loudspeakers—for example, the worsening of the transient response of a certain loudspeaker when the amplitude response was made flatter.

REFERENCES

DiToro, M. J., 1948, Phase and Amplitude Distortion in Linear Networks, *Inst. Radio Eng. Proc.* January, pp. 24–36.

Goldman, S., 1948, *Frequency Analysis, Modulation, and Noise*, McGraw-Hill, New York.

Heyser, R. C., 1967, Acoustical Measurements by Time Delay Spectrometry, *Audio Eng. Soc. J.* **15**:370–382.

Heyser, R. C., 1969a, Loudspeaker Phase Characteristics and Time Delay Distortion: Part II, *Audio Eng. Soc. J.* **17**:130–137.

Heyser, R. C., 1969b, Time Delay Spectrometer, U. S. Patent 3,466,652.

Heyser, R. C., 1971a, Determination of Loudspeaker Signal Arrival Times: Part I, *Audio Eng. Soc. J.* **19**:734–743.

Heyser, R. C., 1971b, Determination of Loudspeaker Signal Arrival Times: Part II, *Audio Eng. Soc. J.* **19**:829–834.

McLachlan, N. W., 1960, *Loudspeakers*, Dover Publications, New York.

Preis, D., 1976, Linear Distortion, *Audio Eng. Soc. J.* **24**:346–367.

Preis, D., 1977, Catalog of Frequency and Transient Responses, *Audio Eng. Soc. J.* **25**:990–1007.

Shorter, D. E. L., 1946, Loudspeaker Transient Response—Its Measurement and Graphical Representation, *Br. Broadcast. Corp. Quart.* **1**:121.

Steinberg, J. C., 1930, Effects of Phase Distortion on Telephone Quality, *Bell Syst. Tech. J.* July, pp. 550–566.

Wheeler, H. A., 1939, Interpretation of Distortion by Paired Ecoes, *Inst. Radio Eng. Proc.* **27**:359–385.

Yenzer, G. R., 1949, Lateral Feedback Disc Recorder, *Audio Eng. Soc. J.* **33**:22–27. (See also H. E. Roys, ed., 1978, *Disc Recording and Reproduction*, Dowden, Hutchinson & Ross, Stroudsburg, Pa., p. 343.)

ANNOTATED BIBLIOGRAPHY

[*Editor's Note:* In Part VI it has seemed desirable to present the annotated bibliography in connection with topics rather than authors, because a given topic is not confined to one author, as for example, in Part I, but is typically treated by all the authors.

In order for the reader to understand more fully the papers presented in Part VI, very specific bibliographic entries are given. Since these entries are not primarily involved with measurement problems, only those pages that shed light on the problem are cited. Most of these entries were generated by the revolution in signal processing that occurred after World War II. This revolution included the widespread use of spectral analysis, the impulse response, the sampling theorem, and then in the 1960s the Fast Fourier Transform (FFT) and low-priced digital processing. It is suggested that the reprinted papers be studied before the reader delves into the annotated bibliography.]

Influence of Phase Distortion on Transient Response

[*Editor's Note:* All of the authors in Part VI are concerned with this subject.]

DiToro, M. J., 1948, Phase and Amplitude Distortion in Linear Networks, *Inst. Radio Eng. Proc.* January, pp. 24–36.
p. 25: DiToro proposes the term *phase bandwidth* to explain the narrowing of the effective bandwidth of an all-pass network.
p. 29: Demonstrates ringing of the impulse response of an all-pass network when phase bandwidth is restricted.

Goldman, S., 1948, *Frequency Analysis, Modulation, and Noise*, McGraw-Hill, New York.
pp. 102–108: Goldman gives the clearest explanation of Wheeler's paired echoes. His Figure 24 gives a graphic illustration of the subtle effect of phase distortion on a monopulse.

Lane, C. E., 1930, Phase Distortion in Telephone Apparatus, *Bell Syst. Tech. J.* **9**:493–521.
pp. 500–503: These pages show experimental distorted transient responses. Both Nyquist and Lane refer the reader to Steinberg.

Steinberg, J. C., 1930, Effects of Phase Distortion on Telephone Quality, *Bell Syst. Tech. J.* **9**:550–566.
pp. 562–563: Steinberg concentrates on the damage done by delay distortion to the intelligibility of speech. He also observes that the primary effect of phase distortion is to narrow the effective bandwidth of the transmission system.

Wheeler, H. A., 1939, Interpretation of Distortion by Paired Echoes, *Inst. Radio Eng. Proc.* **27**:359–385.
p. 359: Wheeler breaks up the distortion into a pre-echo and a post-echo. He promises that the method, an approximation method using the Fourier integral, will quickly specify how much phase distortion can be tolerated.

Advantages of the Laplace Transform Method for Transient Analysis

[*Editor's Note:* Corrington favors the Laplace transform method in Paper 45.]

Gardner, M. F., and J. L. Barnes, 1942, *Transients in Linear Systems,* Wiley, New York.
pp.8–10: See especially Table 1, comparison of four methods.
pp. 99–107: The Fourier method is non-causal for $t < 0$.
pp. 357–358: The Laplace method handles initial conditions better.

Lathi, B. P., 1965, *Signals, Systems, and Communication,* Wiley, New York.
pp. 171–175: The Laplace transform as a logical extension of the Fourier transform.
pp. 208–210: Exp $j\omega t$ can cause trouble with unstable systems, whereas $\exp(\sigma + j\omega)t$ avoids this trouble.
pp. 231–235: Transform-analysis approach versus frequency-analysis approach.

Weber, E., 1954, *Linear Transient Analysis,* vol. I., Wiley, New York.
p. 277: Weber says that real Fourier integrals are usually not very practicable because of the difficulty in integration when poles are present.
p. 284: He now adds that with the help of extensive tables (like Campbell-Foster) and a little ingenuity, you can usually survive with just the real Fourier integral.
p. 289: Shows that the Laplace transform has a built-in damping factor and therefore does not demand any ingenuity.

Advantages of the Fourier Integral Method for Transient Analysis

[*Editor's Note:* Berman and Fincham, and also Heyser, favor the Fourier integral method in Papers 46 and 47.]

Bracewell, R., 1965, *The Fourier Transform and Its Applications,* McGraw-Hill, New York.
p. 220: Says that the essential advantage of the Fourier transform is its physical interpretation as a spectrum and that Laplace transforms are not so interpretable.

Guillemin, E. A., 1931, *Communication Networks,* vol. I, Wiley, New York.
p. 352: Guillemin on normal modes. (Prophetic!)

Guillemin, E. A., 1935, *Communication Networks,* vol. II, Wiley, New York.
pp. 504–556: Guillemin's philosophy of 1935 on the Fourier integral.

Guillemin, E. A., 1955, The Fourier Integral—A Basic Introduction, *Inst. Radio Eng. Trans. Circuit Theory* September, pp. 227–230.
Guillemin's philosophy of 1955. See especially the closing paragraph.

Lathi, B. P., 1965, *Signals, Systems, and Communication,* Wiley, New York.
pp. 234–235: Mentions Guillemin on "black boxes," meaning the sterile mathematical transform-analysis approach versus the more physical frequency-analysis approach.

Schwartz, M., 1959, *Information Transmission, Modulation, and Noise*, McGraw-Hill, New York.
p. 41: Seems to slightly prefer the Fourier transform. Points out that given the pole-zero data, you can always derive the amplitude plus phase response or vice versa.

Choice of Excitation Signal Applied to the Loudspeaker

[*Editor's Note:* Most of the authors in Part VI are concerned with this subject.]

Berman, J. M., and L. R. Fincham, 1977, The Application of Digital Techniques to the Measurement of Loudspeakers, *Audio Eng. Soc. J.* **25**: 370–384.
pp. 371, 375: Why the impulse response should be used.

Heyser, R. C., 1971, Determination of Loudspeaker Signal Arrival Times: Part I, *Audio Eng. Soc. J.* **19**:734–743.
pp. 737–739: Discusses relative advantages of the impulse and the doublet.

McLachlan, N. W., 1960, *Loudspeakers*, Dover Publications, New York.
pp. 332–336: Good pictures of transient response from a repeated impulse.
pp. 336–337: Response from a gated burst, or sine wave train multiplied by a rectangular envelope.

Preis, D., 1976, Linear Distortion, *Audio Eng. Soc. J.* **24**:346–367.
pp. 352–353: Shows experimental impulse responses and tone-burst responses of a number of equalizer-configurations.

Preis, D., 1977, Catalog of Frequency and Transient Responses, *Audio Eng. Soc. J.* **25**:990–1007.
Wonderful pictorial catalog of impulse responses and step responses of networks having garden-variety amplitude and phase distortions (linear).

Schaumberger, A., 1971a, Impulse Measurement Techniques, *Audio Eng. Soc. J.* **19**:101–107.
pp. 105–106: Figures 13–18 show some advantages of the doublet over the impulse.

Schaumberger, A., 1971b, Application of Impulse Measurement Techniques, *Audio Eng. Soc. J.* **19**:664–668.
p. 666: This continues the discussion of the relative merits of the impulse and the differentiated impulse.

Effective Position of the Acoustic Source

[*Editor's Note:* Berman and Fincham, and also Heyser, are concerned with this problem in Papers 46 (pp. 383–384 and Fig. 19) and 47 (p. 33 and Fig. 1).

Heyser, R. C., 1969, Loudspeaker Phase Characteristics and Time Delay: Part II, *Audio Eng. Soc. J.* **17**:130–137.
See p. 136, Summary.

Heyser, R. C., 1971, Determination of Loudspeaker Signal Arrival Times: Part II, *Audio Eng. Soc. J.* **19**:829–834.
See p. 833 on the ensemble of equivalent sources.
Olson, H. F., and F. Massa, 1939. *Applied Acoustics*, 2nd ed., Blakiston, Philadelphia.
pp. 318–319: See especially Massa's discussion of the location of the apparent source of sound within a horn loudspeaker and how it moves toward the throat at high frequencies.

Hilbert Transforms

[*Editor's Note:* This subject, along with minimum-phase networks, is referred to repeatedly by Heyser and by Berman and Fincham in Papers 46 and 47.]

Bode, H. W., 1945, *Network Analysis and Feedback Amplifier Design*, D. Van Nostrand, New York.
pp. 303–336: See especially pp. 305–309 and pp. 318–322 on the relation between magnitude and angle.
Guillemin, E. A., 1957, *Synthesis of Passive Networks*, Wiley, New York.
pp. 296–308: See, for example, Figure 7 on p. 302, showing "loss and phase-lag functions."
Papoulis, A., 1962, *The Fourier Integral and Its Applications*, McGraw-Hill, New York, pp. 198–212.
Pages 210–212 treat the relation between log mag and phase angle of a system function $H(\omega)$.
Tuttle, D. F., 1958, *Network Synthesis*, vol. I, Wiley, New York.
pp. 385–400: See especially p. 393 on the relation between magnitude and angle.

Time-Varying Spectra

[*Editor's Note:* This subject is discussed extensively by Berman and Fincham and by Heyser, in their papers included here (Papers 46 and 47) and elsewhere. Our interest is primarily with the light shed on the measurement of transient distortion.

The time-varying spectra displayed in Shorter (1946) and in Koenig et al. (1946) are concerned with two different problems. Koenig (who is *not* looking for transient response) uses N relatively narrow-band filters that are just broad enough to allow the inputted speech signal in each channel to reach steady state. (See also Flanagan, 1972.)

Shorter (who *is* looking for transient response) uses one broadband filter, namely the full bandwidth of his loudspeaker, for *each* input signal. Thus his inputted tone bursts are modified, during both growth and decay, only by the full bandwidth of his loudspeaker. This is also the approach that Berman (Paper 46, pp. 378, 379) has taken, resulting in his "cumulative spectra," obtained by calculation rather than by direct measurement.

The question of what the ear/brain system is processing during the decay of an acoustic pressure wave, is still unresolved. Some interesting thoughts on the subject are given in Page (1952) and in Heyser (1971b). (See also Heyser [1971a].)

Flanagan, J. L., 1972, *Speech Analysis, Synthesis, and Perception*, 2nd ed., Springer-Verlag, New York, pp. 141–161.
See especially p. 147 and Figure 5.5 showing the output of a filter bank spectrum analyzer. (See also Koenig et al., 1946.)

Heyser, R. C., 1971a, Determination of Loudspeaker Signal Arrival Times: Part I, *Audio Eng. Soc. J.* **19**:734–743.
p. 742: On his hopes for an energy-density microphone.

Heyser, R. C., 1971b, Determination of Loudspeaker Signal Arrival Times: Part II, *Audio Eng. Soc. J.* **19**:829–834.
p. 834: His thoughts on the question of what the ear/brain system is processing during the decay of an acoustic pressure wave.

Koenig, W., H. K. Dunn, and L. Y. Lacy, 1946, The Sound Spectrograph, *Acoust. Soc. Am. J.* **18**:19–49.

Page, C. H., 1952, Instantaneous Power Spectra, *J. Appl. Phys.* **23**:103–106.
p. 103: Some pioneer thinking on the question of what the ear/brain system is processing.

Shorter, D. E. L., 1946, Loudspeaker Transient Response—Its Measurement and Graphical Representation, *Br. Broadcast. Corp. Quart.* **1**:121.

42

Reprinted with permission from pages 522, 523, 524–526, 535–541, and 548 of *Bell Syst. Tech. J.* **9**:522–549 (1930)

Measurement of Phase Distortion *

By H. NYQUIST and S. BRAND

This paper deals with the measurement of phase distortion or delay distortion and is particularly concerned with measurements on telephone circuits. For this purpose, use is made of a quantity defined as "envelope delay," which is the first derivative of the phase shift with respect to frequency. Various methods for measuring this quantity and the principles on which they are based are discussed, the details of the measuring circuits being omitted and sources of further information referred to when possible. Data are included which give the measured envelope delay-frequency characteristics of several kinds of telephone circuits.

AT an early date in the use of long loaded telephone circuits, certain disturbing effects at high frequencies were noticed which have been known as transients.[1] It was found that on such circuits, even when the attenuation was very carefully equalized within the transmitted range, these transient effects still persisted and were made worse. It was realized that these effects were due to phase distortion or delay distortion, that is, the resultant effect of phase shift varying with frequency in a peculiar manner.

[*Editor's Note:* Material has been omitted at this point.]

* Presented at New York Section, A. I. E. E., May 1930.

[1] "Telephone Transmission of Long Cable Circuits," A. B. Clark, *Jour. A. I. E. E.*, Vol. XLII, p. 1, 1923, and *B. S. T. J.*, Vol. II, p. 67, 1923.

It is the purpose of this paper to describe and discuss various methods which have been devised for the measurement of phase distortion. Phase distortion and its effects, as well as methods of correcting for it [5, 6, 7, 8], are considered here only sufficiently for the understanding of these measuring means. It is, of course, necessary that before the correction can be designed, the amount of distortion be known, and that, after the corrective apparatus has been built and applied to the circuit, the overall system be checked to find out how complete the correction has been. The devices described below are for this particular purpose, and before describing them, the fundamental theory upon which they are based will be considered. Some of the principles underlying particular methods of measurement will be considered in the description of the devices themselves.

[*Editor's Note:* Material has been omitted at this point.]

[5] "Phase Distortion and Phase Distortion Correction," Sallie Pero Mead, *B. S. T. J.*, Vol. VII, p. 195, 1928.
[6] "Distortion Correction in Electrical Circuits with Constant Resistance Recurrent Networks," O. J. Zobel, *B. S. T. J.*, Vol. VII, p. 438, 1928.
[7] "Phase Distortion in Telephone Apparatus," a companion paper by C. E. Lane.
[8] "Effects of Phase Distortion on Telephone Quality," a companion paper by J. C. Steinberg.

THEORY UNDERLYING PHASE DISTORTION MEASUREMENT

It should be understood that what is meant here by phase shift is really the insertion phase shift; that is, the phase shift of a system is the change in phase at the receiving terminal due to the insertion of the system under consideration between the generator and the receiving terminal. In the same way, when delays are mentioned, insertion delays are understood unless otherwise specified.

A certain amount of time is required for the transmission of any signal from one place to another; and it has been found that, for a natural reproduction of tone or speech, or the satisfactory transmission of any signal, not only must the attenuation of the various component frequencies be approximately equalized, but also the time of propagation of these same component parts must be nearly the same for different frequencies. This time of propagation is, of course, closely related to the change in phase of the component frequencies during transmission.

In order to have no phase distortion it is necessary that the phase shift be linear with frequency within the frequency range required for transmission.[5, 7] Graphically, this means that if the phase shift is plotted as a function of frequency, the resulting graph will be a straight line within the frequency range under consideration. It is evident then that for such a condition the first derivative of the phase shift with respect to frequency, or the slope of the phase shift-frequency curve, is constant.

The slope or first derivative is closely related to the delay of the envelope. The following statement results from a mathematical consideration of the building up of sinusoidal currents in systems similar to those which we are considering here.[9] *The envelope of the oscillations in response to an e.m.f.* $E \cos \omega t$ *applied at time* $t = 0$ *reaches 50 per cent of its ultimate steady value at time* $t = d\beta/d\omega$ *and its rate of building up is inversely proportional to* $\sqrt{d^2\beta/d\omega^2}$. Various assumptions are made in arriving at this conclusion, but it does not seem necessary to discuss these here except to mention the condition that the attenuation of the system under consideration should be approximately equalized over the frequency range in the neighborhood of the applied frequency.

It is apparent then that this quantity $d\beta/d\omega$ plays a fundamental rôle in determining the delay of a system. Moreover, the use of $d\beta/d\omega$ has an advantage over β in that it is constant for a distortionless system, while β varies with frequency. The quantity $d\beta/d\omega$ will simply be defined here as the "envelope delay" of a system in frequency

[9] "Building Up of Sinusoidal Currents in Long Periodically Loaded Lines," J. R. Carson, *B. S. T. J.*, Vol. III, p. 558, 1924.

ranges where the attenuation is not a function of frequency; that is, the envelope delay in seconds is

$$T = \frac{d\beta}{d\omega} = \frac{dB}{df},$$

where

$\beta =$ the phase shift measured in radians,

$B =$ the phase shift measured in cycles,

$f =$ the frequency measured in cycles per second,

and

$\omega = 2\pi f.$

Hereafter in this paper this notation will be used.

For a distortionless system this quantity is the actual delay of the signal transmitted through the system. However, for a system which introduces phase distortion, the received envelope is usually quite different from the impressed envelope; and the delay of this envelope through the system is then quite indefinite depending upon what particular feature of the envelope is taken for observation. Nevertheless, the quantity defined as envelope delay is perfectly definite for such a system.

The significance of phase shift and envelope delay and the relation between the two is considered at some length in other papers.[6*, 7] The use of phase shift and delay data as a measure of phase distortion is also considered there. Phase shift itself is a rather fundamental quantity and various means of measuring it can be devised when both ends of the system under consideration are available. In this paper, one method of doing this is referred to which has proved very useful in laboratory measurements in the design of apparatus. However, for field measurements on telephone circuits, envelope delay seems to be a more useful quantity with which to work. The derived nature of this quantity makes its measurement somewhat complicated and consequently considerable space is given to methods for this purpose.

The envelope delay is determined from the difference in the steady-state phase shift for a definite interval of frequency. In practical cases finite intervals are used instead of infinitesimally small intervals which would be required for the determination of the derivative, or of the slope of the phase shift-frequency curve. This means then that the measured value is actually the slope of the secant of the curve and is simply an approximate value for the envelope delay, the amount of approximation depending on the size of the interval chosen. The value of envelope delay arrived at in this way will be called T_Δ so that

[6*] l.c. Appendix I.

$$T_\Delta = \frac{\Delta\beta}{\Delta\omega} = \frac{\Delta B}{\Delta f},$$

where $\Delta\beta$ represents a finite difference in β, etc.

The use of steady-state conditions for the measurements of envelope delay * has quite evident advantages practically over a method which would tend to measure the delay of the envelope itself in a transient state.

[*Editor's Note:* Material has been omitted at this point.]

(535)

c. Direct Measurement of Envelope Delay

Here the phase shift of the envelope of a modulated wave is measured under steady-state conditions and this gives a direct measurement of the envelope delay when the measuring set is properly calibrated, inasmuch as the delay of the envelope of the modulated wave is closely related to the differences in phase shift for the component frequencies of the modulated wave transmitted. Before describing the details of the measuring circuits, some of the principles underlying the transmission of simple modulated waves will be considered; and for this purpose envelopes produced by sine wave modulations will be used. It is assumed in this discussion that the modulations in the transmitted current are repeated periodically and that the attenuation of the system used for transmission is completely equalized for all the frequency components.

Consider a 1000-cycle sine wave which is modulated by a 25-cycle sine wave in such a manner that the envelope just reaches zero once per cycle of the modulating wave. This wave is found on analysis to consist of three components, namely, 1000 cycles of two units amplitude, 975 cycles of one unit amplitude, and 1025 cycles of one unit amplitude. At the start, it is somewhat simpler to consider this case with the 1000-cycle component removed. In other words, the current transmitted through the system now consists of 975 and 1025 cycles in equal amounts. This value at the sending end may be conveniently written

$$\sin 975\, \overline{2\pi t} + \sin 1025\, \overline{2\pi t}.$$

The equivalent graphical expression is

$$2 \cos 25\, \overline{2\pi t} \sin 1000\, \overline{2\pi t}.$$

Now suppose that the 975-cycle current suffers a phase change of β_{975} during transmission and that the 1025-cycle current suffers a phase change of β_{1025}, then the analytical expression for the current at the receiving end is

$$\sin (975\, \overline{2\pi t} - \beta_{975}) + \sin (1025\, \overline{2\pi t} - \beta_{1025}).$$

The corresponding graphical expression is

$$2 \cos \left(25\, \overline{2\pi t} - \frac{\beta_{1025} - \beta_{975}}{2} \right) \sin \left(1000\, \overline{2\pi t} - \frac{\beta_{1025} + \beta_{975}}{2} \right).$$

In comparing the graphical expressions for the current at the sending end and the current at the receiving end, it is apparent that the only changes that have taken place are phase shifts of the 1000-cycle carrier wave and of the 25-cycle modulating wave. The phase shift of the 25-cycle modulating wave represents the actual delay of the deformation of the carrier wave. If the circuit is sufficiently long so that this phase shift amounts to one complete cycle, then the corresponding delay equals one period. For any other delay, the phase shift and delay are, of course, proportional. It will be apparent, therefore, that the delay may be represented by the following equation:

$$T_\Delta = \frac{\beta_{1025} - \beta_{975}}{2 \times 25\ (2\pi)} = \frac{\Delta\beta}{\Delta\omega},$$

where T_Δ is expressed in seconds, the numerator in radians and the denominator in radians per second. T_Δ, the value of the delay of this envelope, is according to our previous definition substantially the envelope delay.

This is the simplest form of transmitted current to which the term envelope delay can be applied. This type of wave, being made up of two sinusoidal components of equal magnitude, has the important property that its envelope suffers no distortion regardless of the length and complexity of the circuit as long as it has no non-linear element and as long as the two component frequencies are transmitted with equal attenuation.

Going back now to the original case of the 1000-cycle sine wave modulated by the 25-cycle sine wave where both sidebands and carrier are transmitted, the corresponding graphical expression for this current is

$$2(1 + \cos 25\ \overline{2\pi t}) \sin 1000\ \overline{2\pi t}$$

and the corresponding analytical expression is

$$2 \sin 1000\ \overline{2\pi t} + \sin 975\ \overline{2\pi t} + \sin 1025\ \overline{2\pi t}.$$

Now if the three components suffer phase changes equal to β_{975}, β_{1000}, and β_{1025}, the analytical expression for the wave at the receiving end is

$$2 \sin (1000\ \overline{2\pi t} - \beta_{1000}) + \sin (975\ \overline{2\pi t} - \beta_{975}) + \sin (1025\ \overline{2\pi t} - \beta_{1025});$$

and there is no simple corresponding graphical expression. It will be convenient to consider this wave as being made up of two components, one being the steady component

$$2 \sin (1000\ \overline{2\pi t} - \beta_{1000})$$

and the other being a variable component

$$2 \cos \left(25\ \overline{2\pi t} - \frac{\beta_{1025} - \beta_{975}}{2}\right) \sin \left(1000\ \overline{2\pi t} - \frac{\beta_{1025} + \beta_{975}}{2}\right),$$

which is the same as the total current discussed above. The outstanding complexity in this wave is the presence of a distortion which arises from the fact that the 1000-cycle carrier wave in these two components is not transmitted with the same phase change. The phase change of the 1000-cycle current, making up the steady component, is equal to β_{1000}, whereas the phase change in the variable wave is represented by

$$\frac{\beta_{975} + \beta_{1025}}{2}.$$

In other words, it is the average of the phase changes at 975 and 1025 cycles. Now, if it happens that these two expressions are equal, then there is no distortion. If, however, as is the general case, these expressions are not equal, then there is a distortion which may be very easily exhibited by considering the case where the difference between $(\beta_{1025} + \beta_{975})/2$ and β_{1000} equals 90°. The current which then results is modulated by 50 cycles whereas the original wave was modulated by 25 cycles. Where the difference in question is intermediate in value between 0 and 90°, the detected modulating wave is complex, but has a component equal to 25 cycles. This component gradually gets smaller and disappears completely when 90° is reached. Now, if the circuit is made still longer, the 25-cycle component in the detected modulating wave again makes its appearance but has suffered a discontinuous shift of 180° in passing through the extinguishing point. By the time the phase difference in question has reached 180°, the received wave is distortionless except for the phase shift of 180° in the envelope which is not apparent, or at least is not distinguishable from a delay of one half cycle.

With this distortion in mind, it will now be apparent that the delay suffered by the modulated wave we are considering is no longer a definite quantity. However, it can be made definite for practical purposes by confining attention to the 25-cycle component of the envelope only. The distortion in question consists merely in adding other components to this one but does not shift its phase (except for the discontinuous change spoken of above). Consequently, for practical measuring purposes, if a device is used which eliminates the various harmonics of the 25-cycle current, this wave is perfectly definite for delay measuring purposes excluding the exceptional case where the fundamental component passes through zero.

These two forms of envelopes have been discussed in more or less detail because of the fact that they are the simplest ones for transmission without phase distortion. For this reason they have been used as the basis of the measuring devices which will now be described. The phase shift suffered by the envelope during transmission can be measured by comparison with a standard of the same frequency as the modulating frequency. This will be, of course, a direct measure of the envelope delay.

From the preceding discussion of some of the principles involved in the transmission of modulated waves, it is evident that the phase shift during transmission of the simple envelope considered is equal to one half the difference of the phase shifts of each of the sideband frequencies. This phase shift of the envelope is then the difference in phase shift for a definite frequency interval and is quite convenient for the measurement of the envelope delay. The envelope delay so measured is that for some frequency intermediate between the two sideband frequencies and, although it is not accurately the envelope delay for the carrier frequency, it may be taken as such for all practical cases when the modulating frequency is taken small enough so that the slope of the phase shift-frequency curve for the carrier and the two sidebands may be considered constant.

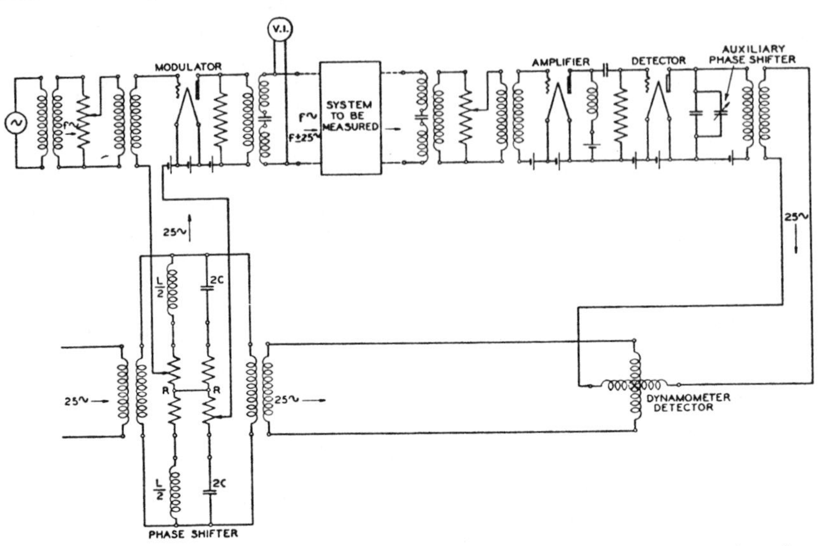

Fig. 7—Arrangement for direct measurement of envelope delay. In the phase shifter shown $R = 50\pi L = 1/50\pi C$.

Fig. 7 shows schematically a circuit for measuring the envelope delay by measuring the phase shift of the envelope.[15] The carrier frequency is modulated with another frequency, 25 cycles for example, and then transmitted through the system to be measured. At the receiving end an ordinary amplifier-detector is used to demodulate the received wave and obtain the modulating frequency. This source can then be compared in phase with a reference frequency which is obtained from the original source. In order to avoid including the effects of the measuring apparatus itself, the change in phase shift so measured through the system under consideration should be compared with a similar measurement made with an artificial resistance line substituted for the system under test. The difference of these two will, of course, be the phase shift suffered in the system by the envelope of the modulated current; and the envelope delay of the system in seconds at the carrier frequency, f, is then given approximately by

$$T_\Delta = \frac{1}{360}\frac{M}{p},$$

[15] U. S. Patent 1,645,618.

where p = the modulating frequency in cycles per second
and M = the phase shift of the envelope of the modulated wave in degrees.

In order to measure the value of M, some method of comparing the phases of various currents must be used. Also it is convenient to have in the measuring circuit a phase shifter or some means of controlling the phase of the modulating frequency.

The value used for the modulating frequency will vary somewhat with the frequency used for measurement and with the conditions under which the measurement is made. Of course, other things being equal, the greater the value of this modulating frequency the greater will be the frequency difference for which the phase shift is measured and the accuracy of the measurement will be correspondingly increased. This is true, however, only when the envelope delay is changing very slowly within this frequency interval. In most cases where the envelope delay is changing quite rapidly with frequency, it is necessary, therefore, to use as small a value for the modulating frequency as will give the required accuracy. In practice, both conditions of measurement will be encountered so that some sort of compromise value should be chosen for a particular measuring set which will do fairly well for its requirements. Various modifications of this circuit for loop and straightaway measurements are given in the patent referred to. Various methods of modulation and detection may be used.

(1) The set-up [16] shown in Fig. 7 has been used extensively for loop measurements on systems, including various telephone circuits and phase correcting networks. The details of the circuit of this set are not given here, but certain phases of its makeup and operation will be discussed. A frequency of 25 cycles from a tuning fork is used for modulation. In measuring the phase shift of the transmitted envelope, a dynamometer detector and phase shifter are used as described in the patent referred to.[15]

When the modulated wave as transmitted over the system is detected, the modulating frequency is obtained. This will, in general, differ in phase from the original modulating frequency. If the detected frequency and the original frequency are now put into the dynamometer detector, the phase of one of these frequencies can be shifted by means of the phase shifter until these two frequencies are 90° out of phase, which is indicated by zero reading of the dynamometer. The amount of phase shift which has been introduced in order to bring about this condition is a measure of the delay of the system being

[16] Compare "Phase Compensation III—Nyquist Method of Measuring Time Delay da/dw," E. K. Sandeman and I. L. Turnbull, *Electric Communication*, Vol. VII, p. 327, 1929.

measured. If the detector were balanced with a zero delay system and, then, rebalanced with the system under question inserted, the difference in these readings as given by the phase shifter would indicate the delay of the system. An integral multiple of π might not be taken care of in this measurement, but this is of little consequence.

For the modulating frequency of 25 cycles, a phase shift of nine degrees in the envelope of the 25-cycle modulation corresponds to a delay of .001 second. For convenience, therefore, the phase shifter used in this set is arranged in steps so that each step corresponds to a phase shift of nine degrees, or a delay of .001 second. In order to read intermediate values of delay, an auxiliary phase shifter, which consists of a variable condenser bridged across the output circuit of the detector tube, is used and calibrated directly in steps of .0001 second.

This particular delay measuring set has been found quite useful in the frequency range of 300 to about 10,000 cycles per second. The absolute value of delay, of course, is not that which is measured, but this can usually be determined from the measured value by adding this measured value to the integral multiple of .04 second, which is suitable for the case in hand.

[*Editor's Note:* Material, including figure 8, has been omitted at this point.]

CONCLUSION

It has not been the intent of this paper to include all the known methods of measuring phase distortion. Various methods for measuring phase shift are, of course, known and these can often be used to indicate phase distortion. In a practical way on telephone circuits, the term defined as envelope delay has certain advantages, and the paper is chiefly concerned with methods of measuring this quantity. In order to avoid including information which is contained elsewhere, the methods have not been given in detail; but references have been given, when possible, to sources where more detailed information can be found.

EDITOR'S NOTE

Page 524: Nyquist's "phase shift" β can usually be replaced by "phase angle" β if we remember that this is the *received* phase angle. The "envelope delay" is defined as $d\beta/d\omega$). The "delay of the envelope" is not defined here, but is merely

$(d\beta / d\omega)_{min}$. This is discussed in Lane's "Phase Distortion in Telephone Apparatus" (*Bell Syst. Tech. J.* **9**:493–521, 1930), in Figure 3 and on pages 498 and 502. A closely related set of terms is: phase velocity, group velocity, and signal-front velocity. These all differ in a dispersive medium. A good discussion is given in Bohm (*Quantum Theory*, Prentice-Hall, New York, 1951, pp. 63–65): V_{ph} is ω / k, V_{gr} is $(\delta\omega/\delta k)$, and $V_{signal\text{-}front}$ needs more words. See also Papoulis (*The Fourier Integral and Its Applications*, McGraw-Hill, New York, 1962, pp. 134–139), and Brillouin (*Wave Propagation in Periodic Structures*, Dover Publications, New York, 1953, pp. 69–80, especially p. 75). See also Kretzmer ("Measuring Phase at Audio and Ultrasonic Frequencies," *Electronics*, October 1949, especially Figure 8) on the phase shift in the testing of loudspeakers.

43

Reprinted with permission from pages 493, 501–502, and 503 of *Bell Syst. Tech. J.* **9**:493–521 (1930)

Phase Distortion in Telephone Apparatus

By C. E. LANE

This paper shows that if, over its transmitting range, the phase shift, B, in radians, of a four-terminal network may be written $B = a_0 + a_1\omega$ ($\omega = 2\pi x$ frequency in cycles per second), there is no phase distortion if $a_0 = N$, N being any integer. However, there is a delay, for any signal, given by $dB/d\omega = a_1$ (seconds). If N is not an integer, there is a delay, a_1, and in addition a distortion, which distortion, generally for speech and music, may be neglected. Typical phase characteristics for lines, filters and all-pass networks are shown. In general over their transmitting range, such phase characteristics which usually are curved, may be regarded as the sum of two characteristics, a straight line having a slope corresponding to the minimum slope of the original and which introduces a delay without distortion and a curved portion to which all of the distortion of the signal may be ascribed. Oscillograms are given showing the distortion for a loaded line and for band filters for a signal which is of the form $y \sin(\omega_0 t + \theta)$ between $t = 0$ and $t = T$ and zero for all other time. A description is given of the means employed for reducing the amount of phase distortion in telephone cable and in low-pass filters in circuits used for program transmission and regular telephone service. Also, phase distortion in repeaters and transformers is described. Brief reference is made to the problem of phase distortion in telegraph, picture transmission, and television circuits.

[*Editor's Note:* Material has been omitted at this point.]

Fig. 9 shows the insertion delay characteristic for a system consisting of four band filters in tandem. The attenuation characteristic is also shown. Fig. 10 shows oscillographs for waves of the above type for f_0 = 260, 300, 480 and 680 cycles per second. Notice that the distortion is much greater where f_0 falls near the edges of the transmitting band, although in every case the wave starts noticeably building up at about .0109 seconds after $t = 0$ for the sent wave. This is the value of $(dB/d\omega)_{\text{min.}}$ in the transmitting range of the filters. In both Figs. 8

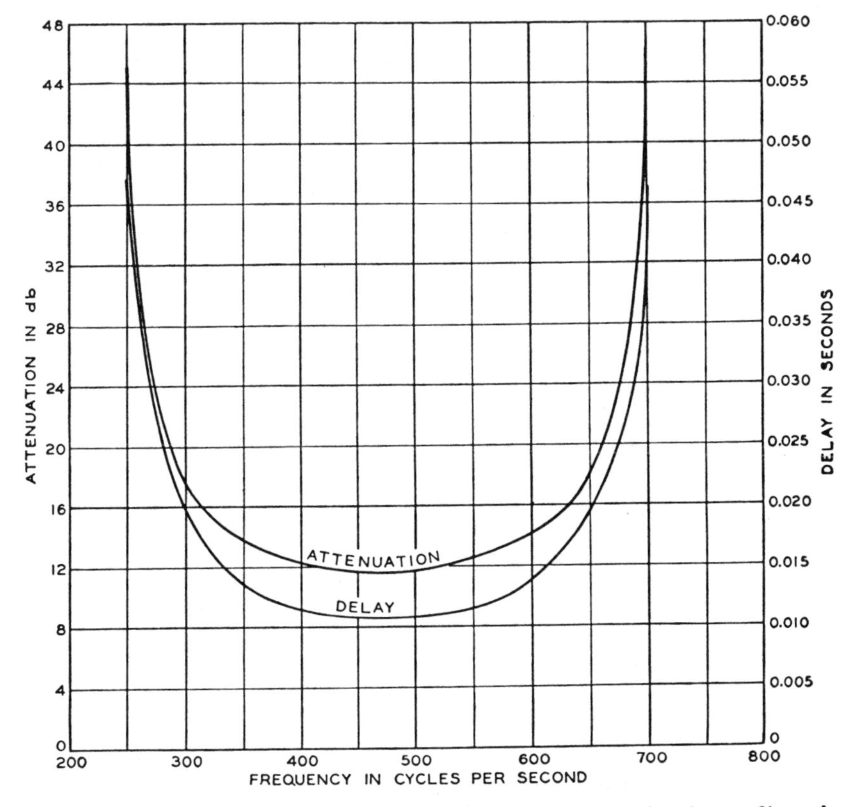

Fig. 9—Insertion delay and attenuation characteristics of 4 band-pass filters in tandem.

and 10 some of the distortion may be ascribed to attenuation although the *elongation effect* is primarily due to phase distortion. It is this elongation effect that is noticeable to the ear in speech and music.

[*Editor's Note:* In the original, material follows this excerpt.]

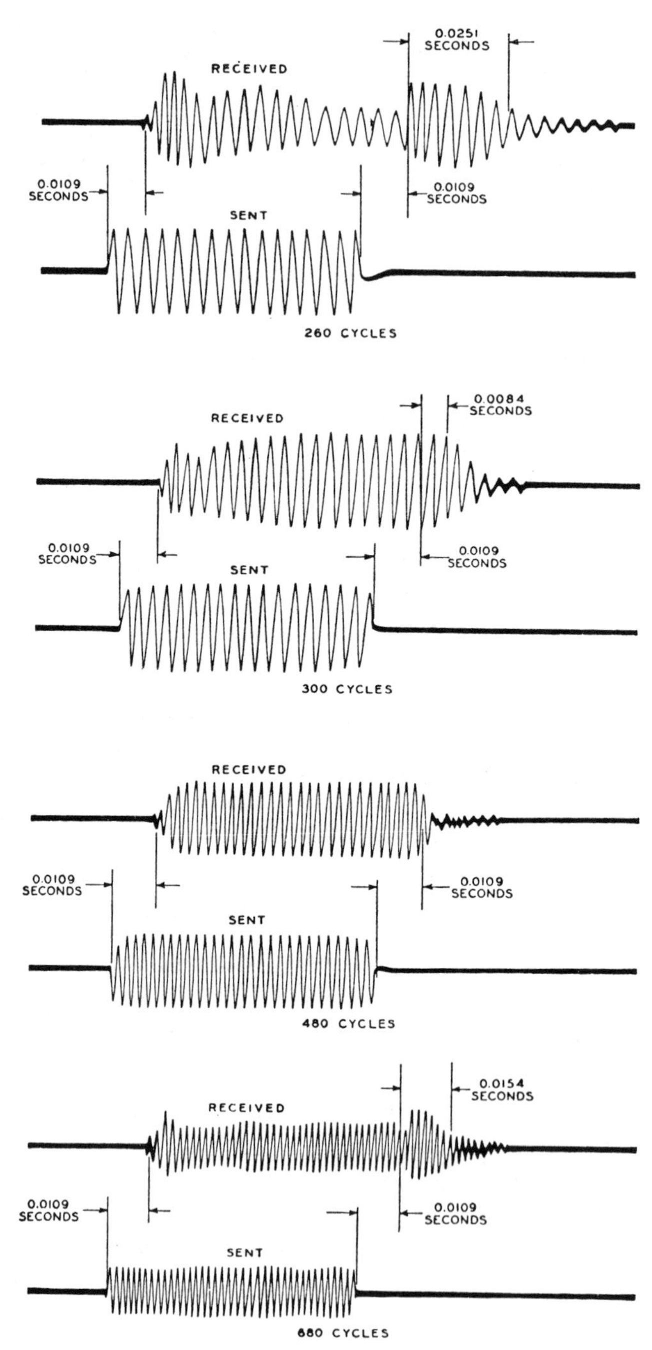

Fig. 10—Distortion resulting from four band filters of Fig. 9 for signals of the form $y \sin \omega_0 t + \theta$ between $t = 0$ and $t = T$ and zero for all other time.

33

Reprinted from pages 115–116 of *Acoust. Soc. Am. J.* **13**:115–123 (1941)

Phase Distortion in Electroacoustic Systems*

FRANCIS M. WIENER

In the course of an investigation of phase distortion in electroacoustical systems the phase shift *vs.* frequency characteristic of a standard miniature condenser transmitter was measured by application of the principle of reciprocity. Measurements were made in a closed cavity and in free space. In analogy to the free field response the phase of the pressure in the undisturbed sound wave was taken as reference. By application of a principle of similitude the phase shift due to diffraction and cavity resonance was determined experimentally and found to be in agreement with theory. The phase characteristics of a number of commercial microphones of representative types were investigated by comparison with the standard condenser microphone. Difficulties were encountered at wave-lengths comparable with the dimensions of the microphones due to the uncertainty of microphone position with respect to the source. Relative response curves were obtained as a supplement and general check on the experimental method. The phase characteristics of a number of electrodynamic loudspeakers are presented.

INTRODUCTION

THE general problem of electrical transmission of speech and music in a single channel from one room to another is complicated and does not lend itself readily to complete theoretical treatment. It may be made manageable with some sacrifice in rigor by considering the problems of transmission and room acoustics separately. Attention is given here to the problem of transmission only, excluding room acoustics from consideration.

A relevant description of the transmission properties of the system is possible by assuming free field conditions at both sending and receiving ends. In an experimental investigation, the microphone at the sending end is excited by a loudspeaker and the sound field of the loudspeaker at the receiving end is explored by a second microphone, free field conditions prevailing at both ends. The transmission system thus created consists of two "electroacousto-electrical" systems, each consisting of a loudspeaker linked acoustically to a microphone, which are connected by purely electrical transmission networks. The system loudspeaker-microphone is equivalent to a passive, linear electric four-pole, if the transducers are linear and reversible and if all mechanical motions are reflected in a single electrical coordinate. In general, the configuration of the equivalent four-pole is not known, but the concepts and theorems of electric network analysis can be applied to the two accessible electrical meshes, notably the concept of transfer impedance and the reciprocity theorem. If the emitted sound waves are not symmetrical in the absence of obstacles,[1] the reciprocity theorem in its simple form cannot be applied and the character of the sources has to be taken especially into account. Application of the Fourier integral theorem yields the well-known conditions for distortionless transmission and shows the sufficiency of the steady-state amplitude and phase transmission characteristics to describe uniquely the behavior of the system for the most general case.

While the phase characteristics of purely electrical networks are well known, no significant amount of data on the phase characteristics of such widely used electroacoustic transducers as microphones and loudspeakers seems to be available in the literature.

In this paper, the phase characteristics of such devices are presented in the form of a survey of existing equipment. Such a survey is necessary before attempting to achieve improved performance.

If the *transfer voltage ratio* of the equivalent four-pole is denoted by

$$e^{-[A(\omega)+jB(\omega)]},$$

* Presented at the 25th Meeting of the Acoustical Society of America, Rochester, New York, May 5, 1941.

[1] Lord Rayleigh, *Theory of Sound* (Macmillan, 1896), Vol. II, §294.

then a plot of $20 \log_{10} e^{-A(\omega)}$ *vs.* frequency is the response characteristic, and a plot of $B(\omega)$ *vs.* frequency is the phase characteristic. In this notation a lagging phase angle is positive. ω is the angular frequency. $B(\omega)$ may be written as

$$B(\omega)=k\omega\pm n\pi+d(\omega), \tag{1}$$

where $d(\omega)$** is the phase distortion,[2] k is a constant, n is an integer. The slope of the phase characteristic, the envelope delay,[3,4] is given by

$$B'(\omega)=k+d'(\omega), \tag{2}$$

where $d'(\omega)$ is a measure of the delay distortion.[3] According to Steinberg,[5] phase distortion results in general in a delay proportional to $B'(\omega)$ and in distortion of the wave form, essentially controlled by the intercepts of the tangents to the phase characteristic on the phase shift axis (intercept distortion).

The amount of tolerable phase distortion in a transmission system for speech and music has to be determined by subjective listening tests. Few data on such tests are available, but all show that delay distortion rather than phase or intercept distortion is most readily noticed by the human ear.[5] Recently published results of listening tests[6] give about 8 milliseconds permissible delay distortion at the high frequencies and about 15 milliseconds at 100 cycles, both referred to 1000 cycles as reference frequency.

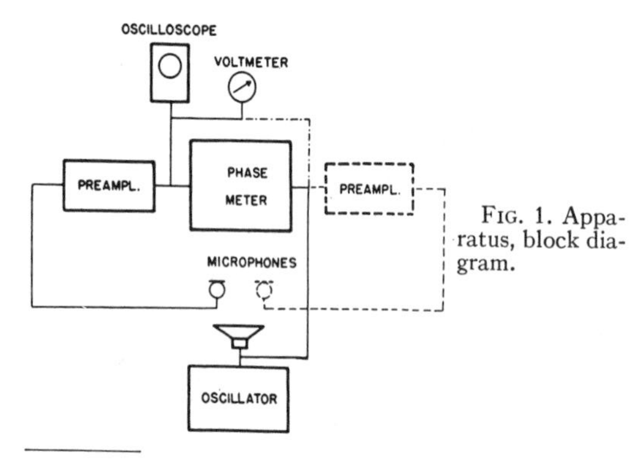

FIG. 1. Apparatus, block diagram.

** $d(\omega)$ is a function of ω other than linear.
[2] S. P. Mead, Bell Sys. Tech. J. **7**, 195 (1928).
[3] C. E. Lane, Bell Sys. Tech. J. **9**, 493 (1930).
[4] R. V. L. Hartley, Bell Sys. Tech. J. **20**, 222 (1941).
[5] J. C. Steinberg, Bell Sys. Tech. J. **9**, 550 (1930).
[6] F. A. Cowan, R. G. McCurdy, and I. E. Lattimer, Bell Sys. Tech. J. **20**, 235 (1941).

Calibration of a Miniature Condenser Microphone

In order to investigate experimentally the phase characteristics of electroacoustic transducers it is necessary to have available a microphone, small in physical size, whose phase shift calibration is known. It is convenient to extend the usual definitions of free field and pressure calibration to include the phase as well as the amplitude.

A miniature W.E. condenser microphone was calibrated in terms of free field pressure in amplitude and phase by application of the reciprocity theorem. The experimental arrangement is shown in Fig. 1. It consists of a phase meter with associated preamplifiers in one or both channels. The phase shift of one microphone with respect to an auxiliary reference voltage or the phase difference between the outputs of two microphones can be measured. An oscilloscope was permanently in use to be on guard against distortion and noise, since filters could not be used.

Using the procedure described by Cook[7] and MacLean,[8] the pressure calibration of the microphone was secured by coupling it to another similar condenser microphone by means of a small pressure chamber. The dimensions of the chamber were small as compared with the wavelength of sound for frequencies below 1000–1500 cycles. The results are shown in Fig. 2, where also a thermophone calibration, made several years ago, is given for comparison. The phase shift is seen to be practically zero for this frequency range. The slight drop in response at low frequencies is most likely due to imperfect sealing of the chamber. Very similar results, shown in Fig. 3, were obtained for the other microphone. Microphone D96436 was used as the reversible transducer. Rayleigh disk and thermophone calibrations are shown for comparison. Systematic deviations between reciprocity and thermophone and Rayleigh disk calibrations exist for both microphones, similar to those reported by Cook.[7] No attempt was made to extend the frequency range by the use of hydrogen.

[7] R. K. Cook, J. Research Nat. Bur. Stand. **25**, 489 (1940); J. Acous. Soc. Am. **12**, 415 (1941); *ibid.* **13**, 81A (1941).
[8] W. R. MacLean, J. Acous. Soc. Am. **12**, 140 (1940).

[*Editor's Note:* In the original, material follows this excerpt.]

Reprinted from *Audio Eng.* **34**:9–13 (August 1950)

TRANSIENT TESTING OF LOUDSPEAKERS

MURLAN S. CORRINGTON*

It is known from the theory of linear dynamic systems of minimum-net-phase-shift type that the amplitude response, the phase characteristic, and the transient response to various applied wave forms are merely equivalent ways of observing the same inherent performance of the circuit.[1,2] Any one of the three can be used to compute either of the other two. The labor of this computation is often great and is seldom attempted. If one had enough experience and were a keen observer, any one test would be adequate.

Many people use sound pressure curves to measure loudspeaker performance, and some laboratories now use transient tests also.[3-16] In the reproduction of speech and music, it is important that the sound start and build up as fast as the original, and when the signal ends the speaker should also stop. Thus when the hammer hits the string of a piano, the sound starts suddenly as the energy of the impact causes the string to vibrate. The loudspeaker must be able to start suddenly also, and when the damper comes down on the string at the end of the note, the loudspeaker must stop suddenly.

One test which is used to observe this performance is to apply a sine wave suddenly to the voice coil of the speaker and to observe the sound pressure with a microphone in front of the speaker on the axis. After a short time the applied sine wave is suddenly reduced to zero. The build-up and decay of the sound pressure is then observed on an oscilloscope. In many cases these results are confusing, because of lack of experience in interpreting the curves. Very little theory has been given in the literature to explain the effect of peaks and valleys in the sound pressure curve on the transient. This paper has three purposes. First there will be presented the theory of ringing of such a circuit when a signal is suddenly applied, and next it will be shown exactly how each peak contributes to the transient. Finally, a conventional 12-inch speaker will be used to show how the cone breaks up into symmetrical and asymmetrical resonant modes, and also the transient performance at each of the selected frequencies.

* *Home Instrument Department, RCA Victor Division, Camden, N. J.*
Paper delivered at the first National Convention of the Audio Engineering Society, New York City, October 29, 1949.

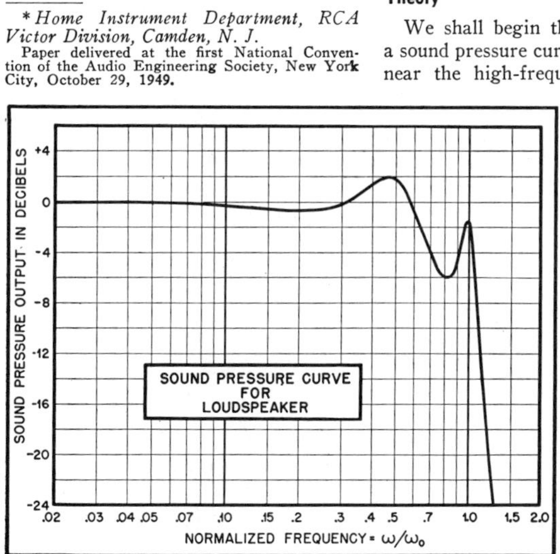

Fig. 1. Sound pressure curve for a loudspeaker.

Theory

We shall begin the theory by taking a sound pressure curve having two peaks near the high-frequency cutoff of the speaker, as shown by *Fig.* 1. The frequency has been normalized to unity at cutoff by dividing by the cutoff frequency ω_0. Thus, if $\omega_0 = 10{,}000$ cps, for example, the first peak occurs near 5000 cps and the second at 10,000 cps. To simplify the interpretation, the low-frequency end is assumed flat. After this simplified case is thoroughly understood, it is easy to generalize to the more complicated curve of an actual speaker. It will be assumed that the loudspeaker is a minimum-net-phase-shift network. This may not be strictly true at high frequencies when cone breakup may cause cancellation at a point, but appears to be true for piston action.

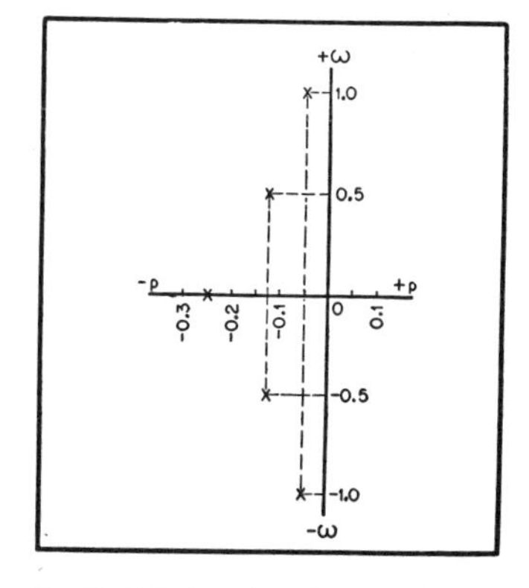

Fig. 2. Pole locations in complex frequency plane for a typical loudspeaker.

The locations of the poles in the complex frequency plane can be determined by noting that the maxima are near frequencies zero, 0.5, and 1.0. The Q's are different, and the pole at zero frequency is low Q, the one at 0.5 is higher, and the one at 1.0 is highest. This is easily decided by noting the width of the resonant peaks. The exact pole location can be found by using a probe in an electrolytic tank, by stretching a rubber sheet over some vertical posts, or by trial and error.[17-19] For the curve of *Fig.* 1, they are located at the points shown by *Fig.* 2. Although the curve is for a loudspeaker, these results apply to any other linear system, such as pickups, electrical networks, microphones, etc.

All of the poles are in the left half of the p-plane. There is one on the axis at $p=-.25$, corresponding to the maximum in the sound-pressure curve at zero frequency. A conjugate pair occurs at $-.125\pm 0.5\,i$ corresponding to the peak at frequency 0.5, and a second conjugate pair occurs at $-0.5\pm i$, corresponding to the high-frequency peak at unity.

Since there are five poles and no zeros, the impedance function is easily found, as shown by equation (*1*).

$$Z=\frac{1}{[p+.25]\,[(p+.125)^2+.25]\,[(p+.05)^2+1]} \qquad (1)$$

The first bracket corresponds to the pole at $p=-.25$. The second bracket is obtained from the conjugate pair of poles at $p=-.125\pm 0.5\,i$ as is readily seen, and the third bracket is obtained from the second conjugate pair of poles at $p=-0.5\pm i$.

The transient response to a unit impulse, which is a spike of current of infinite height, and zero duration, having unit area under the curve, corresponds to the transient generated by a noise pulse. This result can be obtained from the Heaviside expansion theorem,[20] by expanding in partial fractions and applying the inverse Laplace transform term by term,[21] or by evaluating the required contour integrals.[22] It is shown by equation (2).

Transient response to unit impulse

$$= (-.4984 \sin t + 1.1913 \cos t)\, e^{-.05t} + (1.7231 \sin \tfrac{1}{2} t - 4.8112 \cos \tfrac{1}{2} t) \times e^{-.125t} + 3.6199\, e^{-.25t} \qquad (2)$$

The equation shows, in the first line, that the pole at $p=-.05\pm i$ causes ringing at the pole frequency, at $\omega/\omega_0=1$, and has a damping factor corresponding to the distance of the pole from the real frequency axis, $p=-.05$. The higher the Q, the nearer the pole is to the axis. For example, in an ordinary series-loaded parallel-resonant circuit this distance is $-\frac{\omega_0}{2Q}$, where ω_0 is the resonant frequency.

The second line shows the ringing caused by the pole at $p=-.125\pm 0.5\,i$, and the third line is the ringing at zero frequency, which is an ordinary decay curve.

The coefficients are determined by the pole locations and moving any one pole affects all the coefficients.

The curve of the impulse response is shown by *Fig.* 3. It starts slowly from zero, builds up to the first maximum, and then decreases toward zero, oscillating about the zero axis. The two transients of frequency 0.5 and 1.0 are beating together. The components of the transient are shown. The one due to the pole at $-.125\pm 0.5\,i$ is the dominant one, but the one at $-.0.5\pm i$ dies out more slowly. The usual broadening of a pulse by the tuned circuits is evident.

Starting from the impulse response, the response to any other applied wave form can be obtained by applying the convolution or superposition theorem.[20, 21] Another way is to multiply the operational form of the impedance by the operational form of the desired applied wave form, and to find the inverse Laplace transform of the product.[21]

This latter method can be explained more easily in terms of spectra. By means of the Fourier transform or the Laplace transform it is known that the transform of a suddenly-applied unit sine wave $\sin \omega_0 t$ is $\frac{\omega_0}{\omega_0^2-\omega^2}$. This result is equivalent to the amplitude and phase of the spectral density. For a wave of frequency one half the spectral distribution is as shown by *Fig.* 4. The curve rises smoothly from 2 at zero frequency, approaching infinity at frequency one half. At this point the phase shifts 180 deg., and the amplitude then rises smoothly from minus infinity toward zero. If the applied frequency is raised to unity, the shape is similar, as shown. To find the transient response at a given instant, this spectral density is modified in amplitude and phase in accord with the selectivity curve of *Fig.* 1.

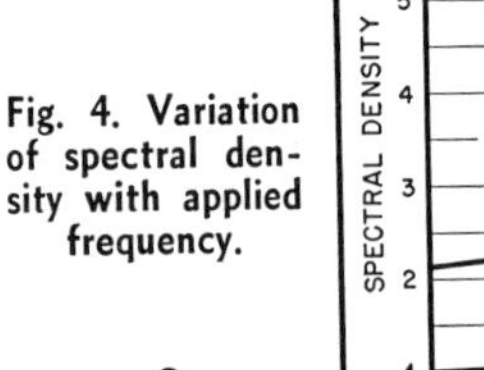

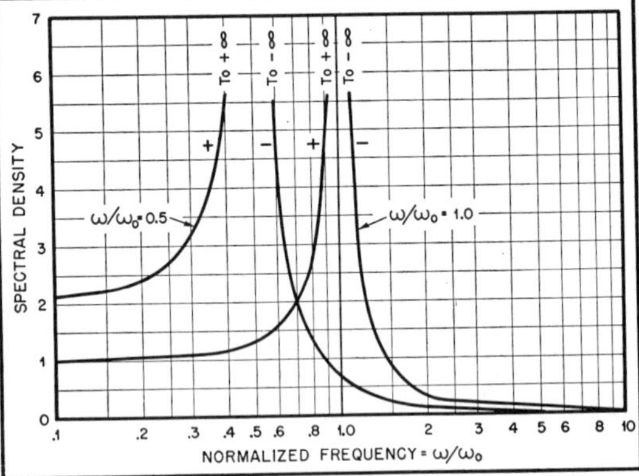

Fig. 4. Variation of spectral density with applied frequency.

It is easy to see that if the points of high spectral density are near to a peak in the response curve of *Fig.* 1, a great deal of energy will be impressed on the network at this frequency. The circuit will ring strongly at the frequency corresponding to the peak.

As an example of this theory, apply the unit sine wave to the two peaks of *Fig.* 1. When this is done, the response is as shown by the following two equations.

Transient response to suddenly applied unit sine wave at $\omega/\omega_0=0.5$

$$=-16.325 \sin \tfrac{1}{2} t - 9.383 \cos \tfrac{1}{2} t + (0.223 \sin t - 0.827 \cos t)\, e^{-.05t} + (19.797 \sin \tfrac{1}{2} t + 4.418 \cos \tfrac{1}{2} t) \times e^{-.125t} + 5.792\, e^{-.25t} \qquad (3)$$

Response at ω/ω_0 *1.0*

$$=10.332 \sin t + 7.039 \cos t - (12.030 \sin t + 4.683 \cos t)\, e^{-.05t} + (3.191 \sin \tfrac{1}{2} t - 5.763 \cos \tfrac{1}{2} t) \times e^{-.125t} + 3.407^{-.25t} \qquad (4)$$

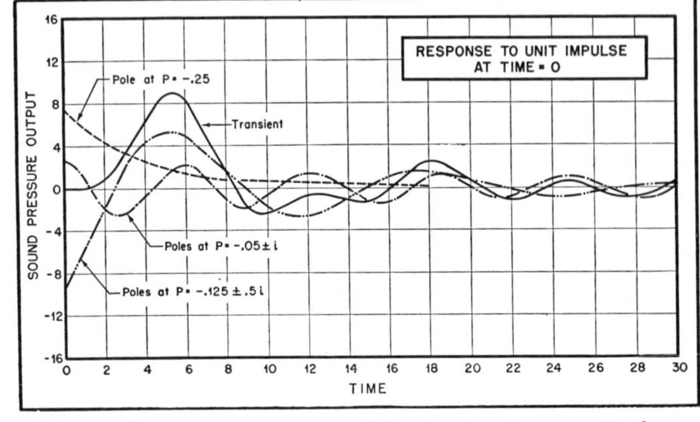

Fig. 3. Transient response to a unit impulse of a sine-wave signal.

Equation (*3*) corresponds to the first peak in the sound pressure curve, and equation (*4*) corresponds to the peak at frequency 1.0. The first line of each equation gives the steady-state term.

It should be noted that the peak nearer in frequency to that of the applied wave causes the stronger ringing, but both are present. As the frequency of the suddenly applied wave moves from the lower peak toward the other, the transient caused by the first pair of poles becomes weaker, and the other transient becomes stronger. The decay curve due to the pole at the origin becomes weaker as the frequency of the applied wave moves away from this pole. The amplitudes of these transient components, for varying applied frequency, are shown by *Fig.* 5. The closer the applied frequency is to a peak in the response curve, the stronger this peak will ring.

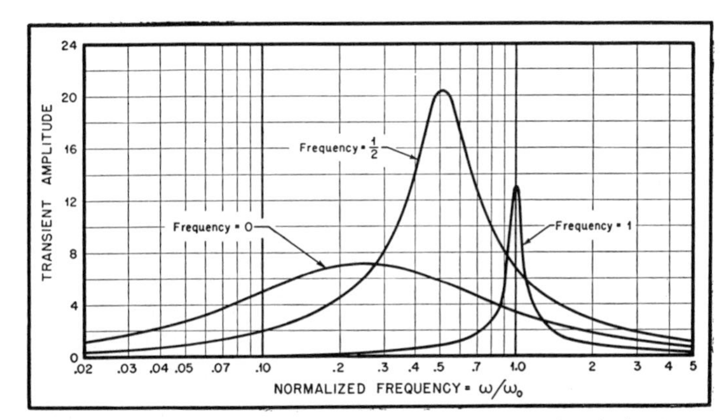

Fig. 5. Variation of transient components with frequency of applied unit-step sine wave.

The response due to a suddenly applied unit sine wave at frequency 0.5, corresponding to equation (*3*), is shown by *Fig.* 6. The transient starts out slowly, due to the time required for the signal to pass through the network, reaches the first maximum, and then oscillates about the zero axis, rapidly approaching the amplitude and phase of the steady-state curve. The transient due to the pole at the same frequency as the applied sine wave is very strong, being nearly as high as the steady-state curve at the first maximum, but it dies out rapidly due to the heavy damping. The transient due to the higher frequency pole is weak, but dies out slowly.

When the frequency of the suddenly applied sine wave is moved up to the second peak of the response curve, the results corresponding to equation (*4*) are as shown by *Fig.* 7. The transient starts out slowly in the positive direction and soon reaches a maximum. Beyond this point the oscillation is about the zero-frequency axis, but of varying amplitude as the two oscillatory components come in and out of phase.

Because of the high Q of this pole, the resulting transient does not approach the final steady state nearly so rapidly as it does at the other pole (see *Fig.* 6).

Experimental Results

The sound pressure curve of a standard 12-inch speaker is shown by *Fig.* 8. The low-frequency resonance occurs at 75 cps, the usual rim resonance occurs in the region between 1200 and 1500 cps, and there is peaking of about 8 db in the upper register. The upper frequencies cut off rather rapidly beyond 6000 cps. The voice coil impedance shows the main resonance at 75 cps. The higher frequency symmetrical modes cause small variations in the impedance curve, but are largely swamped by the rising electrical impedance of the voice coil.

Several points have been marked with the letters *A, B, C, D*, etc. We will now examine the cone breakup and transient response to a suddenly applied sine wave at each of these frequencies. The inextensional radial modes occur at the low frequencies *A, B, C, D,* while the extensional circular modes occur at the high frequencies *F, G, H.*

Figure 9(A) corresponds to point *A* of *Fig.* 8 and was made by sprinkling lycopodium powder on the cone. The frequency was then set near the low-frequency resonance of the cone, which caused the powder to climb up the cone and distribute itself uniformly. The frequency was then raised to 400 cps and left there. The light yellow powder was shaken off of the moving parts of the

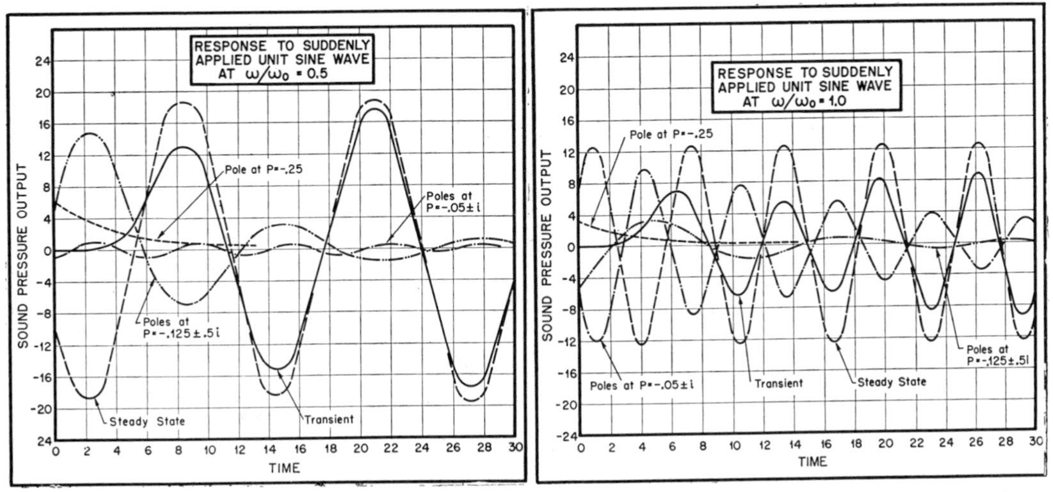

Fig. 6 (left). Response to suddenly applied unit sine wave at $\omega/\omega_0 = 0.5$. Fig. 7 (right). Response to suddenly applied unit sine wave at $\omega/\omega_0 = 1.0$.

cone, and allowed to remain at the nodes. There are thus four radial nodes, uniformly spaced, with the anti-nodes on opposite sides of a given node moving 180 deg. out of phase. This tends to pump air back and forth across a node, and does not result in a peak in the sound pressure output because of this "short circuit" of the air path.

Figure 9(B) shows the response to the suddenly applied sine wave. Since there are no pronounced peaks near 400 cps, the buildup is smooth and fast. The decay is rather uneven as the different resonant frequencies are excited at cutoff and die away at different rates. The frequency corresponding to the main resonance at 400 cps hangs on longer than any other because of the large amount of energy stored in the cone. There is a slow oscillation at 75 cps superimposed corresponding to the resonance peak due to the cone and suspension resonance.

When the frequency is raised to 450 cps, there are eight radial nodes, symmetrically spaced as shown by *Fig.* 10(A). This corresponds to point *B* of *Fig.* 8. There is little effect on the sound pressure curve because the antinodes merely pump air across the nodes.

Since there is no peak near this point, the transient buildup shown by *Fig.* 10(B) is smooth and rapid. When the signal is cut off, the ripple is at the small peak in the response at 480 cps. A slow oscillation at the 75-cps resonance is superimposed. Some high-frequency beating is evident right at cutoff as the higher frequency transients die out rapidly.

The peculiar standing wave pattern of *Fig.* 11(A), corresponding to point *C* of *Fig.* 8, is somewhat unsymmetrical and apparently is a mixture of two normal modes. The one wide node apparently is moving slightly in the center. The corresponding transient response is shown by *Fig.* 11(B). The 75-cps oscillation is superimposed as before, and there is considerable beating between transients just after cutoff.

When the frequency is 940 cps, corresponding to point *D*, the cone vibrates as shown by *Fig.* 12(A). There are two main nodes, but opposite sides are in phase and the rim is not moving strongly. The transient of *Fig.* 12(B) shows that the 75-cps oscillation is much weaker than in the previous cases, since the applied sine wave is much farther from this peak in frequency. The 940-cps note persists for a long time because of the high stored energy and low damping, since the voice coil need not move.

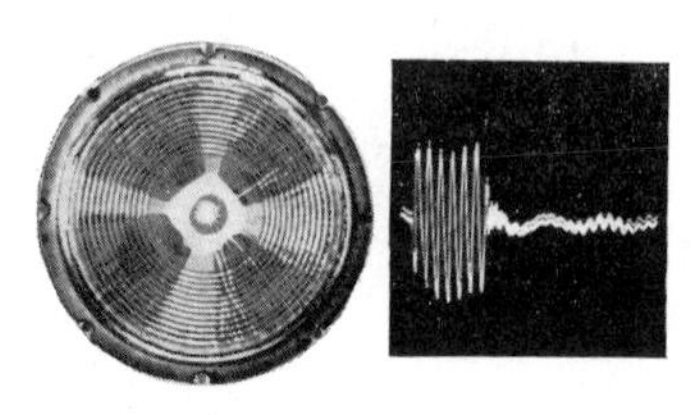

Fig. 9A (left). Cone breakup at 400 cps, with four radial nodes; point A. Fig. 9B (right). Transient response to same signal.

Figure 13(A) is one of the first symmetrical extensional modes and occurs at 1450 cps, at point *E* of *Fig.* 8. The frequency is high because the paper must stretch and this results in high stiffness. The rim is 180 deg. out of phase with the center of the cone. The powder cannot shake off the cap so it appears as a dense white cloud. This resonance is characteristic of all cone speakers where the rim is not properly terminated to eliminate standing waves, and normally shows up as a hole between 1000 and 2000 cps, although it may be somewhat lower in large cones.

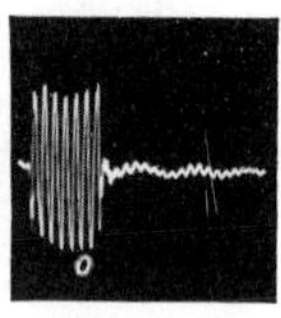

Fig. 10A (left). Cone breakup at 450 cps, with eight radial nodes; point B. Fig. 10B (right). Transient response to same signal.

As shown by *Fig.* 13(B), there is considerable ringing at buildup and decay, because of the dip in the sound pressure curve and because of the high energy storage. The high-frequency oscillation beyond cutoff corresponds to the adjacent peak near 1800 cps.

Figure 14(A) shows the next higher circular mode, corresponding to point *F*, with two nodes. The center cap is moving strongly, a node surrounds it, and there are two more antinodes. There is some evidence of superimposed radial nodes. The buildup and decay of *Fig.* 14(B) are fairly smooth, but there is considerable irregularity as the different transients come in and out of phase.

The next higher symmetrical mode, occurs at 2700 cps and is shown by *Fig.* 15(A). It corresponds to the peak at point *G* in the sound pressure curve. There are 3 circular nodes. The corresponding buildup and decay of *Fig.* 15B shows considerable evidence of the transients, corresponding to the various peaks, coming in and out of phase as they beat together.

The final photograph of *Fig.* 16(A) shows four circular nodes. These are not quite so clearly defined as the preceding ones because the corrugations affect the cone flexibility. The center of the cone is moving strongly. The buildup and decay of *Fig.* 16(B) show the expected interaction between transients at the beginning and end of the wave. Since this frequency is several octaves away from the 75-cps cone resonance, there is very little evidence of 75-cps components in the curve.

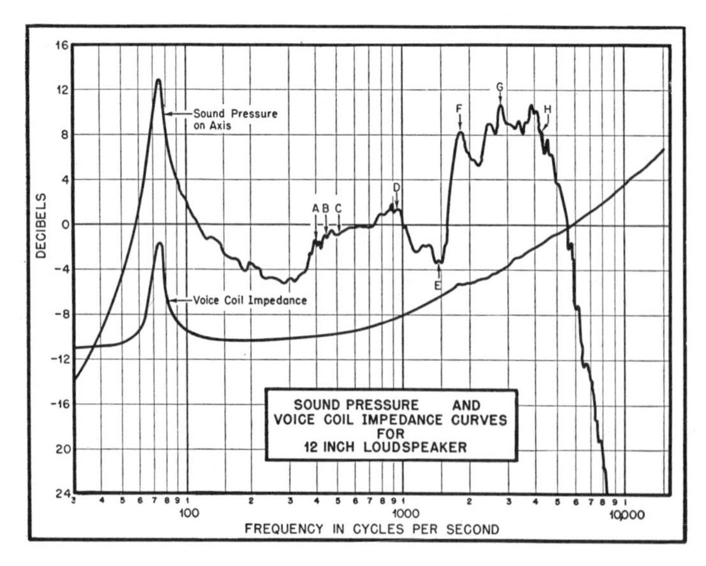

Fig. 8. Sound-pressure and voice-coil impedance curves for a 12-inch loudspeaker.

Conclusions

There are several ways to measure the transient performance of a loud-

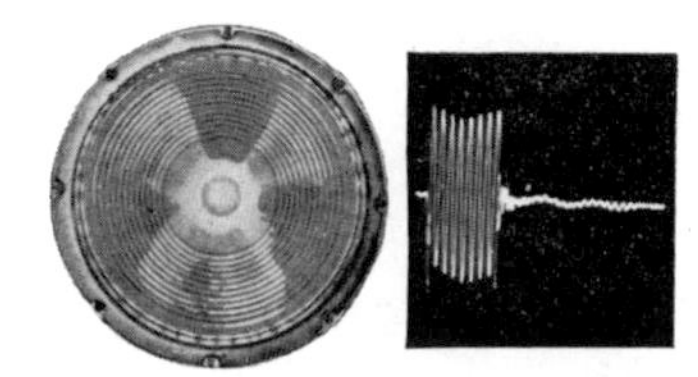

Fig. 11A (left). Cone breakup at 510 cps, with four radial nodes; point C. Fig. 11B (right). Transient response to same signal.

speaker. The unit impulse gives all of the information at one test, but is hard to interpret, because it causes all the peaks to ring strongly simultaneously, in proportion to the area under the curve. The suddenly applied unit sine wave is more selective, since it emphasizes the ringing of the peaks of nearly the same frequency as the applied sine wave.

The resonances in the cone fall into four general classes:

1. Low-frequency resonance—cone moving as piston.
2. Radial modes.
3. Symmetrical modes.
4. Combination modes.

They can be measured by means of

1. Dust patterns.
2. Acoustic probe.
3. Electrostatic or mechanical probe.
4. Stroboscope.

The dust-pattern method is very simple and direct, although the others may give more numerical data.

Whenever a transient wave form is applied to such a system, there will be ringing corresponding to each pole, or

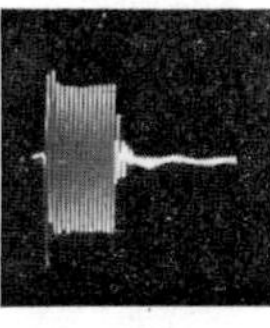

Fig. 12A (left). Cone breakup at 940 cps, with two radial nodes; point D. Fig. 12B (right). Transient response to same signal.

conjugate pair of poles, in the complex frequency plane. The amplitude of each component is determined by the spectrum of the applied wave, and also by the influence of neighboring poles on each other.

Since some of the frequencies generated are not harmonically related to the applied signal, they cause an annoying type of distortion somewhat similar to intermodulation. The ear is very sensitive to this type of inharmonic distortion, and it is thus evident that a high-quality speaker should be as smooth as possible in the sound pressure curve.

Acknowledgment

The equipment used for these tests was built by Roy Fine and Marshall Kidd of this laboratory, and they helped

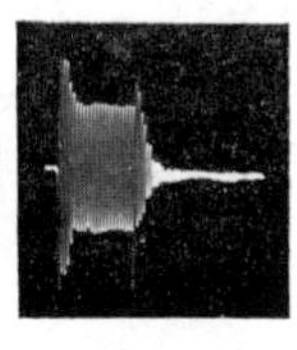

Fig. 13A (left). Cone breakup at 1450 cps, with one circular node; point E. Fig. 13B (right). Transient response to same signal.

obtain the photographs shown. The author wishes to thank them for this invaluable assistance.

References

[1] T. Murakami and Murlan S. Corrington, "Relation between amplitude and phase in electrical networks," *RCA Review,* vol.

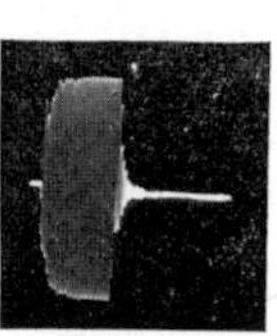

Fig. 14A (left). Cone breakup at 1800 cps, with two circular nodes; point F. Fig. 14B (right). Transient response to same signal.

9, pp. 602–631; Dec. 1948. (Contains an extensive bibliography.)

[2] W. L. Sullivan, "Analysis of systems with known transmission-frequency characteristics by Fourier integrals," *Elec. Eng.,* vol. 61, no. 5, pp. 248–256; May, 1942.

[3] H. Backhaus, "Über Strahlungs- und Richtwirkungseigenschaften von Schallstrahlern," *Zeitschrift für technische Physik,* vol. 9, p. 491; 1928.

[4] H. Neumann, "Ein- und Ausschwingvorgänge an elektro-dynamischen Lautsprechern mit starken Magnetfeldern," *Zeitschrift für technische Physik,* vol. 12, p. 627; 1931.

[5] H. Backhaus, "Über die Bedeutung der Ausgleichvorgänge in der Akustik," *Zeitschuft für technische Physik,* vol. 13, pp. 31–46; 1932.

[6] N. W. McLachlan, "On the symmetrical modes of vibration of truncated conical shells; with applications to loud-speaker diaphragms," *Proc. Phys. Soc.,* vol. 44, pp. 408–420; 1932. Discussion pp 420–425

[7] N. W. McLachlan, "The axial sound-pressure due to diaphragms with nodal lines," *Proc. Phys. Soc.,* vol. 44, pp. 540–545; 1932.

[8] H Benecke, "Über die Schwingungsformen von Konusmembranen," *Zeitschrift für technische Physik,* vol. 13, pp. 481; 1932.

[9] Goswin Schaffstein, "Untersuchungen an Konuslautsprechern," *Hochfrequenztechnik und Elektroakustik,* vol. 45, no. 6, pp. 204–213; June 1935.

[10] J. H. G. Helmbold, "Oszillographische Untersuchungen von Einschwingvorgängen bei Lautsprechern," *Akustische Zeitschrift,* 1937.

[11] W. O. Rogers, "Stroboscope aids dynamic speaker design," *Electronics,* vol. 10, p. 30; Aug. 1937.

[12] P. David, "Dans quelle mesure l'étude d'un haut-parleur in régime permanent, permet-elle de prévoir son comportement en régime transitoire, ?" *l'Onde Électrique,* vol. 17, pp. 309–319; June 1938.

[13] D. E. L. Shorter, "Loudspeaker transient response," *B. B. C. Quarterly,* vol. 1, pp. 121–129; Oct. 1946.

[14] Piero Giorgio Bordoni, "Asymmetrical vibrations of cones," *J. Acous. Soc. Am.,* vol. 19, no. 1, pp. 146–155; January 1947.

[15] G. Guyot, "Étude sur les haut-parleurs en régime transitoire," *Revue Générale de*

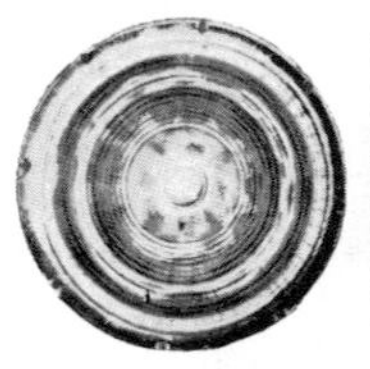

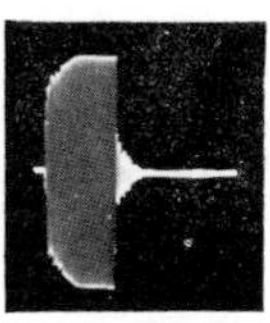

Fig. 15A (left). Cone breakup at 2700 cps, with three circular nodes; point G. Fig. 15B (right). Transient response to same signal.

l'Électricité, vol. 57, pp. 245–253; June 1948.

[16] Robert Campbell, "Sur la vibration d'un haut-parleur elliptique," *Comptes Rendus des séances de l'Academie de Paris,* vol. 228, pp. 970–972; March 21, 1949.

[17] W. W. Hansen and O. C. Lundstrom, "Electrolytic tank impedance-function determination," *Proc. I. R. E.,* vol. 33, pp. 528–534; Aug. 1945.

[18] Richard W. Kenyon, "Network design using electrolytic tanks," *Electronic Industries,* vol. 5, pp. 58–60; March 1946.

[19] A. R. Boothroyd, E. C. Cherry and R. Makar, "An electrolytic tank for the measurement of steady-state response, transient response, and allied properties of networks," *Proc. I. E. E.,* vol. 96, pp. 163–177; May 1949.

[20] Vannevar Bush, *"Operational Circuit Analysis":* John Wiley & Sons, Inc., New York, N. Y., 1929, 1937.

[21] Murray F. Gardner and John L. Barnes, *"Transients in Linear Systems," vol. I;* John Wiley & Sons, Inc., New York, N. Y., 1942.

[22] N. W. McLachlan, *"Complex Variable and Operational Calculus with Technical Applications,"* Cambridge University Press, 1939.

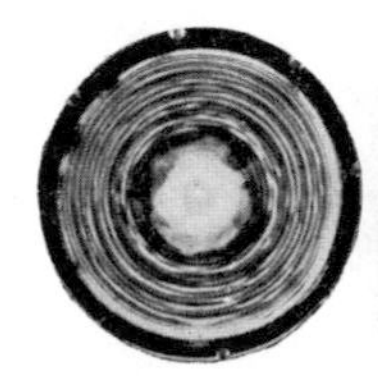

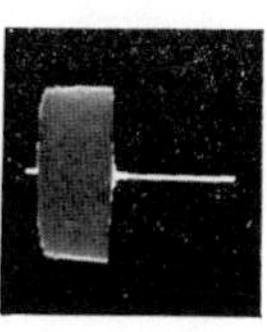

Fig. 16A (left). Cone breakup at 4200 cps, with four circular nodes; point H. Fig. 16B (right). Transient response to same signal.

46

Reprinted from *Audio Eng. Soc. J.* **25**:370–384 (1977)

The Application of Digital Techniques to the Measurement of Loudspeakers

J. M. BERMAN AND L. R. FINCHAM

KEF Electronics Ltd., Tovil, Maidstone, Kent, England

A new approach to the problem of the measurement of loudspeaker characteristics is presented. A digital processor is used to obtain the loudspeaker's transfer function from a direct measurement of the impulse response. The measuring method is discussed in detail along with the forms of display which have so far been used. Some applications are discussed briefly. Finally, the method is shown to be suited to a system approach to the study and development of loudspeakers.

INTRODUCTION: A scientific approach to loudspeaker design and evaluation can only be based on the successful correlation of objective and subjective observations. While we can all too readily gather subjective information on loudspeakers, we have at our disposal only a very limited range of measurements which do not adequately account for the subtle subjective effects found in loudspeakers. Improvements in loudspeakers have not been matched by equivalent improvements in measuring techniques; relatively speaking the subject seems to grow less rather than more scientific, and design still proceeds by a mixture of science, art, and intuition.

An indication of how little loudspeaker measurements have progressed is to be found in the fact that the so-called "frequency-response" curve, which was first used over 50 years ago, is still regarded as the primary means of assessing loudspeaker performance, although it has long been recognized that many important aspects of subjective response cannot be related to the visual appearance of the frequency-response curve.

Many experimenters, observing that speech and music are primarily transient in nature, have turned to transient measurements in the form of tone bursts, step, and pulse responses, but these have always been hampered by signal-to-noise problems and the difficulties of interpretation. Others have recognized that amplitude response is but one part of the frequency response, and there has been considerable interest in the measurement of phase response.

The most significant advance has surely been made by Heyser [1, 2] who, using analog equipment, devised an elegant method of measuring loudspeaker response which overcame many of the previous difficulties. It was the publication of his method in 1969 which provided the inspiration for the digital measuring approach which is described in this paper. Both methods fall into the category of what might be described as the system approach to loudspeaker measurement, in which complete identification of the transfer function is attempted.

Work on the digital testing method was begun in 1971, and interim reports were given in 1973 [3] and 1975 [4]. The method has now been refined and verified over a large number of measurements, and this up-to-date account is based on first-hand knowledge of its use under day-to-day conditions.

THE LINEAR SYSTEM AND ITS MEASUREMENT

The input and output of a linear time-invariant system are always related unambiguously, and this relationship can be expressed as a function of either time or frequency. The time-domain representation takes the form of the response to an ideal impulse, known as the impulse response $h(t)$; the frequency domain counterpart takes the form of the familiar frequency response or transfer function $H(\omega)$, expressed as the real and imaginary parts of a complex function or as the corresponding amplitude and phase characteristics.

The impulse response and frequency response are related mathematically through the Fourier transform, and once either of these is known, the output may be predicted for any arbitrary input. This is calculated in the frequency domain by multiplication of the transfer function with the input expressed as a function of frequency and in the time domain by convolution of the input waveform with the impulse response [5]. These relationships are summarized in Fig. 1.

The advent of the fast Fourier transform (FFT) in the late sixties provided a fast convenient numerical method of transforming from one domain to the other and opened the door to a number of now well-established techniques for obtaining a system transfer function [6, 7]. The ability to measure, digitize, and subsequently manipulate the result numerically permits a wide range of excitation signals to be used, so that one can be chosen to suit the particular system under test.

The great advantage of this approach, which employs a numerical processor, such as a computer, is that once obtained, the system response completely defines the system behavior for *all* possible signals, both transient and steady state alike, and output waveforms can be obtained subsequently through calculation alone.

A loudspeaker, unlike an electrical network, does not have clearly defined output terminals. It creates an acoustic disturbance which radiates into space in a particularly complicated way. Though we would ultimately hope to characterize its behavior entirely, we concern ourselves here with the problem of identifying the behavior for one specific microphone position, and it is this that we refer to in identifying the transfer function of a loudspeaker. It is assumed throughout that the loudspeaker so defined is a linear time-invariant device.

The particular method which has been adopted for identifying system behavior involves a direct measurement of the impulse response. This is not a common approach, but it will be shown that it has significant advantages for the measurement of loudspeakers. In particular, because an impulse is the shortest possible exciting signal for a transfer function determination, the size of the room necessary to perform this measurement is minimized.

DIRECT MEASUREMENT OF LOUDSPEAKER IMPULSE RESPONSE

An impulse response can be obtained by exciting a loudspeaker with a sufficiently narrow pulse. This in itself requires no sophisticated techniques; indeed impulse responses were obtained by McLachlan as long ago as 1930 [8]. Others have since attempted to use the impulse response [9, 10], attracted to the feature that in exciting a system with an impulsive signal, we are effectively investigating its response to all frequencies. Such a narrow pulse, however, cannot always convey sufficient energy to the loudspeaker to give a signal-to-noise ratio that is adequate, if the impulse response is later to be used as a basis for computing the frequency response. Fortunately,

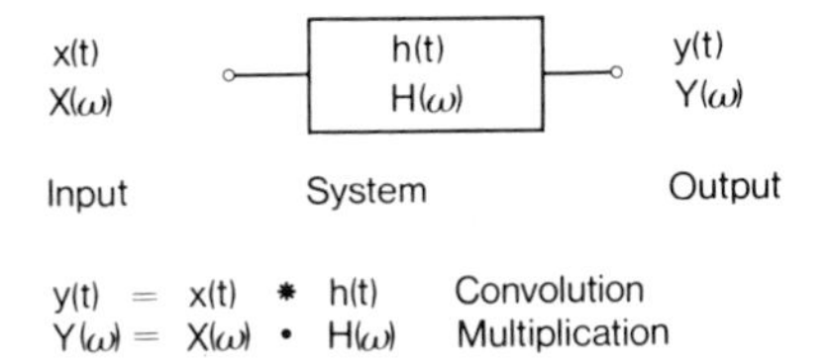

Fig. 1. Symbolic representation of linear system response, expressed in both time and frequency domains.

the limitations of the signal-to-noise ratio can now be overcome by employing a technique known as signal averaging, which is easily implemented when a digital processor is used.

Measuring Method

The measuring chain is shown diagrammatically in Fig. 2. A repetitive short pulse, a single example of which is shown in detail in Fig. 3**a**, is amplified and fed to the loudspeaker, which is positioned close to the center of the measuring room, away from all reflecting surfaces. A microphone is used to detect the consequent acoustic disturbance which is amplified and digitized by the analog-to-digital converter and stored in the computer memory. After each pulse, sound propagates from the loudspeaker and arrives at the microphone after a time delay t_d; it continues out to the walls of the room where it is reflected, and after time t_r the first room reflection returns to the microphone (Fig. 3**b**). The interval between the input pulses is of the order of the room reverberation time, and as each response is captured it is added to a cumulative total of all previous results. The signal is thereby enhanced relative to the background noise, and when the required number of responses have been captured, the cumulative total is divided by the number of responses taken to provide the final result (Fig. 3**c**). This is known as signal averaging. Because the room response is of no interest in this context, all sampled values beyond t_r are set to zero (Fig. 3**d**). The loudspeaker-microphone time delay can also be removed by redefining the time origin (Fig. 3**e**). The loudspeaker impulse response can then be stored permanently on magnetic tape or disc.

Choice of Test Pulse

The primary consideration in selecting a pulse for impulse response measurements is that its spectrum should be flat over the range of frequencies which it is of interest to investigate. It would be convenient to use a pulse of extremely short duration in order to benefit from a spectrum having a greater bandwidth than required and consequently to have no worries about the detailed nature of the waveform. There is so little energy in a short pulse, however, that even with averaging to enhance the signal it is advisable to take a pulse whose energy content is as great as possible, and so a consideration of pulse shape cannot be avoided.

A rectangular pulse is chosen, first because of its close relationship to the ideal impulse, and second because it is so well defined and convenient to obtain in practice. The spectrum of the rectangular pulse has the familiar $(\sin x)/x$ shape shown in Fig. 4. A 10-μs pulse which has its first spectral zero at 100 kHz (Fig. 5) was chosen as a practical compromise between energy and spectral content, while at the same time contributing to the antialiasing function described in the section which follows. This shape of pulse has the additional benefit that it modifies the phase response of the final result only by introducing a small amount of linear phase shift equivalent to the pulse half-width. The practical pulse, of course, begins at $t = 0$,

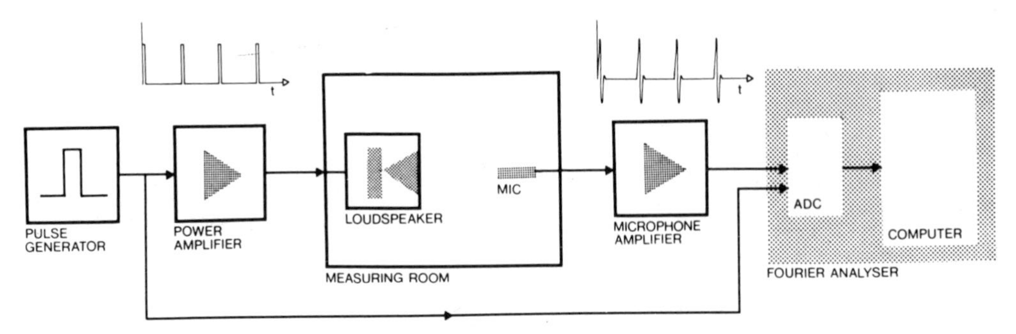

Fig. 2. Block diagram of impulse response measuring chain.

and it is of interest to note that the consequent linear phase shift shown in Fig. 5 appears curved only because of the logarithmic frequency display.

In attempting to provide the loudspeaker with as much energy as possible, the pulse height needs to be maximized, and the demands made of the power amplifier are therefore considerable. The single-sided pulse requires an amplifier with a voltage swing which in audio terms is quite out of proportion to the actual power in the signal. The 60-volt pulse used in practice requires a power amplifier which typically has a continuous rating of 200 watts into 8 ohms.

Sampling Rate and Antialiasing Precautions

In the analog-to-digital conversion process, the analog signal is sampled at equal intervals of time, and the resulting voltage levels are subsequently converted to digital values. If no frequency components greater than half the sampling frequency f_s are present, then the waveform is completely defined by these samples; if not then the components above f_s will "alias" or "fold" back into the frequency range below f_s [11]. This is one of the hazards of analog-to-digital conversion and is commonly safeguarded by employing an antialiasing filter which attenuates the signal sharply above the highest frequency of interest and reduces it to the level of the system noise by half the sampling frequency.

With a suitable choice of sampling rate, however, the antialiasing function for loudspeaker measurements can conveniently be performed by components which are already part of the measuring chain. In the measuring arrangement described in this paper, a half-inch microphone which has a natural cutoff above 40 kHz was chosen. This, along with the rolloff to the first zero of the pulse spectrum, in addition to that of the loudspeaker itself, is generally sufficient to provide adequate antialiasing when a sampling rate of 100 kHz is used. In the rare cases where it is not, the effects can readily be recognized.

Dynamic Range and Signal Averaging

The problem of dynamic range in the impulse-measuring system is highlighted by considering the spectral equivalent of the short pulse, shown in Fig. 5. Because the energy is spread equally among all in-band spectral components, the effective amplitude of each individual component is very much less than that of the testing pulse. It can be shown that, because gain settings of all the equipment are determined by the pulse height, an immediate loss of more than 50 dB of the available dynamic range is inevitable for most loudspeaker measurements.

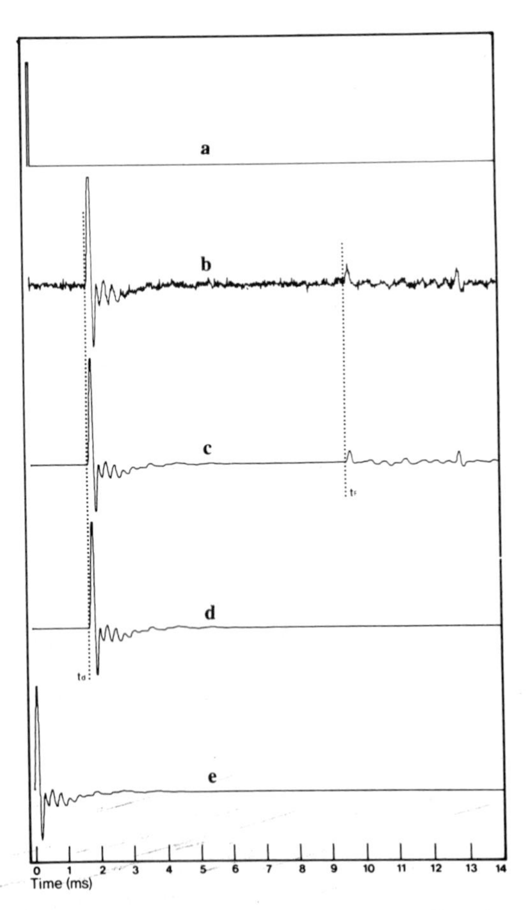

Fig. 3. Impulse response measuring sequence. **a**. Test pulse; **b**. Single-shot response; **c**. Signal-averaged response; **d**. Response after removal of room reflections; **e**. Impulse response, corrected for loudspeaker–microphone time delay.

In order to increase the signal-to-noise ratio, the input pulse is made to be repetitive, and the resulting microphone responses are stored and averaged by the computer. The pulse repetition rate is made to be as high as possible, consistent with residual sound from one impulse response making a negligible contribution to the next. The impulse response is successively reinforced, whereas random noise is not, and this gives a theoretical improvement in the signal-to-noise ratio of a factor of $\sqrt{N}$, where N is the number of averages. Signal averaging, however, is not the universal panacea which it appears to be at first sight. The theoretical improvement is only approached if the noise is entirely random. If periodic or impulsive components are present, the improvement is very much less, and because the $\sqrt{N}$ relationship is a law of diminishing returns, a quiet room is required if high-quality results are to be obtained (acoustic noise is rarely random in nature). For this reason it is essential to minimize any form of mains (line-frequency) hum or interference throughout the measuring equipment. A particularly severe effect occurs if there are periodic components coherent with the pulse repetition rate, for the noise is then reinforced in exactly the same way as the signal. It must also be remembered that signal averaging relies on the repeated coincidence of the measured signal. Variations in the time of arrival of the acoustic signal, which result from wind or thermal air currents, generally preclude outdoor measurements unless weather conditions are exceptionally calm and stable.

Finally, care should be taken over the manner in which the averaging is realized computationally. Loss of precision from, for instance, a simple integer realization with insufficient word length, can result in severe distortion of the averaged waveform.

The amount of averaging required depends entirely on the particular situation. In the limit, averaging can be continued to give a signal-to-noise ratio as great as the precision of the analog-to-digital converter will allow. The results for the present paper were obtained with 64 averages, which represents an 18-dB improvement in the signal-to-noise ratio.

Measuring Environment

The room in which measurements are made must be large enough so that the response of the loudspeaker dies down to a negligible level before the first room reflection arrives at the microphone. A room having a minimum dimension of 3 meters, for example, permits the capture of impulse responses up to 6 ms long for a measuring distance of 1 meter. This has been found adequate for the reliable characterization of loudspeakers down to a frequency of 200 Hz. For measurements extending to lower frequencies, a correspondingly larger room is required.

The total measurement time for a given number of averages depends on the pulse repetition rate, which in turn depends upon the reverberation time of the room at the frequencies of interest. While the environment need not be anechoic, a low reverberation time is advisable if the measurement time is to be kept reasonably short. The 64 averages needed for the results given in this paper take less than 30 seconds.

To obtain accurate impulse responses, the measuring conditions need far more consideration than would normally be given for steady-state measurements. Reflections from quite small objects can cause disturbances which are readily identified in the impulse response. Microphone stands, loudspeaker stands which are larger than the base of the loudspeaker under test, and other reflecting surfaces should be avoided. Even the microphone cable should be well covered with acoustic absorbent. Such reflections have only a small effect on the frequency response, but complicate the interpretation of the impulse response considerably. An example of the extremely high resolution inherent in this technique is given in Fig. 6. The pulse at 1.7 ms is a result of the reflection of the tweeter pulse in the 0.5-inch (12-mm) microphone capsule which is reflected again in the loudspeaker baffle; for such measurements the microphone needs to be angled at some 10° to avoid the effect. It is clear from these results that transient measurements are rarely taken with sufficient precautions to ensure meaningful results. It is tacitly assumed that

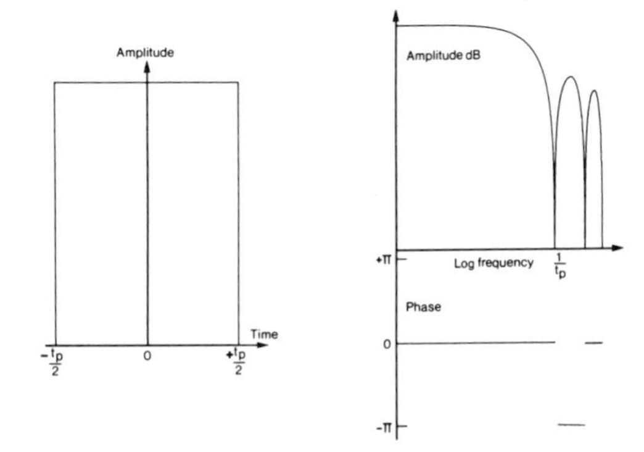

Fig. 4. Rectangular pulse (left) and its frequency spectrum.

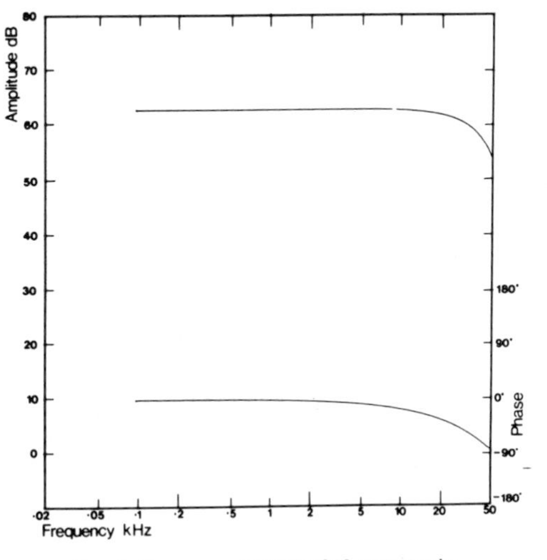

Fig. 5. Frequency spectrum of 10-μs test pulse.

conditions which are sufficient for a good steady-state measurement are sufficient for a transient one, but this is certainly not the case. As a corollary, it will be appreciated that reflections of this order are to be found in so-called "anechoic" chambers from metal parts such as floor supports and even from the absorbing material itself. So an anechoic chamber, while providing an excellent environment in terms of reverberation time, is extremely wasteful for impulse response measurements, in that only the internal chamber volume can be used.

Computation and Analog-to-Digital Conversion

A detailed discussion of computational techniques is not appropriate here, but consideration must be given to computational accuracy. A fast Fourier transform implemented in 16-bit arithmetic has been used, which has the ability continually to optimize its dynamic range. The computational errors are arranged to be equally distributed throughout the transform to give an effective accuracy of 12 bits. To take the maximum advantage of this FFT algorithm a 12-bit analog-to-digital converter is used which provides a peak-to-peak signal-to-noise ratio of 72 dB and a corresponding dynamic range in the frequency domain of over 80 dB.

EXPERIMENTAL VERIFICATION

The assumption of linearity raises the most serious potential objection to the impulse testing technique; most people feel intuitively that such a violent excitation must send the transducer into its nonlinear operating region. A pulse of 10 μs, however, is by its very nature so short that, even with an amplitude of 100 volts, the mechanical parts of the loudspeaker hardly move at all. Comparison of impulse and frequency responses, obtained with pulses of very different amplitudes, have shown no evidence of differences which could be attributed to nonlinear effects. Similarly, the polarity of the pulse used has been found to be of no consequence.

A comparison with traditional analog measurements may be made with results taken outdoors, because these are truly anechoic if the loudspeaker is sufficiently far from the ground. Fig. 7 compares results taken outdoors at a height of 10 meters above the ground with an equivalent digital measurement. For convenience of comparison, the digital result has been plotted on recorder paper with a grid identical to that used for the analog result. The impulse response was also used to calculate a tone burst and square wave response, and these are shown in Fig. 8 along with the equivalent analog measurements. The experience of intensive use of this measuring system alongside traditional analog measuring equipment has given complete confidence in the method.

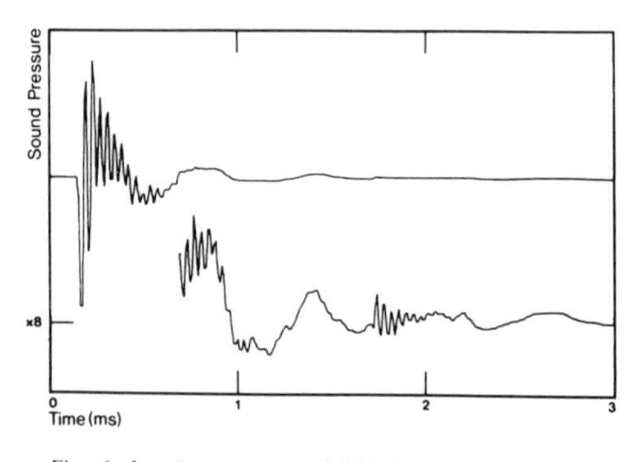

Fig. 6. Impulse response of high-frequency unit showing reflection from microphone capsule.

SIGNAL PROCESSING AND PRESENTATION OF RESULTS

One of the great advantages of this digital approach to measurement is the ability to calculate, from one single measurement, the response of the loudspeaker to any possible excitation and to alter the form of display at will. It is interesting to note that whereas the analog results of

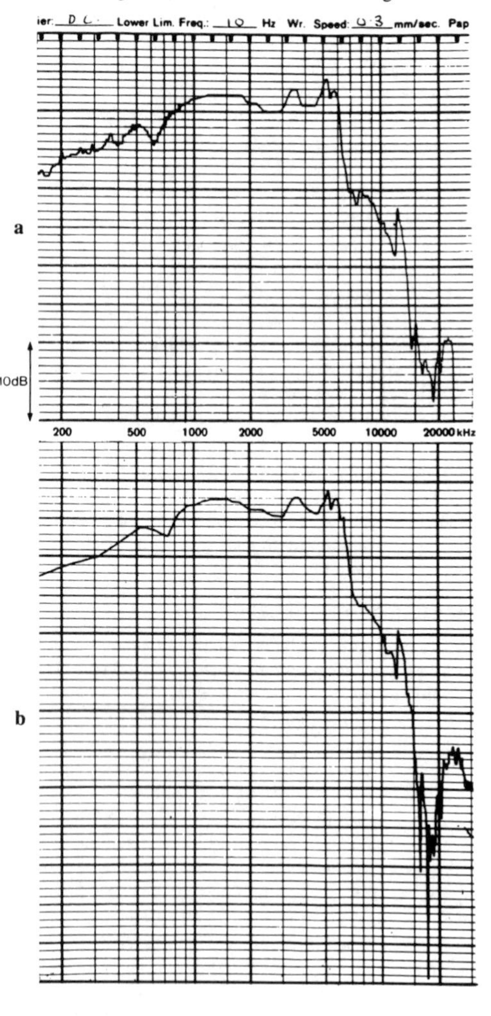

Fig. 7. Comparative amplitude responses of 200-mm drive unit in 45-liter enclosure. **a**. With swept sine-wave measurement outdoors; **b**. Equivalent digital measurement.

the previous section exist only as curves on recorder paper, the digital measurement, which is stored on magnetic disc and paper tape in impulse response form, can readily be recalled. Even now, many months after the measurement, we can continue to answer questions about different aspects of the performance.

The particular manner in which the impulse response data is processed and displayed, has been arrived at in the course of a search for optimum presentations of the information, but it in no way precludes alternatives.

Impulse Response

A surprising amount of information can be obtained from the impulse response alone. The increased dynamic range provided by signal averaging permits a detailed study of the response which was not previously possible with single-shot methods.

A threefold form of display has therefore been adopted in which the tail of the impulse response is magnified, first 8 times and then 64 times. This happens to be convenient for computation, since these magnifications are powers of 2. It gives, in three displays, an optimum view of most loudspeaker impulse responses and offers a visual dynamic range of over 50 dB. The information in the tail is entirely repeatable and can give considerable insight into a loudspeaker's behavior. Whereas loudspeakers of different types produce widely different results, impulse responses of units of the same type bear a close family resemblance. The results from two nominally identical units are shown in Fig. 9, demonstrating family likeness throughout the decay.

Frequency Response

The frequency response embodies both the amplitude-frequency and the associated phase-frequency response and is obtained from the impulse response by numeric transformation using the FFT.

The loudspeaker impulse response may be completely isolated from that of the room in which it is measured by setting to zero all values in the sampled waveform which occur from the start of the first room reflection. The Fourier transform of the resulting response is then a truly anechoic frequency response of the loudspeaker, even though the measurement is made in a semireverberant room.

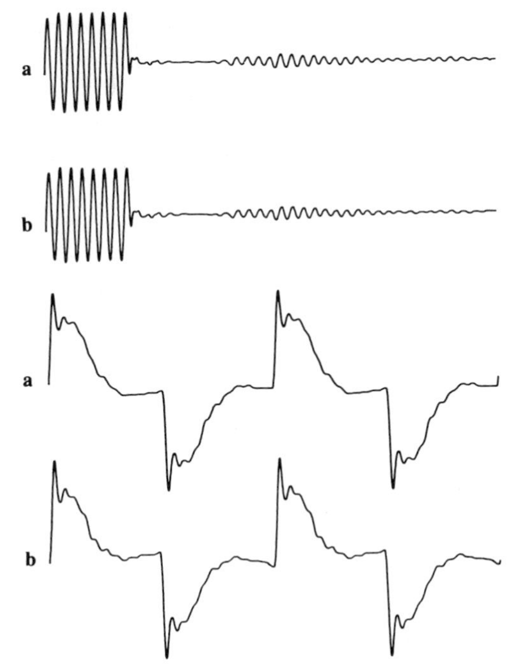

Fig. 8. Comparison of 4-kHz tone burst (above) and 1-kHz square-wave responses for the loudspeaker of Fig. 7. **a**. Measured directly; **b**. Computed from the impulse response.

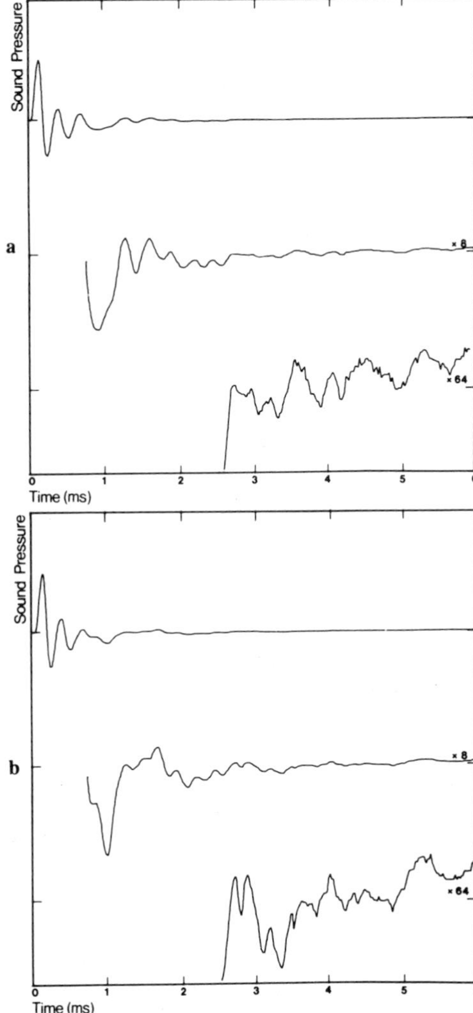

Fig. 9. Measured impulse responses of two nominally identical 110-mm drive units in a 7-liter closed box.

In the same way that the impulse response is represented by a set of samples, the frequency response, after operation of the FFT, is similarly represented by a set of discrete values. Appendix I deals with the relationship between these two domains. The number of samples required in the first instance is determined by the sampling rate, which has already been discussed, and the size of the room used for measurement, along with the additional consideration that the FFT algorithm requires samples which are a power of two in number (see Appendix I). All results given in this paper were obtained with a sampling rate of 100 kHz and a 10-ms length of signal capture resulting in 1024 samples. The 512 frequency components which result are consequently spaced approximately 100 Hz apart (see Appendix I).

If the impulse response has died away sufficiently before the first room reflection, then the frequency response of the device may be completely characterized; if not, some information will be lost resulting in a smoothed frequency response. In either case, the discrete values produced by the FFT do completely define the available knowledge about the response shape in the same way that samples of a time waveform define the continuous signal, so we need only interpolate correctly to construct a continuous curve. In practice, straight line interpolation is found to be adequate, although frequency response curves obtained in this way sometimes appear angular at low frequencies. If straight-line interpolation is used and the visual resolution is thought to be insufficient, more points can be obtained by adding zero values to the tail of the impulse response and thereby increasing the length of the data transformed. In the present case, for example, 4096 points would give a point spacing of 25 Hz. This procedure was adopted for the result shown in Fig. 7 **b**. It should be noted, however, that such a procedure only improves the visual presentation; it does not increase the information content.

In order to display the phase response in a meaningful way, the linear phase shift associated with the time taken for the sound to travel between loudspeaker and microphone (as evidenced by multiple phase rotations in Fig. 10**a**) has to be removed. It is not sufficient simply to calculate the time delay from a measurement of the physical distance between microphone and baffle. The effective acoustic center of a drive unit lies closer to the plane of the voice coil than to that of the baffle on which it is mounted. With analog phase measurements this represents a real difficulty, whereas with the impulse response method the time at which the response begins can be identified within one or two sampling points, and the necessary shifting of the time origin can easily be made. The compensated phase response is shown in Fig. 10 **b**.

Hilbert Transform

It has become common practice to refer to the amplitude response alone as the frequency response of a loudspeaker, whereas strictly speaking the frequency response, as already discussed, embodies both an amplitude and a phase characteristic. In certain cases the amplitude response is sufficient for complete characterization, so that the behavior can be deduced without recourse to the phase response. It happens that most system functions in linear network theory fall into this classification, known as minimum-phase functions, and the tradition which arose of reglecting phase response was subsequently carried over into the field of acoustics. It is of more than academic interest to know if a loudspeaker falls into the minimum-phase category. If it does, then the transient response may be improved by equalization of the amplitude response without regard for the phase response; if not, the same procedure may have quite the reverse effect and degrade the transient performance [2, p. 32].

The mathematical operation which relates the amplitude and phase of a minimum-phase device is known as the Hilbert transform. This is implemented in computational form so that the "minimum-phase" phase response can be

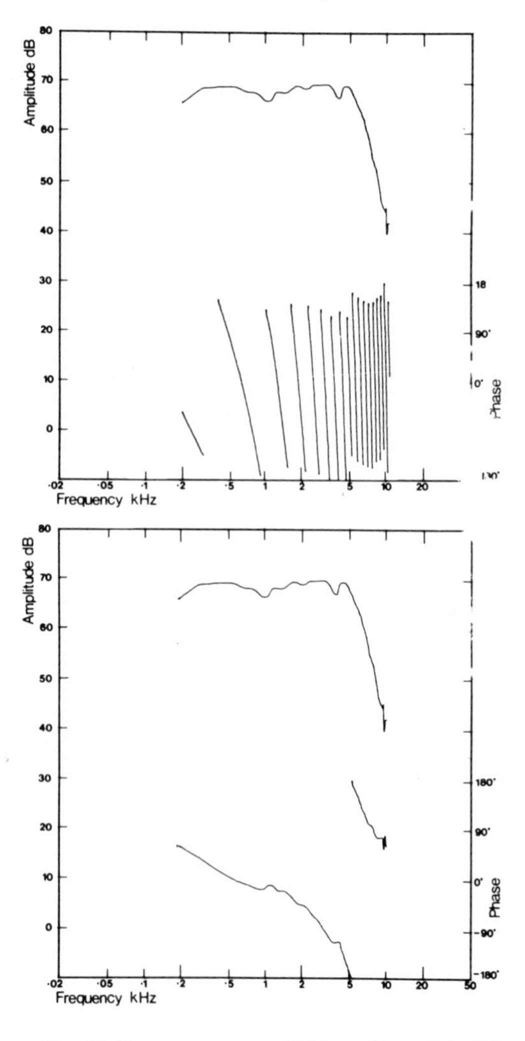

Fig. 10. Frequency response of 110-mm drive unit in 150-liter enclosure measured on axis at a distance of 25 cm. **a**. Before time-delay compensation. **b**. After time-delay compensation.

computed from the measured amplitude response and compared with the actual phase response. Fig. 11 shows the measured response of a twin-cone loudspeaker along with the calculated Hilbert transform, in which non-minimum phase shift behavior is clearly exhibited.

The Hilbert transform has been found to be of use in another important way. The exact acoustic position of a drive unit in a multiway system may be determined by finding how much delay must be removed to make the phase response coincide with the "minimum-phase" phase response; by definition, the minimum-phase response must have no linear phase shift in excess of that inherent in the system itself.

Cumulative Spectra

Impulse response and frequency response are statements of system behavior in the two mutually exclusive domains of time and frequency. In both of these the entire system behavior is visually presented but not, it seems, in an optimum form. Because speech and music are perceived as pitch varying with time, there is good reason to look for alternative methods of displaying the system information in a form which may correlate more closely with subjective response.

In 1948, Shorter pointed out that transient effects in loudspeakers, which were subjectively significant, often contributed so little to the steady-state response that objective identification of the faults was impossible [12]. He introduced the notion of delayed response curves for looking into this matter, and developed an experimental technique based on tone-burst responses to obtain such curves. Although this technique was shown to be of value, the difficulties of instrumentation prevented its coming into general use.

It is now possible, by simple computation and use of the FFT, to produce delayed response curves directly from the impulse response. The three-dimensional perspective display of these multiple curves is given the term "cumulative spectra." Appendix II explains the method of calculation and confirms the relationship of this form of display to the tone-burst response. It is as though we were able simultaneously to excite the loudspeaker with all possible tone-bursts and to record the amplitude of each decaying frequency as a function of time.

The examples shown in Figs. 12–15 demonstrate two of the forms that this display can take. A linear amplitude scale is found to be useful for assessing the early part of the decay and a logarithmic amplitude scale for a more searching analysis of low-level information. A logarithmic frequency scale has been found to cause confusion of the display in the high-frequency region and a linear frequency representation has consequently been adopted.

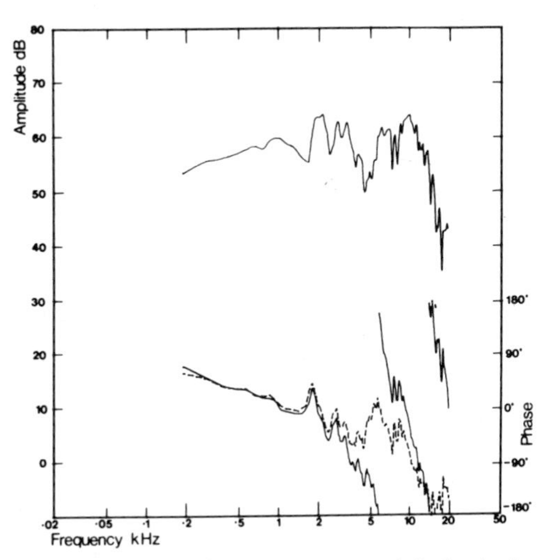

Fig. 11. Example of nonminimum-phase behavior in the frequency response of 200-mm twin-cone drive unit in 150-liter closed box. Dashed line indicates the Hilbert transform of the amplitude response.

APPLICATION OF THE METHOD

Specific applications of these digital techniques deserve detailed reports. For the present, this paper is concerned first with indicating in general terms how these techniques can usefully be applied to the development of loudspeakers, and second with demonstrating some practical results which have not previously appeared in the literature.

Drive Units

It is in the area of individual drive units that the forms of display discussed in this paper are used together to the greatest advantage. It is, of course, a most important area, because drive units are the fundamental building blocks of any loudspeaker system.

In contrast to the impulse responses of Fig. 9, which demonstrate nominally identical units, Figs. 12 and 13 show the results from two very different units. A linear amplitude cumulative-spectra display suffices to view the gross differences in the initial decay of these two loudspeakers, and to perceive that the faults in the unit with an irregular steady-state response are largely resonant in nature.

Figs. 14 and 15, on the other hand, demonstrate that small changes to a unit, while making only a small difference to the frequency response, can make dramatic changes to both the impulse response and the cumulative spectra. The two units are different only in respect of their voice-coil mass. The impulse response of Fig. 15 shows clear evidence of ringing, but even with this and the frequency response together, it is difficult to ascertain the exact frequency of this oscillation. In fact it becomes clear from the cumulative spectra that this ringing, which is evident as a ridge in the later spectra, emanates from a dip in the steady-state response and early spectra; not at all what might have been expected. These drive units were mounted in the same enclosure for the purpose of measurement, and the elevation at 1 ms which extends over the whole frequency range is confirmed to be a reflection from the rear of the enclosure. This could not have been identified from either time or frequency responses alone.

The Hilbert transform has been used to determine whether units are minimum-phase devices, and has led to

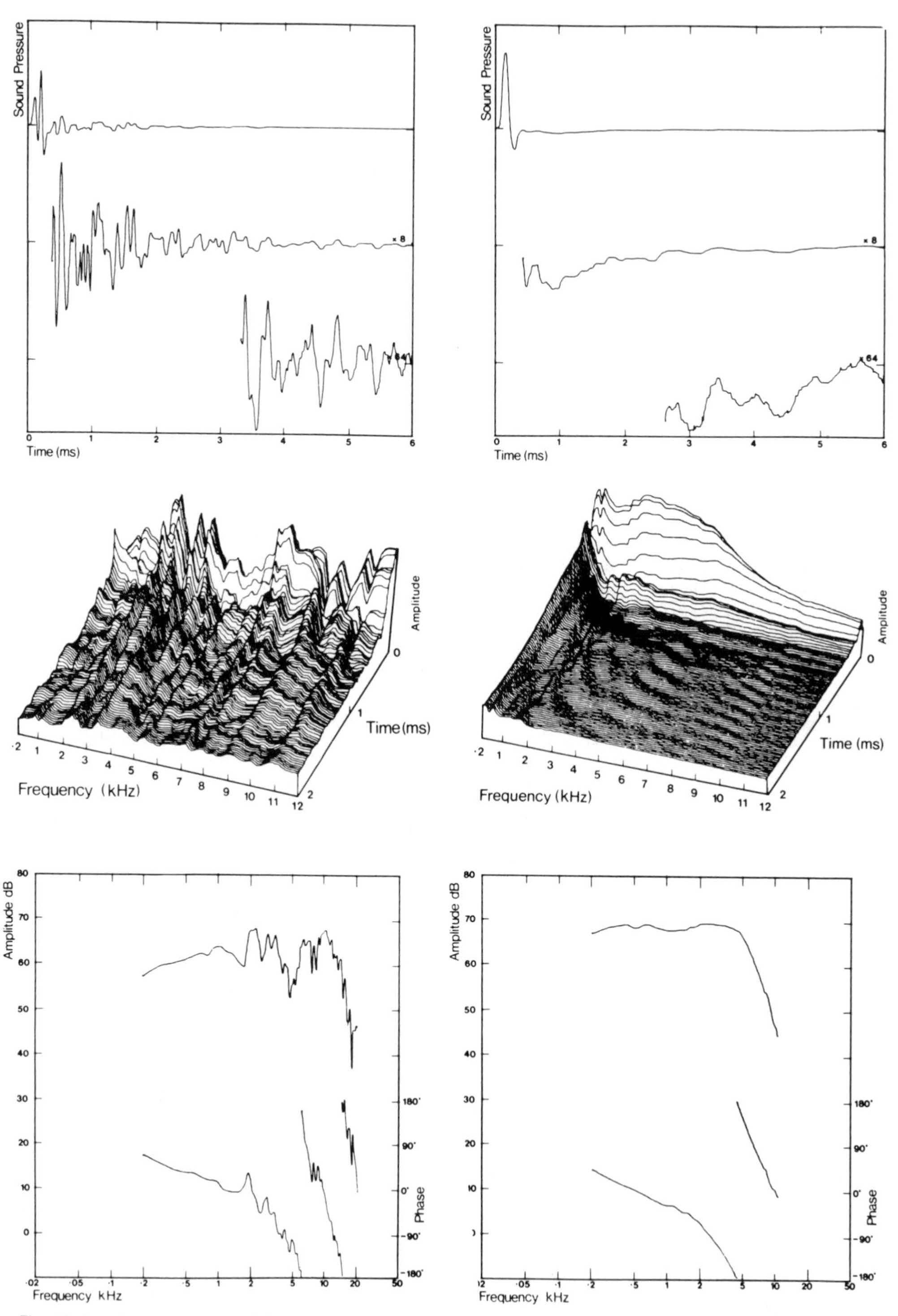

Fig. 12. Impulse response, cumulative decay spectra, and frequency response of 200-mm twin-cone drive unit in 150-liter closed box.

Fig. 13. Impulse response, cumulative decay spectra, and frequency response of 110-mm development drive unit in 150-liter closed box.

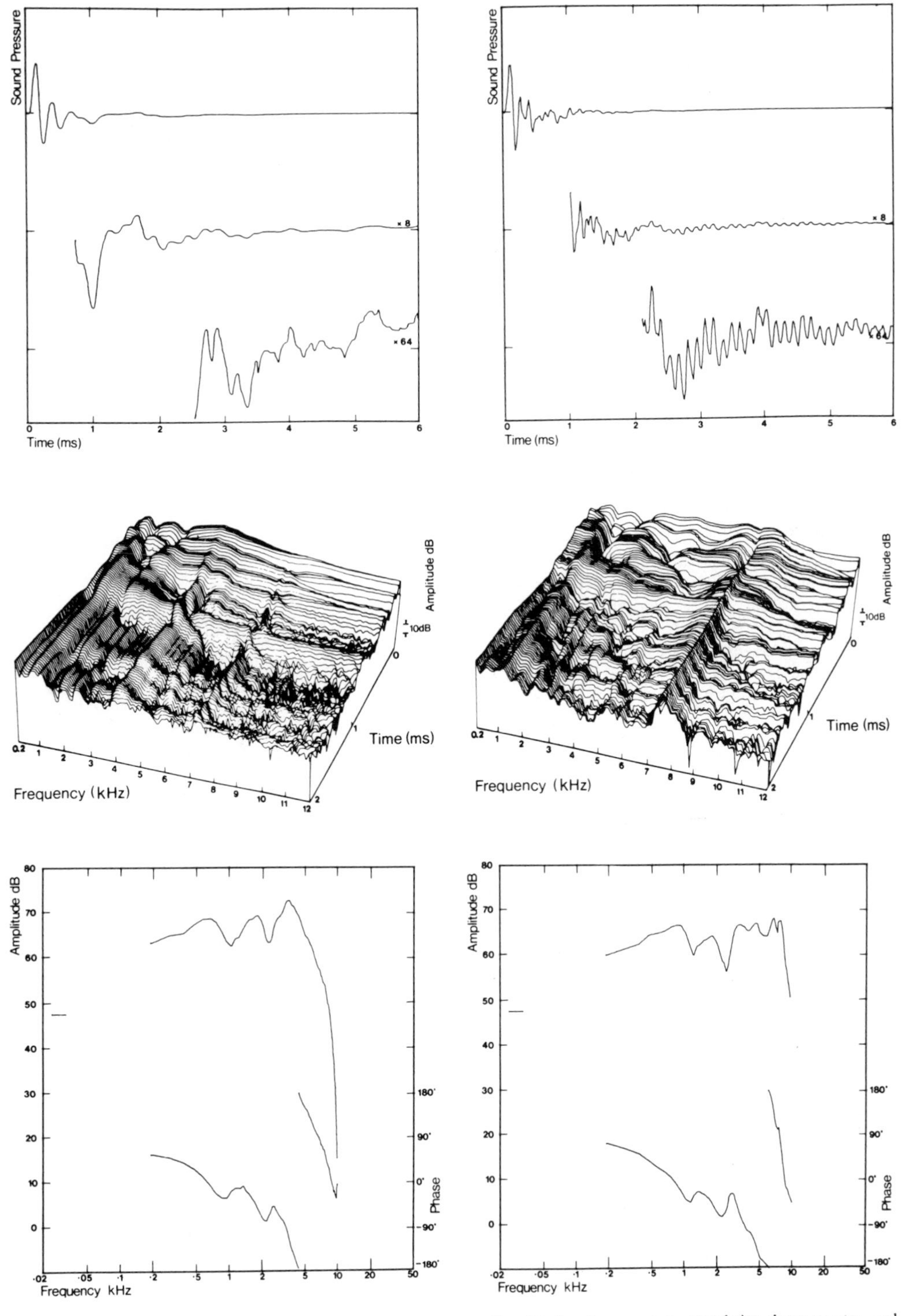

Fig. 14. Impulse response, cumulative decay spectra, and frequency response of 110-mm drive unit in 7-liter closed box.

Fig. 15. Impulse response, cumulative decay spectra, and frequency response of 110-mm drive unit with lightweight voice coil in 7-liter closed box.

the surprising result that the nonminimum-phase loudspeaker shown in Fig. 11 demonstrates a rare exception to the general rule. Almost every unit that has been measured has turned out to be of the minimum-phase variety, and a typical example of the agreement obtained is shown in Fig. 16.

Enclosures

The impulse response alone provides a useful way of separating the effect of a loudspeaker drive unit from the enclosure in which it is housed.

While the early part of the impulse response is related predominantly to the drive unit, the latter part is dominated by sound radiated from the enclosure walls and retransmitted through the loudspeaker cone after reflection within the enclosure. Each enclosure exhibits its own characteristic "signature" waveform, which is substantially independent of the drive unit. Fig. 17 shows one application of this, in which the effect of panel damping is demonstrated. The technique can, of course, equally be applied to the investigation of cabinet construction and lining materials.

Loudspeaker Systems

It is interesting to look at the manner in which individual drive units and crossovers combine to make a loudspeaker system. Fig. 18 shows the usual two-way arrangement in which a tweeter is mounted in the enclosure above the low-frequency unit and the measurement taken on the high-frequency unit axis. The effective acoustic position of the bass unit is further from the microphone than that of the high-frequency unit, and the measurements of curves A and B in Fig. 19, in which the impulse responses have been shifted to the time origin of the tweeter, show this clearly. When added together they produce the impulse response of Fig. 19**c** in which the disturbance after the first doublet, often interpreted as "ringing," is simply the delayed doublet generated by the bass unit. The frequency response of the complete system is shown in Fig. 20 along with the corresponding Hilbert transform. It will be seen that this is far from being a minimum-phase system, yet it is entirely typical of the response produced by multiway loudspeaker systems which have their units mounted in the same plane.

DISCUSSION

The paper has described a method for obtaining the transfer function of a loudspeaker by direct measurement of its impulse response. This is a complete description of

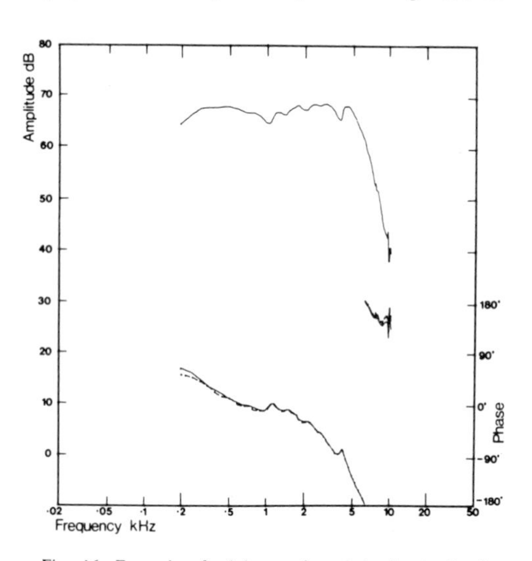

Fig. 16. Example of minimum-phase behavior in the frequency response of 110-mm drive unit in 150-liter closed box. Dashed line shows Hilbert transform of amplitude response.

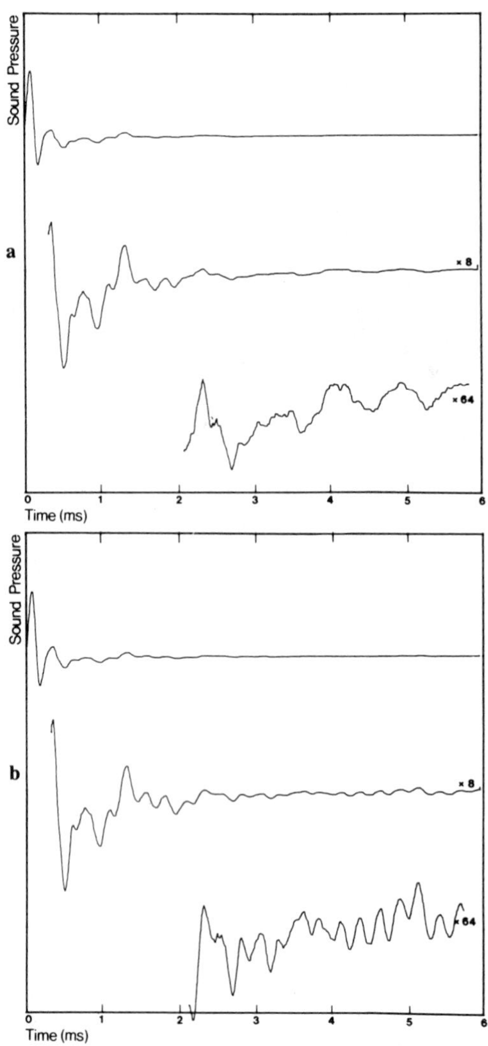

Fig. 17. Impulse responses of 110-mm drive unit in 7-liter, 12-mm chipboard closed box. **a**. Without panel damping. **b**. With panel damping.

the loudspeaker's behavior in its particular relation to the measuring microphone. Once the impulse response has been captured, the response of any steady-state or transient signal can be calculated and no further measurements are necessary. The method has been completely verified against equivalent analog measurements where such measurements have been possible. Some applications have been mentioned briefly to indicate how the particular displays of impulse and frequency response, Hilbert transform, and cumulative spectra can usefully be employed. However, these do not preclude other forms of display; indeed it is the flexibility of display which offers the future possibility of presenting the information in a form more closely related to subjective experience.

There is, without doubt, room for further improvement and modification of the measuring technique. No mention has been made of linear distortion introduced by parts of the measuring chain, but clearly the amplitude and phase response of the microphone, amplifiers, etc., will be superimposed on those of the loudspeaker under test. It is fortunate that so many measurements have been found to be minimum phase in nature because this in turn has confirmed that the measuring chain itself must be a minimum-phase system. Consequently, inaccuracies in the measuring chain can easily be compensated for by numeric manipulation, having consideration for the amplitude response alone, after the loudspeaker has been measured. Following from this, the pulse could be pre-

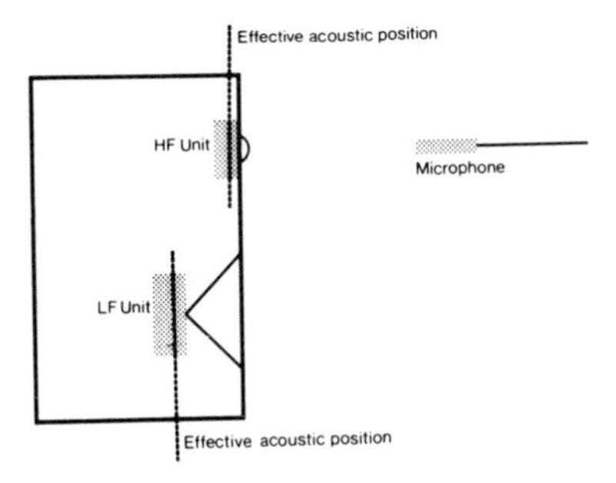

Fig. 18. Two-way loudspeaker system showing effective acoustic position of units relative to customary measuring microphone position.

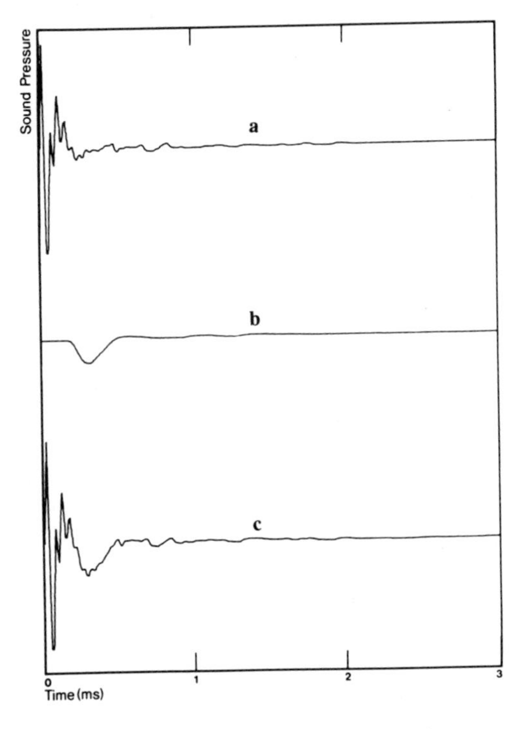

Fig. 19. Impulse responses of a two-way loudspeaker system measured on the tweeter axis at 1-meter distance. **a**. High-frequency unit and crossover; **b**. Low-frequency unit and crossover; **c**. Complete system.

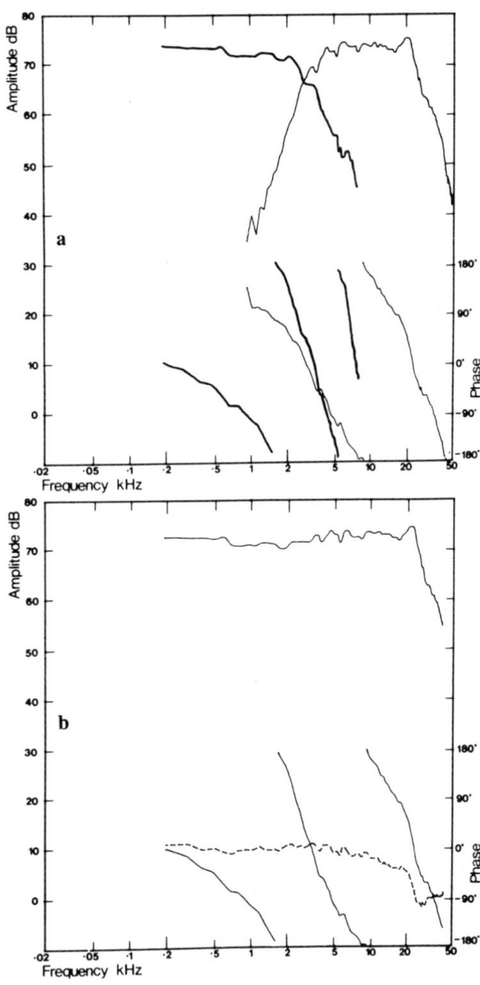

Fig. 20. Frequency responses of a two-way loudspeaker system. **a**. Low- and high-frequency units separately. **b**. Complete system. Dashed line shows Hilbert transform of amplitude response.

shaped and the measured result modified accordingly; such a technique would allow the signal to be optimized to take account of, for example, a less than ideal measuring environment.

Details of the equipment used for the results presented in this paper are given in Appendix III. The computer-based Fourier analyser may appear overspecified for the measuring technique inasmuch as the impulse response capture and subsequent Fourier transformation can now be performed with much more modest equipment. The measuring method, however, is just one part of a complete philosophy, rooted in linear system theory, which treats the component parts of a multiway loudspeaker as subsystems of a complete linear system. The impulse response provides a means of identifying each component part, and the associated ability to store each result, and to recall it for computation when required, facilitates the construction of the system response from the component parts alone. Once the drive units have been measured, for instance, loudspeaker system development can proceed by computer-aided design of crossovers without recourse to physical construction. Such manipulation requires elaborate facilities for handling the results after measurement, so a versatile processor is essential if this digital approach to loudspeaker measurement is to be used to full advantage.

APPENDIX I

The Fast Fourier Transform (FFT)

The FFT provides a highly efficient means of computing the reversible discrete Fourier transform (DFT) of a time series [13]–[15]. The coefficients of the DFT are calculated iteratively, which reduces the total number of arithmetic operations from N^2 to $2N \log_2 N$, so that the computation time of most signal analysis transforms is reduced from minutes to fractions of a second. It is, however, a condition of the iterative algorithm that the number of samples in the time series be exactly a power of 2.

N time-domain values give rise to an identical number in the frequency domain, which can be written in place of the time values during computation. The frequency domain is arranged in the form of $N/2 + 1$ real values from direct current to half the sampling frequency and $N/2 - 1$ related imaginary values. This takes advantage of the fact that the imaginary part is an odd function of frequency and consequently must always have a value of zero at direct current and at half the sampling frequency. Conversion from real and imaginary to amplitude and phase involves only a change of coordinates, which is of course reversible. Whereas time samples are separated by the sampling period $\Delta t = 1/f_s$, frequency values are $\Delta f = 1/t$ apart, where t is the total length of the initial time series. These relationships are summarized in Fig. 21, which shows a 16-point transform.

APPENDIX II

Cumulative Spectra

Consider the system, whose impulse response is $h(t)$, subjected to the sudden onset of a generalized sinusoidal oscillation $e^{j\omega t}$; the input excitation is then $U(t)\, e^{j\omega t}$, where $U(t)$ is a step function. The system output $y(t)$ at any instant t_1 can then be found by evaluating the convolution integral:

$$y(t) = \int_{-\infty}^{+\infty} U(t-\tau)\, e^{jw(t-\tau)}\, h(\tau)\, d\tau, \text{ at } t = t_1. \tag{1}$$

This may be expressed as

$$y(t_1) = C \int_{-\infty}^{+\infty} H(\tau)\, e^{-jw\tau}\, d\tau \tag{2}$$

where $C = e^{jwt_1}$ and is constant and

$$H(\tau) = U(t_1 - \tau)\, h(\tau). \tag{3}$$

Eq. (2) may be recognized simply as the Fourier integral evaluation of $H(\tau)$ for the frequency ω. However, $H(\tau)$ is the product of the system impulse response $h(\tau)$, illustrated in Fig. 22**a**, and the delayed step function $U(\tau - t_1)$ reversed in time to become $U(t_2 - \tau)$, as shown in Fig. 22**b**.

It follows from this that at any given time t_1 after the onset of the sinusoidal oscillation, the instantaneous amplitude and phase of the response to all possible frequency excitations may be obtained by evaluating the Fourier transform of the step-function-windowed impulse response. By successively evaluating the Fourier transform for different values of t_1, we can build up a three-dimensional picture of the tone-burst response. Similarly, the decay of the output when a sinusoidal excitation abruptly ceases is calculated by multiplying the impulse response by the delayed step function $U(\tau - t_1)$ shown in Fig. 22**c**, and again performing successive Fourier transformation.

APPENDIX III

EQUIPMENT DETAILS

Lyons bipolar pulse generator: PG 71N
SAE power amplifier: Mark IIICM

Bruel & Kjaer half-inch microphone type 4133
Bruel & Kjaer microphone amplifier type 2606

Hewlett-Packard Fourier analyzer type 5451B (see Fig. 23).

ACKNOWLEDGMENT

The authors would like to acknowledge all those who have contributed to this project, in particular Mr. R. V. Leedham, Senior Lecturer in Electrical Engineering at the University of Bradford.

REFERENCES

[1] R. C. Heyser, "Acoustical Measurements by Time Delay Spectrometry," *J. Audio Eng. Soc.*, vol. 15, p. 370 (1967).

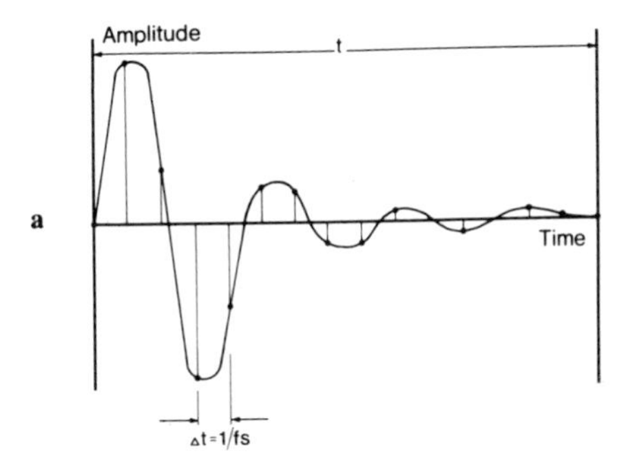

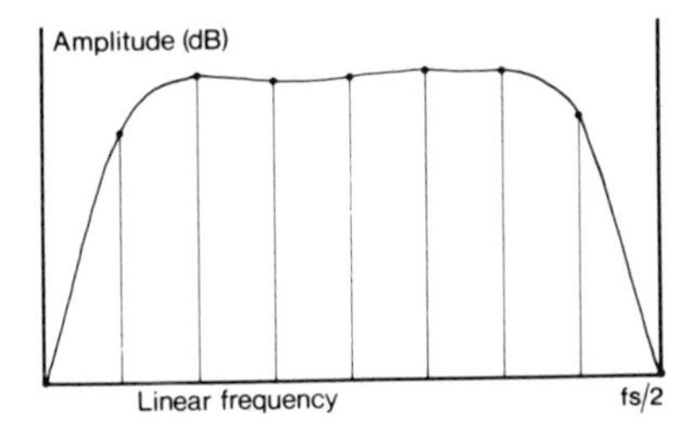

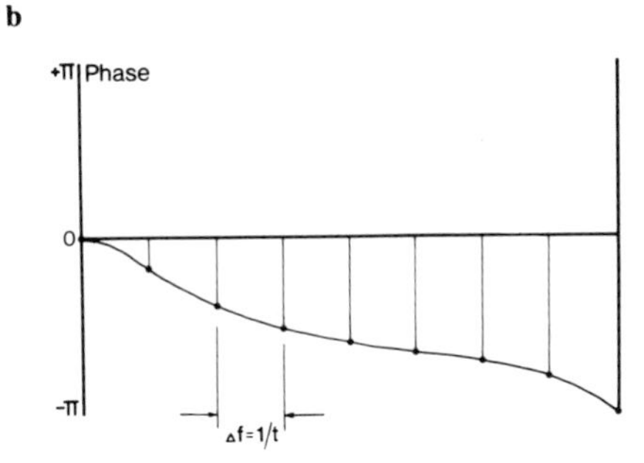

Fig. 21. Illustration of data structure in fast Fourier transform computations. **a**. Time-domain signal and its representation by discrete samples. **b**. Frequency-domain spectrum, presented as amplitude and phase, showing discrete frequency sample points and corresponding continuous transform.

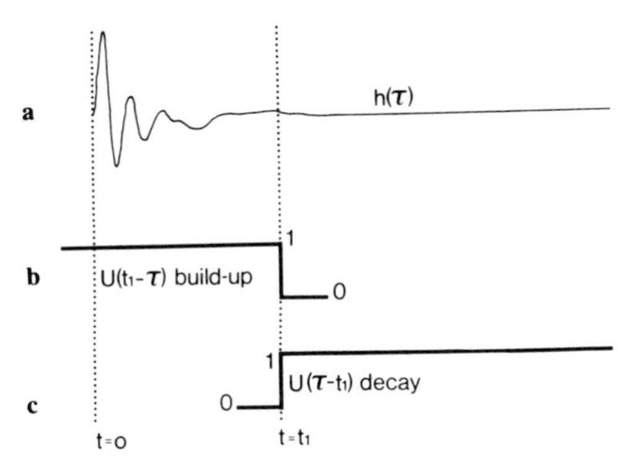

Fig. 22. Window functions used in cumulative-spectra calculations. **a**. Loudspeaker impulse response; **b**. Reversed delayed unit step; **c**. Delayed unit step.

Fig. 23. Hewlett-Packard Fourier analyzer type 5451B used for measuring impulse response of loudspeakers.

[2] R. C. Heyser, "Loudspeaker Phase Characteristics and Time Delay Distortion, Pts. I and II," *J. Audio Eng. Soc.*, vol. 17, p. 30 (1969); p. 130 (1969).

[3] L. R. Fincham and R. V. Leedham, "Loudspeaker Evaluation Using Digital Fourier Analysis," presented to British Section of the Audio Engineering Society at the IEE, London, England, Feb. 1973.

[4] J. M. Berman, "Loudspeaker Evaluation Using Digital Techniques," presented March 4, 1975, at the 50th Convention of the Audio Engineering Society, London, England.

[5] A. Papoulis, *The Fourier Integral and Its Applications* (McGraw-Hill Electronic Series, New York, 1962).

[6] P. R. Roth, "Effective Measurements Using Digital Signal Analysis," *IEEE Spectrum*, vol. 8 (Apr. 1971).

[7] R. G. White, "Evaluation of the Dynamic Characteristics of Structures by Transient Testing," *J. Sound Vib.*, vol. 15, no. 2, pp. 147-161 (1971).

[8] N. W. McLachlan, *Loudspeakers. Theory Performance, Testing and Design* (Dover, New York, 1960), pp. 332-336.

[9] P. Chapelle, "Les Essais Impulsionnels de Hautparleurs," *Electroacoust.*, vol. 17, p. 35 (1970).

[10] A. Schaumberger, "Impulse Measurement Techniques for Quality Determination in Hi-Fi Equipment, with Special Emphasis on Loudspeakers," *J. Audio Eng. Soc.*, vol. 19, p. 101 (1971).

[11] R. Bracewell, *The Fourier Transform and Its Applications* (McGraw Hill, New York, 1965), p. 189.

[12] D. E. L. Shorter, "Loudspeaker Transient Response—Its Measurement and Graphical Representation," *B. B. C. Quart.*, vol. 1, p. 121 (1946).

[13] W. T. Cochran *et al.*, "What Is the Fast Fourier Transform?" *IEEE Trans. Audio Electroacoust.*, vol. AU-15, p. 45 (1967).
[14] A. V. Oppenheim and R. W. Schafer, "Digital Signal Processing," (Wiley, New York, 1975).
[15] G. D. Bergland, "A Guided Tour of the Fast Fourier Transform," *IEEE Spectrum*, vol. 6, p. 41 (July 1969).

47

Reprinted from *Audio Eng. Soc. J.* **17**:30–41 (1969)

Loudspeaker Phase Characteristics and Time Delay Distortion: Part 1

RICHARD C. HEYSER

Jet Propulsion Laboratory, California Institute of Technology, Pasadena, California

A technique is described for measurement of the complete frequency response of a loudspeaker, including amplitude and phase. A concept of time delay is introduced which provides a physical description of the effect of phase and amplitude variations as a frequency-dependent spatial smearing of the effective acoustic position of a loudspeaker.

INTRODUCTION In a previous paper a derivation was given for a new acoustic testing technique which allowed anechoic measurements to be taken in a normally reverberant environment [1]. The quantity measured was shown analytically to be the complex Fourier transform of the system impulse response of any signal with a fixed time delay. When applied to loudspeakers, this means that beside the conventionally measured pressure amplitude spectrum there is a pressure phase spectrum, which to the best information available to this author has not been as well investigated. In measuring both the amplitude and phase spectrum of some common loudspeaker types it was immediately evident that some peculiarities in the behavior of phase were not apparent from an inspection of amplitude alone. Since the measurement technique allows a subtraction of time delay incurred between the electrical stimulus applied to the loudspeaker terminals and the pressure wave incident upon the test microphone, it was apparent from the beginning that there was in many cases a frequency-dependent time delay in excess of that caused by the travelling of the pressure wave over the known distance from voice coil to microphone diaphragm. Attempts at understanding this delay in terms of an equivalent electronic transfer function were not satisfactory: the only concept of circuit time delay which might prove useful is that of group delay, also called envelope delay, and it was not evident exactly how one might go about application of this concept, particularly when open literature definitions are accompanied with disquieting phrases such as ". . . when the amplitude does not change rapidly with frequency . . ." and ". . . if there is no absorption . . .". In fact it is rather disturbing that a substitute more in alignment with nature is not used in those cases when this so-called time delay becomes negative, a condition found to be quite common in loudspeaker measurements. For this reason it became necessary to derive a time delay more in keeping with concepts of causality as well as to investigate the amplitude and phase relationships in loudspeakers. The resulting concept of a spatial "spreading out" of the effective acoustic position of a loudspeaker behind its physical position would appear to be a more unified approach to time delay in such a complicated network. This paper is a documentation of some characteristics which have been measured on typical loudspeakers, as well as the derivation of a concept of time delay useful for interpreting these characteristics. The ultimate purpose is to provide audio engineers working on loudspeaker design a criterion by which a loudspeaker may be equalized to provide a more perfect response than would be possible by the use of pressure response measurements alone.

THE LOUDSPEAKER AS A NETWORK ELEMENT

In considering the role of a loudspeaker in the reproduction of sound, it would appear logical to describe this device as a network element. Although a transducer of electrical to acoustic energy, a loudspeaker may dissipate power, store energy, modify frequency response, introduce distortion, and in general impart the same aberrations in its duty as any conventional electronic network. Unlike a conventional network, however, a loudspeaker interfaces with the spatial medium of a human interpreter and may possess characteristics in this medium, such as polar response and a time-delayed phenomenon, which are unlike a normal electronic network element. While recognition of these latter effects exists, measurement has proven cumbersome.

In the discussion to follow a loudspeaker will be considered as a general network element. This element will have some transfer function relating the output sound pressure wave to the applied electrical stimulus. The output of the loudspeaker will be considered to be its free-field response. Where severe interaction with its environment is desired, as for example in the case of a corner horn or wall-mounted dipole, then that portion of the environment will be included. No simplification to an equivalent circuit is sought; indeed, the system is assumed to be so complicated that the only knowledge one has is a direct acoustic response measurement.

DEFINITION OF FREQUENCY RESPONSE

The transfer function of a loudspeaker will be assumed independent of signal level. This considerably simplifies the analysis and is justified when one considers that deviations from linearity of amplitude are much less than deviations from uniform frequency response. This means that, as in linear circuit theory, one can now define the transfer function to be a function of frequency. Our concern then rests with determining the frequency response of a loudspeaker.

The frequency response of a loudspeaker will be defined as the complex Fourier transform of its response to an impulse of electrical energy. If one considers the magnitude of the frequency dependent pressure response, this definition coincides with the practice of using a slow sinewave sweep in an anechoic chamber and plotting the pressure response as a function of frequency [17]. This definition, however, also includes the phase angle as a frequency-dependent term. Thus if a loudspeaker has an amplitude response $A(\omega)$ and a phase response $\phi(\omega)$ the loudspeaker response $S(\omega)$ is

$$S(\omega) = A(\omega)e^{i\phi(\omega)}. \tag{1}$$

It quite frequently happens that the amplitude response is better characterized in decibels as a logarithmic function, $a(\omega)$, which leads to the simplified form

$$S(\omega) = e^{a(\omega)}e^{i\phi(\omega)}. \tag{2}$$

The frequency response of a loudspeaker will then consist of two plots. The plot of magnitude of sound pressure in decibels as a function of frequency, $a(\omega)$, will simply be called amplitude. The plot of phase angle of sound pressure as a function of frequency, $\phi(\omega)$, will be called phase. These taken together will completely characterize the linear loudspeaker frequency response by the relation

$$\ln S(\omega) = a(\omega) + i\phi(\omega). \tag{3}$$

Equation (3) allows us to state that the amplitude and phase are the real and imaginary components of a function of a complex variable. This complex function is thus a more complete definition of what is meant by the frequency response of a loudspeaker. From this we see that measuring amplitude alone may not be sufficient to specify total speaker behavior.

THE LOUDSPEAKER AS A MINIMUM PHASE NETWORK

The measurement of loudspeaker pressure response is not a new art, and substantial definitive data has existed for over 40 years [17]. With a few very rare exceptions these measurements have been of the amplitude response [23, 24, 25]. Indeed, it is common practice to call this the frequency response of a loudspeaker.

It is of more than casual concern to investigate the conditions under which the measurement of amplitude response is sufficient to characterize the complete behavior of a loudspeaker. Obviously, whenever this is the case, the measurement of phase is academic, and a conventional amplitude response measurement should continue as a mainstay of data. Referring to Eq. (3) one may inquire into the conditions under which a prescribed magnitude function results in a definite phase angle function and conversely.

For a complex function such as expressed in Eq. (3) there will exist a unique relationship between the real and imaginary parts, determined by a contour integral along the frequency axis and around the right half-plane, if the logarithm is analytic in this right half-plane [2]. A network which meets this requirement is called a minimum phase network. Thus, if a loudspeaker is a minimum phase network, the measurement of either phase or amplitude is sufficient to characterize the frequency response completely.

When one has a minimum phase network the amplitude and phase are Hilbert transforms of each other related by [2, 3, 5, 6]

$$a(\omega) = \frac{1}{\pi} P \int_{-\infty}^{\infty} \frac{\phi(x)}{\omega - x} dx \tag{4}$$

$$-\phi(\omega) = \frac{1}{\pi} P \int_{-\infty}^{\infty} \frac{a(x)}{\omega - x} dx, \tag{5}$$

where P indicates that the principal part of the integral is to be taken at the pole ω. These integrals are the counterpart of Cauchy's differential relations on the frequency axis and are a sufficient condition to ensure that the loudspeaker does not have an output prior to an input signal, that is, $f(t) = 0$ for $t < 0$. While it may appear absurd to even concern oneself with the obvious fact that a minimum phase network cannot predict the output, it is quite important to the concept of group delay for a minimum phase function. It can be shown that group delay is not coincident with signal delay for a minimum phase network [26].

Given that one seeks confirmation of minimum phase behavior in a loudspeaker, how can one determine when this condition is actually achieved? It is unfortunate that one cannot determine this from either amplitude or phase alone. This follows either from an analysis of Eqs. (4) and (5) or the Cauchy-Riemann equations (Eq. A2 in Appendix A) or from a consideration of the factors needed to say that there are no zeros in the right half-plane. Thus it would seem that it is necessary to measure both amplitude and phase in order to determine if a measurement of amplitude alone would have been sufficient. On its face value, this is a convincing argument for measuring phase.

Hard on the heels of such an analysis comes the query as to why one would be concerned whether a loudspeaker is minimum phase or not. The answer comes from a very important property of minimum phase networks which may be paraphrased [4] as: If a loudspeaker is a minimum phase network then it can be characterized as a ladder network composed of resistors, capacitors, and inductors, and there will always exist a complementing network which will correct the loudspeaker frequency distortion as closely as one chooses. Thus, if one assumes linearity, a minimum phase loudspeaker could be completely compensated to become a distortionless transducer.

Consider what this means if one does not have a minimum phase loudspeaker. Assume that a pressure response (amplitude) measurement is made on a loudspeaker in an anechoic chamber. The speaker will of course have peaks and dips in its response. One might naturally assume that by diligent work it would be possible to level out the peaks and fill in the dips by a combination of electrical and mechanical means to produce a loudspeaker with a considerably smoother measured response. However, if one does this with a non-minimum phase loudspeaker, even if it is non-minimum phase over only a portion of the spectrum, one will have a transducer with a smoother $a(\omega)$ but quite probably a considerably distorted $\phi(\omega)$, having inadvertently created a complicated all-pass lattice network. As shown in Appendix B and elsewhere [26], this is equivalent to a dispersive medium. When program material is fed into this "equalized" loudspeaker the odds are that the sound heard by a listener will be considerably more unpleasant than that coming through the loudspeaker without benefit of such equalization.

The unhappy consequence of this type of experience will be an assumption that the anechoic chamber response bears little relationship to the quality of reproduction, and that perhaps the best loudspeakers were indeed made in a previous generation. If, on the other hand, one knew what portions of the frequency spectrum were minimum phase, equalization of that portion *would* improve the response, provided one did not attempt equalization on the remaining non-minimum phase portions. Thus if one is concerned with improving the quality of sound reproduction by means of passive or active equalization one is immediately interested in phase response of a loudspeaker and determination of minimum phase criteria.

TIME DELAY IN A LOUDSPEAKER

A network is said to be dispersive if the phase velocity is frequency-dependent, which means that the phase plot is not a linear function of frequency [11]. In pursuing the mathematical analysis of network transfer functions one eventually arrives at considerations of the effect of dispersion on time delay of a signal through a network. Perhaps the best that can be said of such considerations is that they are seldom lucid and almost never appear applicable to the problem at hand. One must nonetheless recognize that group delay (envelope delay) and phase delay are legitimate manifestations of the perturbations produced on a signal by a network. In this section we shall investigate some general considerations of time delay and apply them to the loudspeaker.

There is, of course, a very real time delay incurred by an acoustic pressure wave as it travels from loudspeaker to listener. The velocity of sound in air at a fixed temperature is constant, at least to the precision required for this discussion. Furthermore, this velocity is not a function of frequency. If a pressure wave travels a given distance at a given velocity it takes a period of time expressed as the quotient of distance to velocity.

When one considers the loudspeaker one can no longer assume a non-dispersive behavior. Appendix B contains a derivation of the effect of a dispersive medium (without absorption) on the transfer function of a network. It is shown that if the time lag is a function of frequency the effect will be an excess phase lag in the transducer transfer function. A very important result of this is that one should look first at the phase terms when considering time delay. Equations (4) and (5) indicate that, at least for a minimum phase network, either phase or amplitude give the same information. Bode [4] has shown that for a minimum phase network the phase lag, and hence to some measure time lag, of the low-frequency portions of the spectrum are governed by the amplitude function at high frequencies. Since it is known that the amplitude response of a loudspeaker must fall off at some sufficiently high frequency, there is some justification to believe that there will be an additional time delay in a loudspeaker due to the rolloff of high-frequency response. Stated another way, the acoustic position of a loudspeaker should, on the average, lie behind its physical position by an amount that is some inverse function of its high-frequency cutoff. The acoustic position of a woofer will be further behind the physical transducer than that of a tweeter. This important fact is quite frequently overlooked by engineers who consider that spatial alignment of voice coils is sufficient to provide equal-time path signals from multirange loudspeakers.

Having thus considered air path delay and time lag in the loudspeaker due to high frequency cutoff, one comes to the seemingly nebulous concept of dispersive lag in the transducer. Two types of so-called time delays have been defined for the purpose of expressing the distortion of a signal passing through a medium. These are phase delay and group delay, also called envelope delay [7-8]. Phase delay expressed in seconds is a measure of the amount by which a sinusoid disturbance of fixed frequency is shifted in phase after passing through a network. This conception is applicable only for a sinewave and only after total equilibrium is achieved, and is consequently of no use in considerations of the realistic problem of aperiodic disturbances. Therefore, we will not consider phase delay any further.

Group delay, also expressed in seconds, is an attempt at expression of the relative time shift of signal frequency components adjacent to a reference frequency. When used in this manner and applied to a medium with a substantial time delay (in the classic sense of distance traversed at a finite velocity) group delay is of benefit in distortion analysis. Group delay has caused a considerable confusion among engineers who have attempted to stretch the definition beyond the frequency range within which it is valid. Group delay is expressible very simply as the slope of the phase-frequency distribution; thus for a network with the transfer function of Eq. (2),

$$t_{Group} = -d\phi(\omega)/d\omega. \qquad (6)$$

Group delay, although expressed in the dimension of time, is not a satisfactory substitute for the engineering concept of time delay which one intuitively feels must be present in a medium. There is a strong desire to ascribe a causal relation between an applied stimulus to a loudspeaker, for example, and the emergent pressure wave which as a premise, must have some time lag. There are, however, enough examples of total inapplicability—so called anomalous dispersion—to create suspicion regarding the primacy of group delay.

It can be shown that there is indeed a time delay phenomenon in a network more in alignment with engineering experience [26]. This delay is not necessarily single-valued, but may be multiple-valued or possibly a time distribution. An engineering interpretation which can be put on the time delay of a network may be secured by investigating the behavior of a given frequency component of an input signal as it finally emerges in the output. It is true that a sudden change in a parameter at the input will in effect create a broad spectrum around the particular frequency the delay of which we wish to characterize. However, this does not invalidate the premise that there will still exist a spectral component of this frequency, and one may legitimately ask what happens to that spectral component. It is shown that it may take some finite time for the component to appear at the output and that for a general network there may be many components of the same frequency arriving with different time delays.

This multiplicity of delayed outputs at a given frequency means that in the case of a loudspeaker one could also think in terms of a number of loudspeakers arrayed in space behind the physical transducer in such a way that the air-path delay of each produces the appropriate value of delay. Each frequency in the reproduced spectrum will then possess some unique spatial distribution of the equivalent loudspeakers. The emergent sound pressure wave will be a perfect replica of electrical signal only if these equivalent loudspeakers merge into one position for all frequencies since only then will there be no frequency-dependent amplitude or phase terms and all signal components will arrive at the same time [26].

Figure 1a is a highly schematic attempt at illustrating the phenomenon of loudspeaker time delay. A single loudspeaker is assumed, with a physical position in space indicated by the solid line. The effective position of the source of sound as a function of frequency is shown by speaker symbols which lie behind the physical location of the actual speaker by an amount determined by the delay time and velocity of sound in air. There may be many such equivalent perfect speakers distributed in space with more or less energy from each. The average position of the distribution of these equivalent speakers is

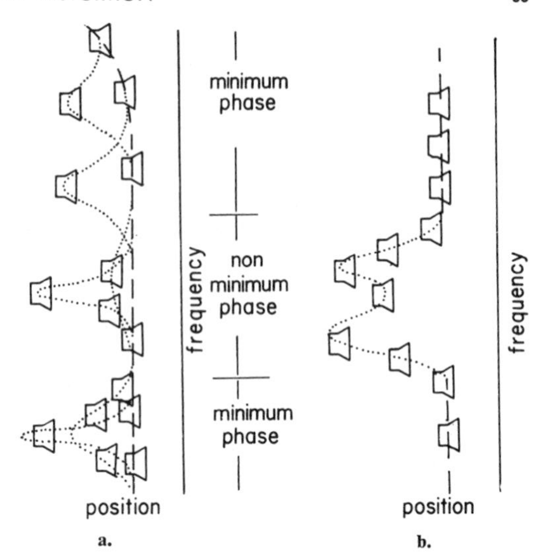

Fig. 1. **a.** Symbolic representation of the effect of frequency-dependent amplitude and phase variations of a single loudspeaker as equivalent to an assemblage of perfect loudspeakers distributed in space behind the physical position of the single loudspeaker. Each of the equivalent perfect loudspeakers has a flat amplitude and phase pressure response but assumes a frequency-dependent position indicated by the **dotted lines**, and in general differs in amount and polarity of energy radiated. The physical position of the actual radiator is indicated by the **solid line** while the average acoustic position determined by high-frequency cutoff lies behind this at the position of the **dashed line. b.** Symbolic representation of the acoustic effect of a response equalization based entirely on correcting the amplitude variations of the speaker of Fig. 1a. The frequencies at which the unequalized speaker is minimum phase may be corrected to perfect response indicated by a space-fixed perfect loudspeaker, but excess time delay results for non-minimum phase frequencies.

indicated by the dashed line, and corresponds to the delay attributed to high-frequency rolloff. Those frequencies at which the original loudspeaker is minimum phase are shown. Note that at no frequency will an equivalent speaker be found in front of the physical speaker. This is a result of the obvious fact that a loudspeaker can have no response prior to an input.

In Fig. 1b the result of amplitude equalization alone is symbolized. The equivalent speakers coalesce into one for the minimum-phase frequencies but spread out for all others.

TECHNIQUE OF MEASUREMENT

Analysis of the technique of time delay spectrometry has shown that the intermediate frequency amplifier contains the complex response of Eq. (2) with the frequency parameter ω replaced by a time parameter at [1]. This complex spectrum is convolved with a modifying term which is composed in turn of the convolution of the impulse response $i(t)$ of the intermediate frequency amplifier with the sweeping window function $w(t)$. The derivation of this relation is presented in a previous paper [1] and is sufficiently lengthy that it will not be reproduced here.

Equation (37) of that paper shows that if a loudspeaker is the subject of this test, the time function one finds in the intermediate frequency amplifier is $o(t)$, where

$$o(t) = (Gain\ Factor)[i(t) \otimes w(t)] \otimes S(at). \quad (7)$$

The effect of the convoluting, or folding integrals, indicated by the symbol $\otimes$ is a scanning and smoothing of the response $S(at)$. This means that if one observes appropriate precautions in sweep rate a, the intermediate frequency amplifier contains a signal which may be interpreted as

$$o(t) = Smoothed\ S(at). \quad (8)$$

Expressing this differently, one might say that the desired spectrum (Eq. 2) is contained in the equipment but all "sharp edges" have been smoothed off. The amount of smoothing is a function of the bandwidth of the equipment and the rapidity of the sweep. What is important is that there are no surprises or genuinely false patterns created by time delay spectrometry. If the loudspeaker contains a peak in response, then this peak will show in the analysis.

Having established the fact that time delay spectrometry will yield the complex transfer function of Eq. (2), it remains to see how both amplitude and phase may be extracted. It is assumed that the previous paper [1] will be used as reference for the basic technique. Figure 2 is a simplified block diagram of a time delay spectrometry configuration capable of measuring both amplitude and phase of a loudspeaker response.

In Fig. 2 a crystal oscillator is used as a source of stable fixed frequency. A countdown circuit derives a rate of one pulse per second which triggers an extremely linear sawtooth generator. This sawtooth provides a horizontal display to an oscilloscope as well as drive to a voltage-controlled sweep oscillator. The crystal oscillator is also used as a frequency source for two frequency synthesizers. A fixed frequency synthesizer converts the crystal frequency to the frequency of the intermediate frequency amplifier for the purpose of providing a reference to the phase detector. A tunable frequency synthesizer with digital frequency control is used to provide a precise offset frequency to down-convert the sweeping oscillator to the audio range; the filtered audio-range sweeping tone is then used to drive the speaker under test. A delay line is used between the sweeping oscillator and the microphone-channel balanced mixer for the purpose of precise cancellation of the time lag of the down-conversion lowpass filter. The intermediate-frequency amplifier containing the up-converted microphone spectrum feeds both a limiter-phase detector channel and a logarithmic amplifier-amplitude demodulator channel. The output of either the amplitude or phase detectors may be selected for display on the vertical axis of the oscilloscope which has the sawtooth horizontal drive.

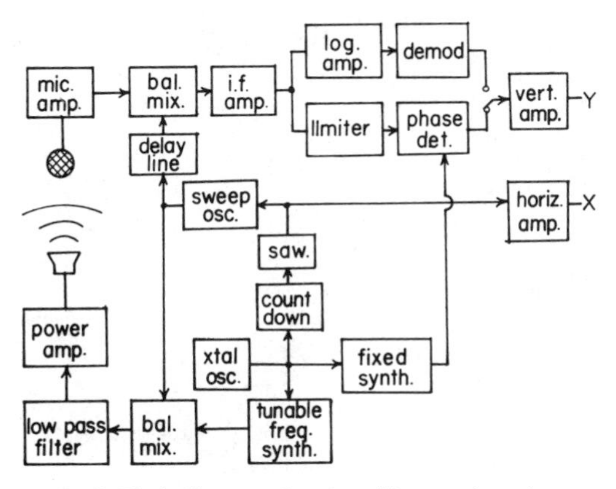

Fig. 2. Block diagram of a time delay spectrometry configuration capable of measuring both amplitude and phase frequency response of a loudspeaker.

It will be recognized that the demodulation of phase information has necessitated a considerable increase in circuit complexity over that required for amplitude alone. Particularly important is the fact that subtraction of the free air-path time lag and a stable display of phase not only requires a fixed offset frequency precisely divisible by the frequency of the sawtooth sweep but also necessitates a coherent phase-stable locked loop involving the time lag of the measurement air path.

As an illustration of the precision required, assume that one is measuring a response from dc to 20 kHz in a one second sweep and that an intermediate frequency of 100 kHz is being used. An offset of one part in 100,000 of the oscillator with respect to the phase standard will produce 360° of phase drift in one second. With a high-persistence phosphor, standard with spectrum analyzers, a distinct image blur will occur with a drift of 3.6° per sweep, yet this requires the oscillators to be offset by no more than one part in ten million. The use of a sampling attachment with an x-y plotter might require two orders of magnitude of stability over this value, i.e., oscillators which if multiplied up to 1,000 MHz would differ by no more than one Hz. This requires a self-consistent set of frequencies provided by obtaining all frequencies from a single oscillator.

Distance-measuring capability is similarly impressive. If one is measuring the phase response in the assumed 20 kHz band, one has a wavelength of approximately 650 mil at the highest frequency. A physical offset in distance of only 6 mil between loudspeaker and microphone will produce a 3.6° phase change at 20 kHz. This corresponds to a ½ μsec time difference in path length.

While the phase measurement provides an enormously sensitive measure of time delay, it should not be implied that it is at all difficult to obtain good data. It is unusual when there is sufficient motion of the air path or transducers to upset a phase measurement, and such an effect is visible within the time of two sweeps of the display.

In making measurements on a loudspeaker, the microphone is positioned at the desired speaker polar angle and at a distance consistent with the spectrum sweep rate and closest reflecting object. With the loudspeaker energized by the sweeping tone, the closest integer offset frequency is dialed which corresponds to the relation

$$F_0 = (X/c)(dF/dt) \quad (9)$$

where F_0 is the synthesizer offset in Hz, X is the distance from speaker to microphone in feet, c is the velocity of sound in feet per second, and the sweep rate is measured in Hz per second.

The amplitude response should be at or near its peak

value and require no further adjustment. The displayed phase response will quite generally be tilted with respect to the frequency axis at a slope which is proportional to the deviation of the offset oscillator from the proper value. An offset discrepancy of only two Hz will accumulate 720° of additional shift in a one second sweep. The proper offset frequency is easily found by observing the phase display and using that value which produces the most nearly horizontal phase plot. If the value is between two integer frequencies with a one second sweep, it may be helpful to slightly reposition the microphone toward or away from the speaker. Having found the required offset frequency it is more than likely that the display will pass through 360° at one or more frequencies, producing phase ambiguities. The phase of the offset synthesizer must then be slipped until the most nearly optimum display results. This is most readily done by dialing an offset which is 0.1 Hz above the integer Hz value, which will cause the phase display to move vertically by 36° per sweep. When the proper pattern is obtained the 0.1 Hz digit is removed, leaving a stationary pattern of phase *vs* frequency.

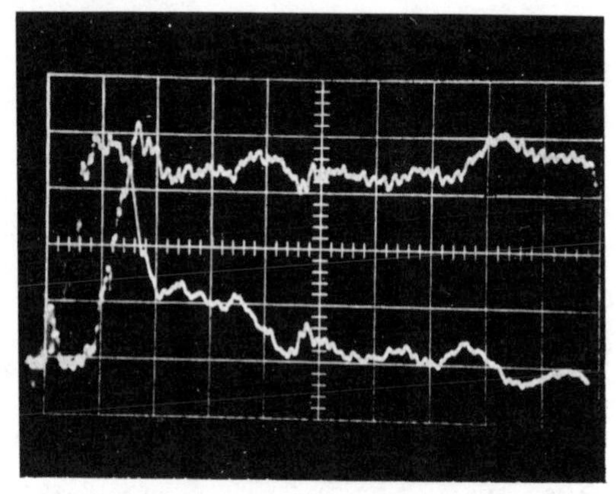

Fig. 3. Simultaneous amplitude (**upper**) and phase (**lower**) response of an inexpensive horn-loaded compression tweeter. Response is shown from dc to 10 kHz at 1 kHz per horizontal division. Amplitude is 10 dB per division and phase 60° per division.

MEASUREMENTS ON TYPICAL LOUDSPEAKERS

In the analysis of typical loudspeaker amplitude and phase characteristics it is desirable that a simultaneous display of both functions be made so that variations in either parameter at any given frequency may be compared. The oscillographs of this section are therefore dual-trace plots with the following characteristics: 1. The horizontal axis represents increasing frequency to the right with a linear scale; 2. Amplitude is the upper plot with logarithmic response and increasing signal as an upward deflection; 3. Phase is the lower plot with phase lead as an upward deflection. These characteristics, with the possible exception of a linear frequency scale, are among the standard conventions of network analysis.

Figure 3 is an example of an inexpensive horn-loaded compression tweeter. The plot extends from zero to 10 kHz with a frequency deflection factor of 1 kHz per cm (one cm is represented by a major division). The amplitude deflection factor is 10 dB per cm and the phase is 60° per cm. By referring to Fig. A-2 in the appendix it can be seen that this speaker is of the minimum phase type. The peaks at 1.6 kHz, 4 kHz, 7 kHz, and 8 kHz, and the dips at 4.6 kHz and 7.3 kHz have the appropriate phase fluctuation. Of particular interest is the peak at 1.6 kHz, identified by the phase change as a primary resonance in the driver occurring above horn cutoff.

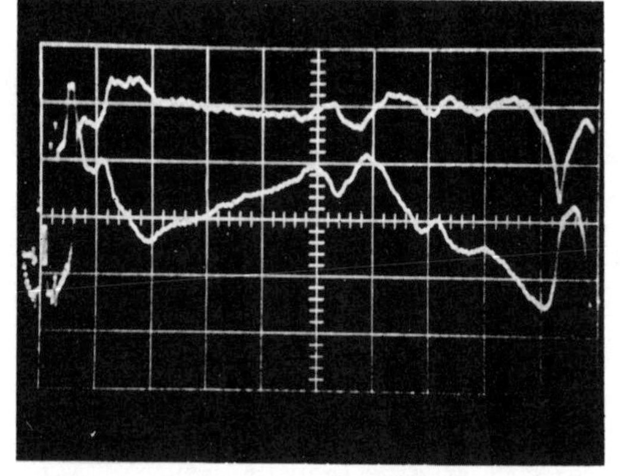

Fig. 4. Amplitude (**upper**) and phase (**lower**) response of midrange horn-loaded driver. Coordinates identical to those of Fig. 3.

Figure 4 is an example of a midrange horn-loaded speaker of the non-minimum phase type. Small variations are of minimum phase, including the dip at 9.4 kHz, but inspection of the phase identifies the fact that the speaker has a frequency-dependent spatial location. The acoustic position of the farthest forward equivalent speaker as determined by the offset frequency in the spectrometer is approximately 0.75 in. behind the position of the phase plug. This is in general agreement with the delay one would expect with a 10 kHz cutoff. From 2 kHz to 5.5 kHz the phase slope with independence of amplitude indicates a position which might be called Position 1. From 5.5 kHz to around 9.4 kHz the acoustic position

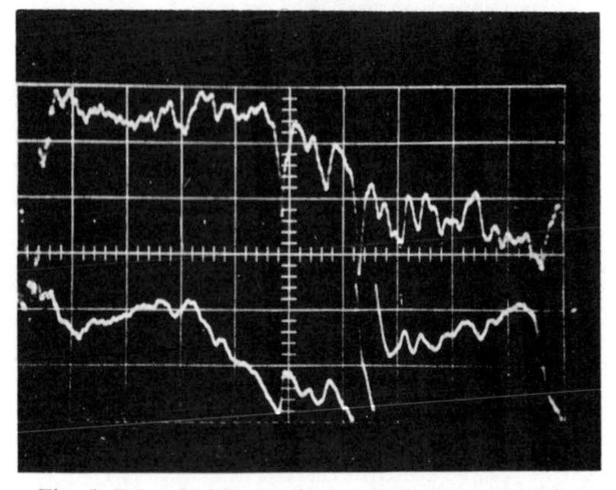

Fig. 5. DC to 20 kHz amplitude and phase response of the speaker of Fig. 4. Amplitude is 10 dB per division and phase 120° per division.

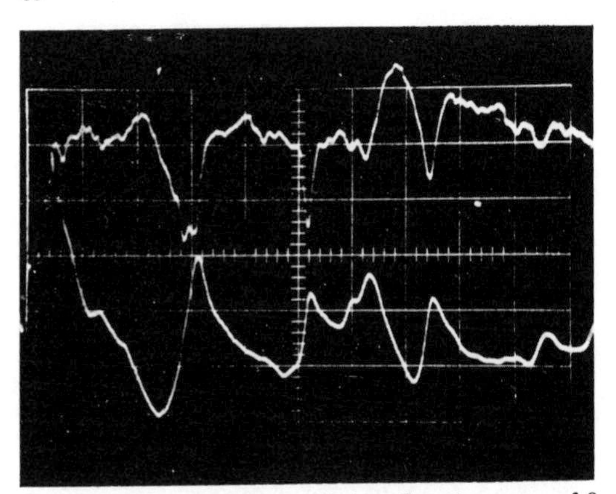

Fig. 6. Amplitude (**upper**) and phase (**lower**) response of 8 in. loudspeaker mounted in a reflex cabinet. Frequency scale is from dc to 2 kHz at 200 Hz per division. Amplitude is 5 dB per division and phase 30° per division.

can be seen to be behind Position 1 by almost 2 in. This is because the phase slope difference between the two regions differs by about 60° per kHz, which yields a time difference of 1/6 msec. The phase behavior between 1.1 kHz and 2 kHz in conjunction with the broad peak indicates a minimum phase resonance and not necessarily a change in location. This is further bolstered by observing that the phase slope from 800 Hz to 1.1 kHz is that of 2 kHz to 5.5 kHz. The high negative phase slope at the cutoff and the phase peak at 500 Hz is the minimum-phase behavior to be expected of the rapid drop in amplitude. The interpretation of the region around 1.5 kHz is that of a multi-pole resonance similar to a band-pass filter.

Figure 5 is the response from dc to 20 kHz for the speaker of Fig. 4. The behavior pattern shows definite non-minimum phase between 12 kHz and 14 kHz, with minimum phase elsewhere.

Figure 6 shows the on-axis behavior from dc to 2 kHz, at 200 Hz per cm, of an 8 in. speaker mounted in a reflex cabinet. The dip at 600 Hz coincides with the undamped back-wall reflection, and from the phase characteristic can be seen to be of the minimum phase type. This oscillograph was made with a five times expansion of a 10 kHz sweep in order that the space-equivalent bandwidth could be made small enough for measurement in a small room. For this reason the smoothing bandwidth is 70 Hz and the phase plot will appear slightly to the left of its proper frequency due to smoothing. By making the appropriate correction it can be seen that each of the peaks and dips is of the minimum phase type shown in Fig. A-2. The strong local fluctuations, such as the dips at 600 Hz, 1100 Hz, and 1500 Hz and the peaks at 400 Hz and 1350 Hz, produce phase fluctuations which are superimposed on the overall phase characteristic due to the low-frequency cutoff. This phase variation due to low-frequency cutoff is characteristically a high negative phase slope in the cutoff region and a smooth extension into the region of normal response.

The low-frequency behavior, which is strong due to the system response zero at dc is shown also in Fig. 7. This is a response measurement at 1 kHz per cm made from −5 kHz to +5 kHz passing through zero. It is a 15° off-axis response of an unenclosed paper whizzer-cone loudspeaker of the type normally used for replacement purposes in automobile radios. The required even and odd frequency characteristic of amplitude and phase is quite pronounced. For this loudspeaker the phase change is 180° at zero frequency, and does not commence its transition until within 200 Hz of zero. The apparent phase breakup below 100 Hz is due to strong low-frequency disturbances in the measuring room which captured the limiter when the loudspeaker signal dropped below their spectral distribution. This oscillograph demonstrates that the phase at zero frequency for which there is no loudspeaker output, may be obtained by centering the time delay spectrometry display at zero frequency and observing the point of symmetry as one approaches zero from both directions. This loudspeaker may be seen to have a minimum phase dip at 3.5 kHz and a possible minimum phase peak at 2.2 kHz, while the remainder of the spectrum is non-minimum phase. This is not unexpected since the response was obtained off-axis and the diffraction and reflection around the whizzer are substantial. Note in particular the absorption dips around 3.5 kHz which do not have substantial phase variations.

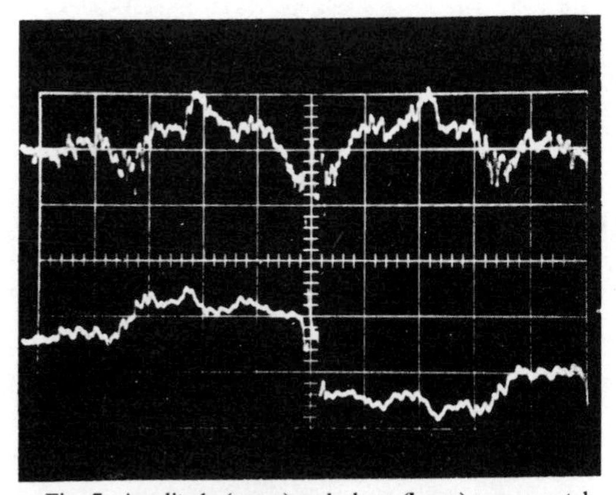

Fig. 7. Amplitude (**upper**) and phase (**lower**) response taken 15° off-axis on an unenclosed whizzer-cone speaker. Frequency scale is 1 kHz per division and extends from —5 kHz through dc to +5 kHz. The even frequency symmetry of amplitude and odd frequency symmetry of phase necessary for causality is readily discerned. Amplitude is 10 dB per division and phase is 120° per division.

Figure 8 shows the response of another horn-loaded compression tweeter. The scale factors for frequency, amplitude and phase are 1 kHz, 10 dB, and 30° per cm respectively. The spectrum encompasses dc to 10 kHz. With the possible exception of the region around 8.4 kHz the response is definitely minimum phase.

Figure 9 shows the response from dc to 20 kHz of a quality midrange electrostatic loudspeaker. This is also of minimum phase characteristic throughout the spectrum. The large apparent phase jump at 18 kHz is a phase detector transition through the equivalent 360° point.

An extremely difficult loudspeaker to measure, due to severe environmental dependence, is the full-range corner horn. The inclusion of the necessary walls and floor rather effectively nullifies a normally anechoic environ-

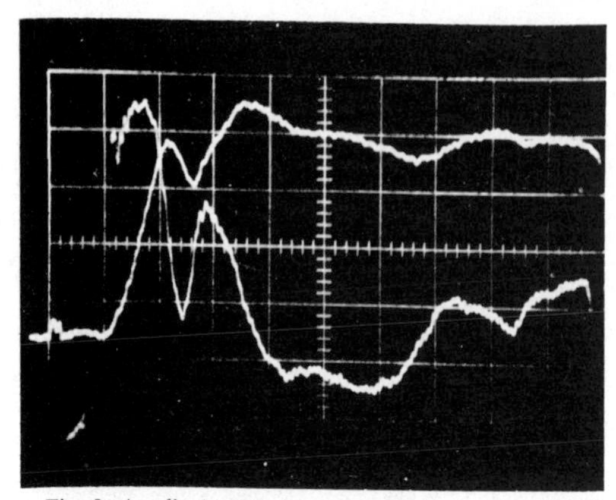

Fig. 8. Amplitude and phase response from dc to 10 kHz at 1 kHz per division of a moderately expensive horn-loaded compression tweeter. Amplitude is 10 dB per division and phase 30° per division.

ment. Figure 10a is for a measurement taken 3 ft on-axis in front of a medium-size corner horn enclosure. The frequency range covered is dc to 1 kHz at 100 Hz per cm. The amplitude scale is 10 dB per cm while the phase scale is 60° per cm. The response dip at 260 Hz was determined to be a genuine loudspeaker aberration and not a chance room reflection by the simple expedient of moving the microphone physically around and noting response. The basic response is seen to be reasonably uniform from about 70 Hz to well beyond 1 kHz with the exception of a strong minimum phase dip at 260 Hz. Figures 4, 8, and 10a compared near cutoff reveal a rather similar behavior pattern. According to the analysis of Appendix A the response dip at 260 Hz is minimum phase and removable. Figure 10b is the result of a simple inductance-capacitance peaking circuit placed between the power amplifier and loudspeaker terminals. The scale factors and magnitudes of Fig. 10b are identical to those of Fig. 10a to show a direct comparison before and after removing the minimum phase response dip. Because the network represents a loss, investigation of Fig. 10b reveals that the overall transfer gain was reduced by about

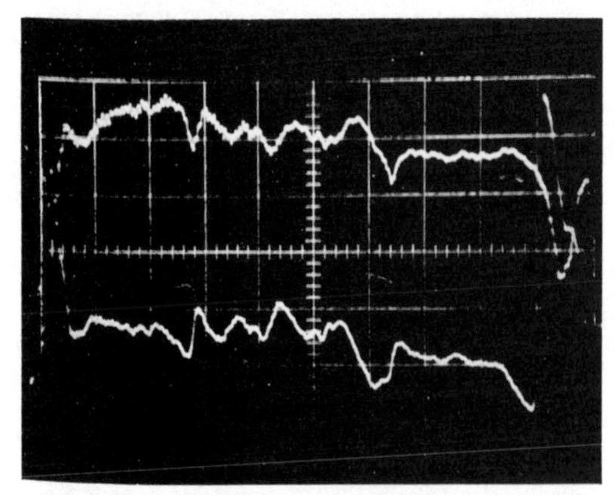

Fig. 9. DC to 20 kHz response at 2 kHz per division of a midrange electrostatic loudspeaker. Amplitude is 10 dB per division and phase is 60° per division.

20 dB with an additional 60° of incurred midrange phase lag. However, the response dip of Fig. 10a estimated at close to 20 dB has been effectively removed. The remaining response dip at 150 Hz may be seen to be an independent response aberration by comparing with Fig. 10a.

Investigation of the equalized response of Fig. 10b and other portions not illustrated revealed a smooth amplitude and phase response at frequencies which were harmonically related to 260 Hz. Thus, if room reflections did not constitute substantial energy at the microphone location, the squarewave response with a 260 Hz fundamental should have been improved. Figure 10c is the response to such a squarewave of the configuration of Figs. 10a and 10b. The upper trace is the response after equalization. The lower trace, made as a second exposure with the equalizer removed and system drive reduced accordingly, shows the unequalized response of Fig. 10a. There is no question that the transient behavior is improved. The squarewave frequency of Fig. 10c was chosen to show the improvement due to the response null removed, and was not deliberately modified for a more pleasing waveform after equalization. In fact, a substantial range of squarewave frequencies from about 70 Hz to 300 Hz show a distinct waveform improvement for the equalized speaker even with obvious room resonances.

CONCLUSION

This paper represents a preliminary report on loudspeaker frequency response measurement. An attempt has been made to provide a more rigorous approach to understanding the role that the neglected partner, phase, plays in the resultant performance of a loudspeaker. The measurement of phase as a spectral distribution and its correlation with response in the time domain has not received the attention which has been devoted to amplitude. There is, in fact, a substantial void in open literature discussion of phase distributions, which it is hoped will be partially filled by this paper. While there is rather complete agreement about the effect of peaks and dips on the response of a loudspeaker, one can usually expect animated discussion on the subject of phase variations. Part of the reason is the difficulty in instrumenting a phase measurement since phase is intimately related to time of occurrence. Thus measurements on magnetic recorders, disk recorders, loudspeakers and microphones are normally restricted to amplitude characteristics. Where lack of reverberation allows definite measurement of a source, measurements in the time domain by impulse testing or otherwise are used as a supplement to amplitude response measurements in the frequency domain [20]. By utilizing a different measurement technique it has been demonstrated here that loudspeaker phase response measurements may be made with the same facility and validity as amplitude response ones.

Having thus secured the capability of simultaneous measurement of amplitude and phase spectra, it is necessary to demonstrate a reasonable need for this capability. Looked at another way, any prior analysis which dictated a valid need for phase information would most certainly have precipitated a measurement technique. It

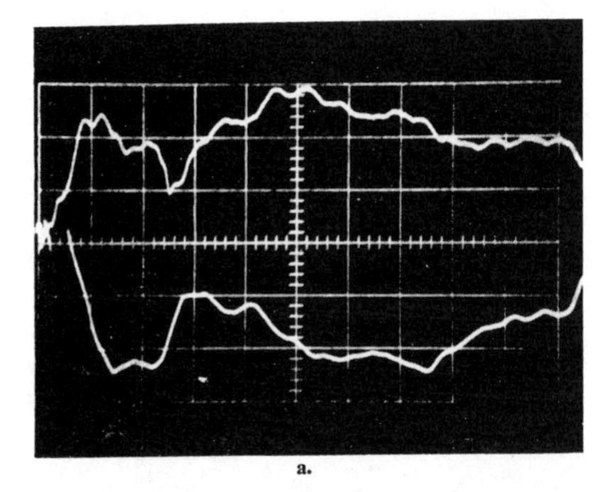

a.

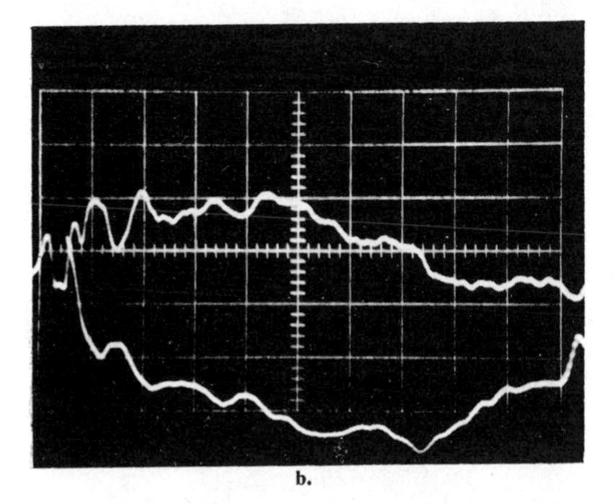

b.

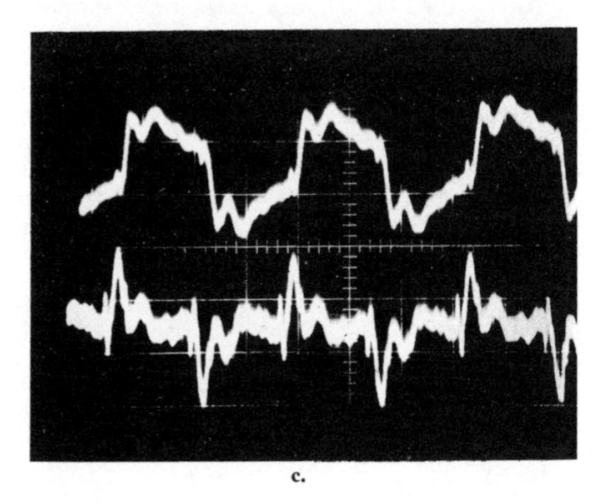

c.

Fig. 10. **a.** Three foot on-axis measurement of a corner horn loudspeaker system. Frequency scale is 100 Hz per division and covers dc to 1 kHz. Amplitude (**upper**) is 10 dB per division and phase (**lower**) is 60° per division. **b.** Result of equalization of the single major response dip of Fig. 10 a. **c. Upper** trace is the acoustic response to a square-wave of the equalized loudspeaker of Fig. 10b. **Lower** trace is the acoustic response of the unequalized loudspeaker of Fig. 10a. Vertical scale is uncalibrated pressure response and horizontal scale is 1 msec per division.

was required, then, to analyze the conditions under which it is necessary to have both amplitude and phase characteristics before one can say one knows everything about the frequency behavior. This led to electronic circuit analysis, where both spectra are normally considered, and a consideration of that class of networks known as minimum phase. The answer was not found completely in circuit analysis, since the very complex behavior of a loudspeaker quite frequently overtaxes the simplified concept of circuit time delay. The appendices to this paper help to bridge the gap in using circuit concepts for analysis of loudspeakers when one has both amplitude and phase spectra. Appendix A derives a simple graphical relationship between amplitude and phase which allows one to state whether a network under analysis is of the minimum phase type. This derivation was necessary since circuit analysis normally proceeds from a knowledge of the equations governing behavior, whereas loudspeaker characteristics are a product of a measurement on a system for which one has incomplete analysis.

Appendix B is a derivation of the inter-domain Fourier transformation relationship in the case of a dispersive medium without absorption. This is, of course, precisely what one finds with a loudspeaker if all minimum phase aberrations are removed. Surprisingly, this is also a common delay situation in many other branches of wave mechanics, yet the transform of a fixed delay of time or frequency (a special case of a non-dispersive medium) is the only form found in widely respected literature.

A head-on attack on the concept of time delay in a dispersive medium with absorption, which is a general characterization of loudspeakers, seems required. The concept of time delay in such a medium does not lack publication; however, such analysis is generally so specific that an engineer quite understandably hesitates in applying it to a general problem. An attack on this problem, discussed in a paper originally planned as a third Appendix to the present one [26], proceeds from the premise that a real-world system is causal; the output follows logically from the input and cannot predict the input. In pursuing this premise we were willing to accept a multiple-valued time behavior of spectral components. As a result, we found that there is indeed a concept of time delay of a system which is causal and makes engineering sense. Furthermore, there is a natural analogy between frequency and time which allows an engineer to draw from knowledge in one domain to add to comprehension in the other domain, and thus to relate modulation theory to time delay incurred by a complex frequency transfer function. Although a unique concept of time delay, the derivation involved is shown to be consistent with the work of Rayleigh [15], Brillouin, [16] and MacColl [9]. An important byproduct of this analysis is an explanation for the confusion created by attempts at using group delay in an absorptive medium. It is shown that the classic concept of group delay is not applicable to a minimum phase medium, and hence to any causal medium with absorption.

Thus, it is possible to identify the effect of amplitude and phase variations as equivalent to what would exist if the actual loudspeaker were replaced by a large number of perfect loudspeakers spread out in space in a frequency-dependent manner. With this approach to the interrelation of amplitude and phase, one can state from measured performance whether a given loudspeaker may be improved by electrical or mechanical means. Since it makes no difference whether the response characteristic is traceable to the loudspeaker, enclosure, or immediate environment, if the resultant behavior is minimum phase one knows that it can be improved.

Finally, some amplitude and phase spectra of typical loudspeakers have been included. The product of any mathematical analysis of phase and time delay would be questionable if it could not be applied to the practical physical problems of an audio engineer. It has been demonstrated by example that a loudspeaker may not be minimum phase. In the case where a loudspeaker is found to be minimum phase, a simple example was included to demonstrate that such minimum-phase response deficiencies can be removed by simple networks.

As more familiarity is gained with the phase response of loudspeakers it is likely that such measurement will become more commonplace. By relating the amplitude and phase characteristics to a seldom considered form of distortion, time delay, and by providing a visual means of determining whether this time delay distortion is removable by relatively simple means, it is hoped that a tool has been provided which will lead to improved quality of sound reproduction.

APPENDIX A

Relations Between Amplitude and Phase for Minimum Phase Network

Assume the transfer function of a network is

$$H(s) = A(s)e^{i\phi(s)} = e^{\alpha(s)+i\phi(s)} \tag{A1}$$

where $\alpha(s) = \ln A(s)$ and $s = \sigma + i\omega$. This will be defined as the transfer function of a minimum phase network if $\alpha(s)$ and $\phi(s)$ are uniquely related each to the other. If either $\alpha(s)$ or $\phi(s)$ are known, then everything is known about the network. This implies that there are no zeros or poles of the transfer function in the right-half s plane, since the derivatives must not only exist along the $i\omega$ axis for all values of the frequency ω but are related by the Cauchy-Riemann equations within and on the boundary of the right-half s plane [2,6]:

$$\frac{\partial\alpha(s)}{\partial\sigma} = \frac{\partial\phi(s)}{\partial\omega}, \quad \frac{\partial\alpha(s)}{\partial\omega} = -\frac{\partial\phi(s)}{\partial\sigma}. \tag{A2}$$

Taking further derivatives,

$$\frac{\partial}{\partial\sigma}\left(\frac{\partial\alpha(s)}{\partial\sigma}\right) = \frac{\partial}{\partial\sigma}\left(\frac{\partial\phi(s)}{\partial\omega}\right), \tag{A3}$$

$$\frac{\partial}{\partial\omega}\left(\frac{\partial\alpha(s)}{\partial\omega}\right) = -\frac{\partial}{\partial\omega}\left(\frac{\partial\phi(s)}{\partial\sigma}\right) \tag{A4}$$

$$= -\frac{\partial}{\partial\sigma}\left(\frac{\partial\phi(s)}{\partial\omega}\right),$$

from which it follows that

$$\frac{\partial^2\alpha(s)}{\partial\omega^2} = -\frac{\partial^2\alpha(s)}{\partial\sigma^2} \tag{A5}$$

and

$$\frac{\partial^2\phi(s)}{\partial\omega^2} = -\frac{\partial^2\phi(s)}{\partial\sigma^2}. \tag{A6}$$

If one assumes that the complex s plane is a topological map with $\alpha(s)$ as the elevation, the situation is as plotted in Fig. A-1 for the simple one-pole one-zero

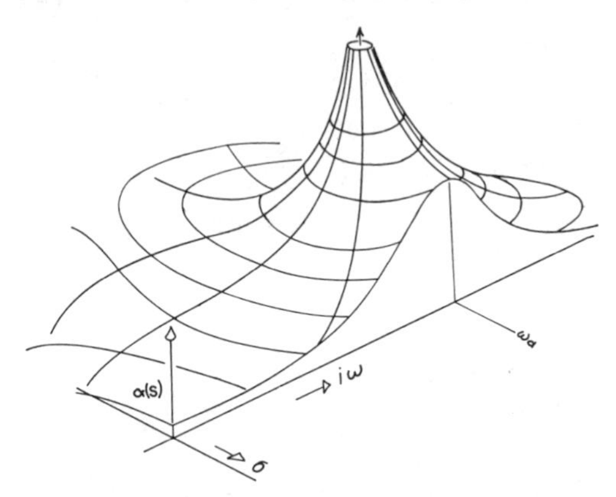

Fig. A-1. Representation of the topological plot of amplitude response in the complex s-plane. A section is taken along the $i\omega$ axis which corresponds to the curve normally considered as the amplitude response. The behavior near a pole in the left half s-plane is characterized and lines of steepest descent and of equipotential are shown.

function. For such a plot of $\alpha(s)$ the lines of steepest descent for $\alpha(s)$ are equipotential lines for $\phi(s)$. The lines of descent for $\alpha(s)$ can originate only from those points at which $\alpha(s)$ is positive or negative infinite, since no finite maxima or minima occur and there can only be saddle points where

$$dH(s)/ds = 0. \tag{A7}$$

Because there are no singularities in the right half-plane one can then state that for a point on the imaginary axis (where $\sigma = 0$ and the real-world concept of transfer function exists) when at or near a frequency of closest approach to a singularity such as ω_a in Fig. A-1,

$$\partial\alpha(s)/\partial\sigma \text{ is a maximum negative value.} \tag{A8}$$

From Eq. A2, then, at this frequency

$$\partial\phi(s)/\partial\omega \text{ is a maximum negative value,} \tag{A9}$$

or, a point of inflection exists for $\phi(s)$,

$$\partial^2\phi(s)/\partial\omega^2 = 0. \tag{A10}$$

At this point any penetration into the s plane along the direction of the σ axis must be along an equipotential line of phase and consequently a maximum rate of change of phase with respect to σ, since it follows from Eqs. A6 and A10 that

$$\partial^2\phi(s)/\partial\sigma^2 = 0. \tag{A11}$$

This means that

$$-\frac{\partial}{\partial\sigma}\left(\frac{\partial\phi(s)}{\partial\omega}\right) = \frac{\partial^2\alpha(s)}{\partial\omega^2} \tag{A12}$$

$$= \text{maximum negative.}$$

This leads to the very simple rule for a minimum phase

network: At a local maximum or minimum in the transfer function, a frequency of maximum curvature of amplitude corresponds to a point of inflection of phase.

It may also be seen that the skew symmetry of Eqs. A2 allow a similar rule stating that a maximum curvature of phase corresponds to a point of inflection of amplitude.

Also, from inspecting the polarity of the functions it is evident that if one considers the direction of increasing frequency, when $\partial^2 a(\omega)/\partial\omega^2$ is a maximum positive then $\partial\phi(\omega)/\partial\omega$ is a maximum positive, while when $\partial^2\phi(\omega)/\partial\omega^2$ is a maximum positive then $\partial a(\omega)/\partial\omega$ is a maximum negative.

Note that these relationships should be held precisely only if the frequency under analysis is that which is closest to the singularity (pole or zero in transfer function) which produces the high rate of change of curvature. Perturbations on either side of the frequency of the singularity will adhere less to this pattern the further one proceeds from the singularity. Figure A-2 shows several simple cases of amplitude and phase functions which one can use to identify minimum-phase networks. Points of maximum curvature are shown by the dashed circle around which the curve might be thought of as bent. Points of inflection are shown by horizontal dashes. Note that the frequency scale is linear, amplitude is plotted logarithmically in dB with increasing gain as a positive quantity, and phase is plotted as an angle with the standard convention of phase lag as a negative angle.

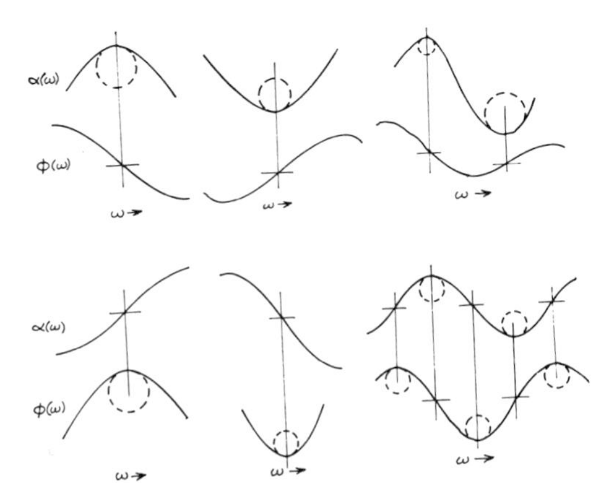

Fig. A-2. Characteristic amplitude and phase plots which may be used to identify minimum phase behavior. Points of maximum curvature are shown by **circles** around which the curve may be considered as bent, and associate points of inflection indicated by **horizontal lines.**

APPENDIX B

Fourier Transform of Frequency-Dependent Delay

A medium in which a time-dependent disturbance propagates is said to be dispersive if the velocity in the medium, or time of traverse, varies with the frequency of the time dependence. If there are many frequency constituents in the initial form of the disturbance, then some time later these constituents are dispersed. The observed waveform at a fixed "down-stream" location will no longer be identical to the initial disturbance, and the resultant waveform will be a distortion of the original waveform even if no reduction in amplitude of these constituents has occurred.

Assume that the time delay for each frequency ω has been determined to be

$$T(\omega) = T_0 + \tau(\omega) \tag{B1}$$

where T_0 is a fixed delay in seconds and $\tau(\omega)$ is a dispersive time delay. Since we are interested in frequency dependent factors, the frequency phase dependence becomes

$$\Theta(\omega) = \int_0^\omega T(\omega)\,d\omega = T_0\omega + \int_0^\omega \tau(\omega)\,d\omega. \tag{B2}$$

This is the phase dependence of each frequency component comprising the dispersed output waveform. The expansion of all such components in a Fourier integral will yield the resultant time function at the output. Thus, for the case of an otherwise perfect transmission system, an input signal with a time dependence $g(t)$ and a frequency transform

$$G(\omega) = \int_{-\infty}^{\infty} g(t)e^{-i\omega t}dt \tag{B3}$$

will produce an output from the dispersive medium of $f(t)$, where

$$f(t) = (1/2\pi)\int_{-\infty}^{\infty} G(\omega)e^{-i\Theta(\omega)}e^{i\omega t}d\omega. \tag{B4}$$

The negative sign is used for time delay. This leads to the important finding that:

If $G(\omega)$ is the transform of $g(t)$, then

$$G(\omega)\cdot e^{\pm i\int_0^\omega T(\omega)d\omega}$$

is the transform of

$$g[t \pm T(\omega)].$$

By the same reasoning, if $g(t)$ is the transform of $G(\omega)$ then

$$g(t)\cdot e^{\pm i\int_0^t \Omega(t)dt}$$

is the transform of

$$G[\omega \mp \Omega(t)].$$

Note that the commonly encountered transform relations for a fixed delay constitute the special case of non-dispersive time delay.

REFERENCES

1. R. C. Heyser, "Acoustical Measurements by Time Delay Spectrometry," *J. Audio Eng. Soc.* **15**, 370 (1967).
2. C. H. Page, *Physical Mathematics* (D. Van Nostrand Co., Princeton, 1955), p. 220.
3. S. J. Mason and H. J. Zimmerman, *Electronic Circuits, Signals, and Systems* (Wiley, New York, 1960), p. 350.
4. H. W. Bode, *Network Analysis and Feedback Amplifier Design* (D. Van Nostrand Co., New York, 1955), p. 312.
5. Bateman Manuscript Project, *Tables of Integral Transforms* (McGraw Hill Book Co., New York, 1954), p. 239.

6. E. A. Guillemin, *The Mathematics of Circuit Analysis* (M.I.T. Press, Cambridge, 1965); p. 330.
7. W. R. Bennett and J. R. Davey, *Data Transmission* (McGraw Hill Book Co., New York, 1965), p. 84.
8. R. V. L. Hartley, "Steady State Delay as Related to Aperiodic Signals," *B.S.T.J.* **20**, 222 (1941).
9. S. Goldman, *Frequency Analysis Modulation and Noise* (McGraw Hill Book Co., New York, 1948), p. 102.
10. G. A. Campbell and R. M. Foster, *Fourier Integrals* (D. Van Nostrand Co., Princeton, 1961), p. 7.
11. M. J. Lighthill, "Group Velocity," *J. Inst. Math. Applics. I* (Academic Press, London, 1965), p. 1.
12. S. P. Mead, "Phase Distortion and Phase Distortion Correction," *B.S.T.J.* **7**, 195 (1928).
13. P. R. Geffee, *Simplified Modern Filter Design* (J. F. Rider, New York, 1963), p. 73.
14. S. Hellerstein, "Synthesis of All-Pass Delay Equalizers," *IRE Trans. on Circuit Theory* **PGCT-8**, No. 3, 215 (1961).
15. Rayleigh, *Theory of Sound* (Dover Press, London, 1945), Vol. 1, p. 301.
16. L. Brillouin, *Wave Propagation and Group Velocity* (Academic Press, London, 1960).
17. H. F. Olson, *Acoustical Engineering* (D. Van Nostrand Co., Princeton, 1960).
18. A. Papoulis, *The Fourier Integral and Its Applications* (McGraw-Hill Book Co., New York, 1962), p. 148.
19. D. Slepian and H. O. Pollak, "Prolate Spheroidal Wave Functions, Fourier Analysis and Uncertainty," *B.S.T.J.* **40**, 353 (1961).
20. E. D. Sunde, "Pulse Transmission by A. M., F. M., and P.M. In The Presence of Phase Distortion," *B.S.T.J* **40**, 353 (1961).
21. G. N. Watson, "The Limits of Applicability of the Principle of Stationary Phase," *Proc. Cambridge Phil. Soc.* **19**, 49 (1918).
22. C. Eckart, "The Approximate Solution of One-Dimensional Wave Equations," *Rev. Mod. Phys.* **20**, 399 (1948).
23. C. A. Ewaskio and O. K. Mawardi, "Electroacoustic Phase Shift in Loudspeakers," *J. Acoust. Soc. Am.* **22**, 444 (1950).
24. F. M. Weiner, "Phase Distortion in Electroacoustic Systems," *J. Acoust. Soc. Am.* **13**, 115 (1941).
25. F. M. Weiner, "Phase Characteristics of Condenser Microphones," *J. Acoust. Soc. Am.* **20**, L707 (1948).
26. R. C. Heyser, "Group Delay, Excess Delay, and Overall Time Delay," to be published.

AUTHOR CITATION INDEX

SUBJECT INDEX

About the Editor

HARRY B. MILLER is a designer of transducers and arrays at the Naval Underwater Systems Center, New London, Connecticut. During the 1960s he headed the Electroacoustics Laboratory at General Dynamics/Electronics in Rochester, New York. He was head of the Acoustic Engineering Department at the Brush Development Company in Cleveland, Ohio, during the 1950s.

Throughout his career Mr. Miller has been responsible for a wide variety of design projects in applied acoustics involving ultrasonic nondestructive testing, hearing-aid transducers, underwater projectors and hydrophones, electrically steered arrays, pulsed modulation, magnetic and disc recording, and modern signal processing. Much of this work is documented in his patents and journal articles.

Mr. Miller is a senior member of the Institute of Electrical and Electronics Engineers, a Fellow of the Acoustical Society of America, and a former chairman of that society's Technical Committee on Engineering Acoustics. He is currently writing a series of monographs on transducer design.